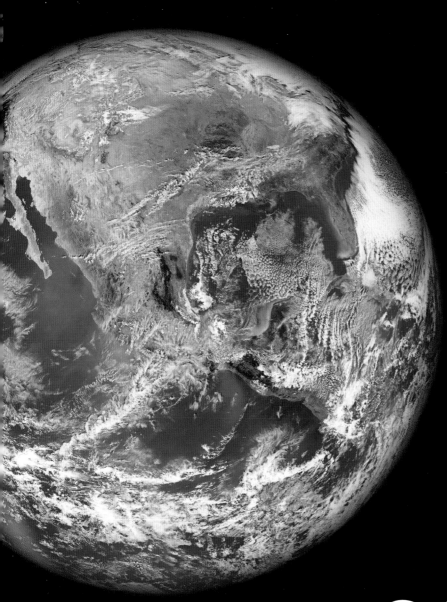

SPACE
ATLAS

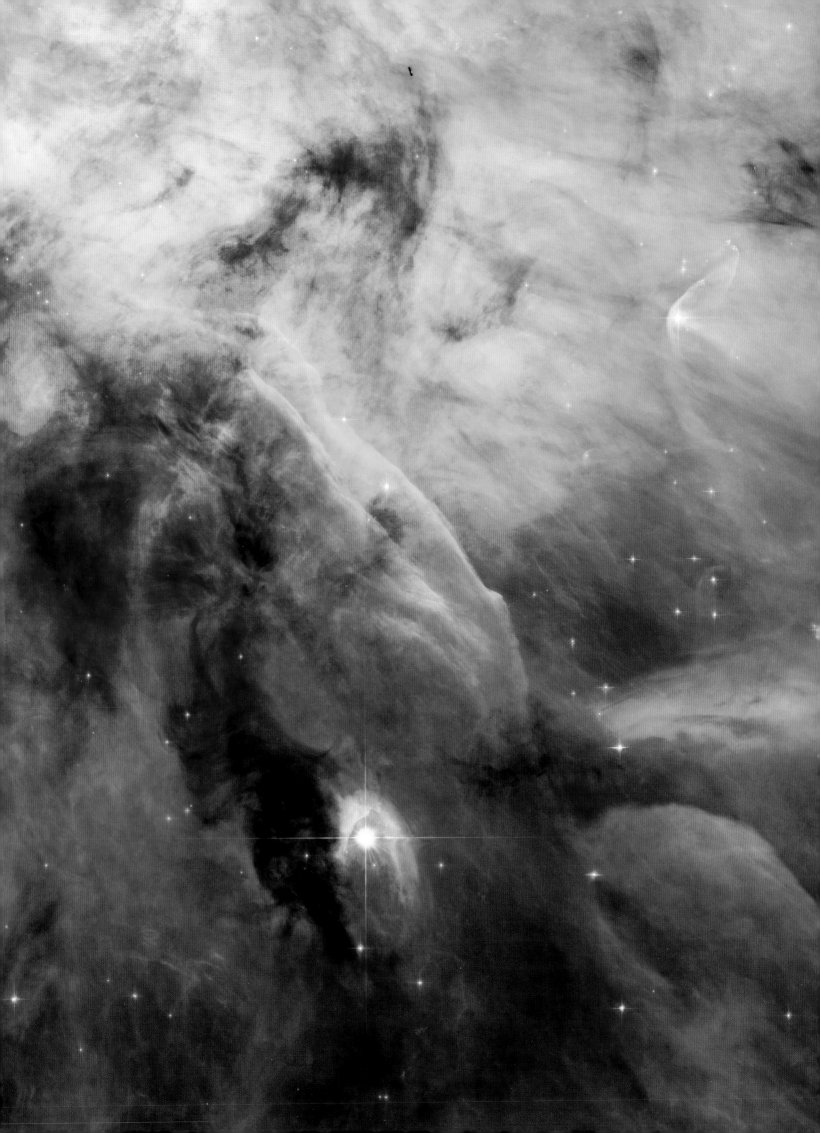

SPACE
ATLAS

MAPPING
THE UNIVERSE
AND BEYOND

JAMES TREFIL

FOREWORD BY BUZZ ALDRIN

NATIONAL GEOGRAPHIC

WASHINGTON, D.C.

CONTENTS

PREVIOUS PAGES: (Page 1) Western Hemisphere of planet Earth. (Pages 2–3) Orion Nebula, with star LP Orion is at lower left. THESE PAGES: An image compiled from both visible-light and x-ray sources reveals the bubble-shaped remnants of supernova SNR 0509-67.5, located in the Large Magellanic Cloud galaxy. Heated material is shown in green and blue; the pink shell represents the shock wave of the supernova moving through the gas.

THE SOLAR SYSTEM

Queen of the solar system, Saturn is famed for its spectacular ring system. This image, compiled in natural light from the Cassini orbiter, shows the complex structure of the rings and gaps and the sharp shadows they throw across the gas giant's atmosphere.

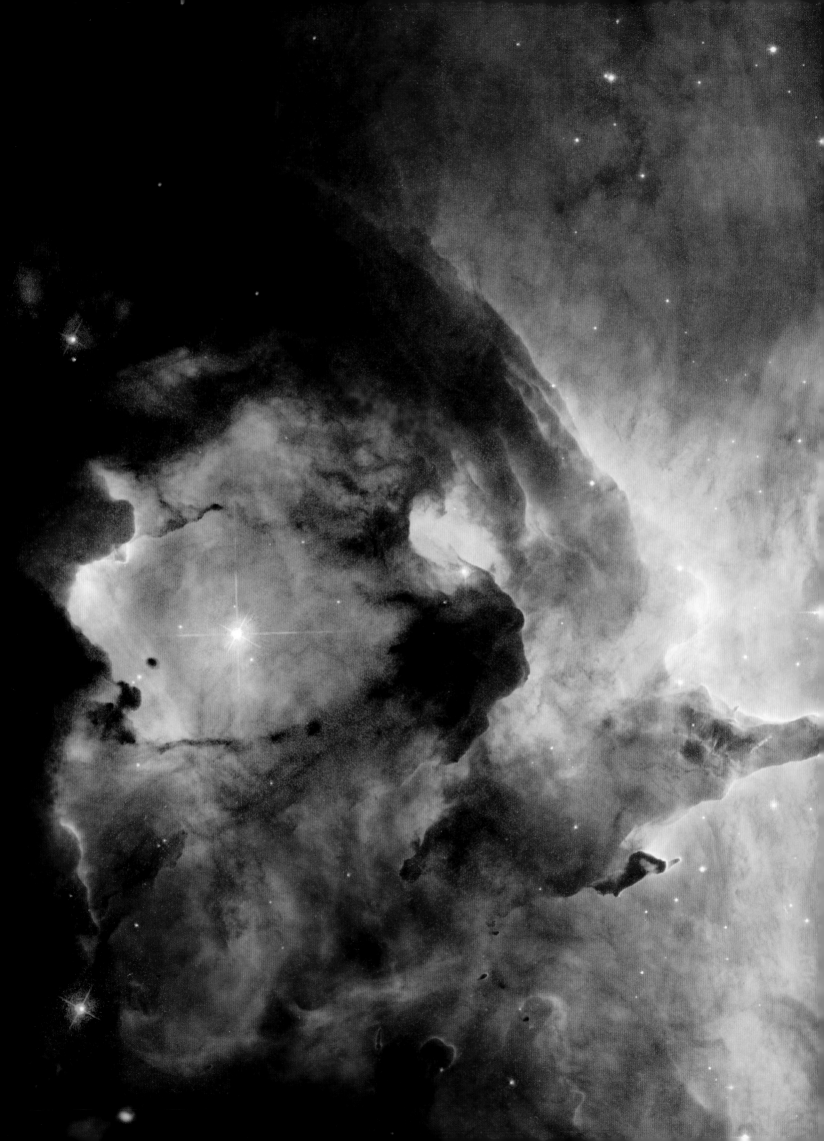

THE GALAXY

Brilliant stars and gaseous, star-forming regions make up the nebula NGC 6357. The brightest star in the image above is actually a double star, Pismis 24-1, with an enormous combined mass about 200 times that of our sun.

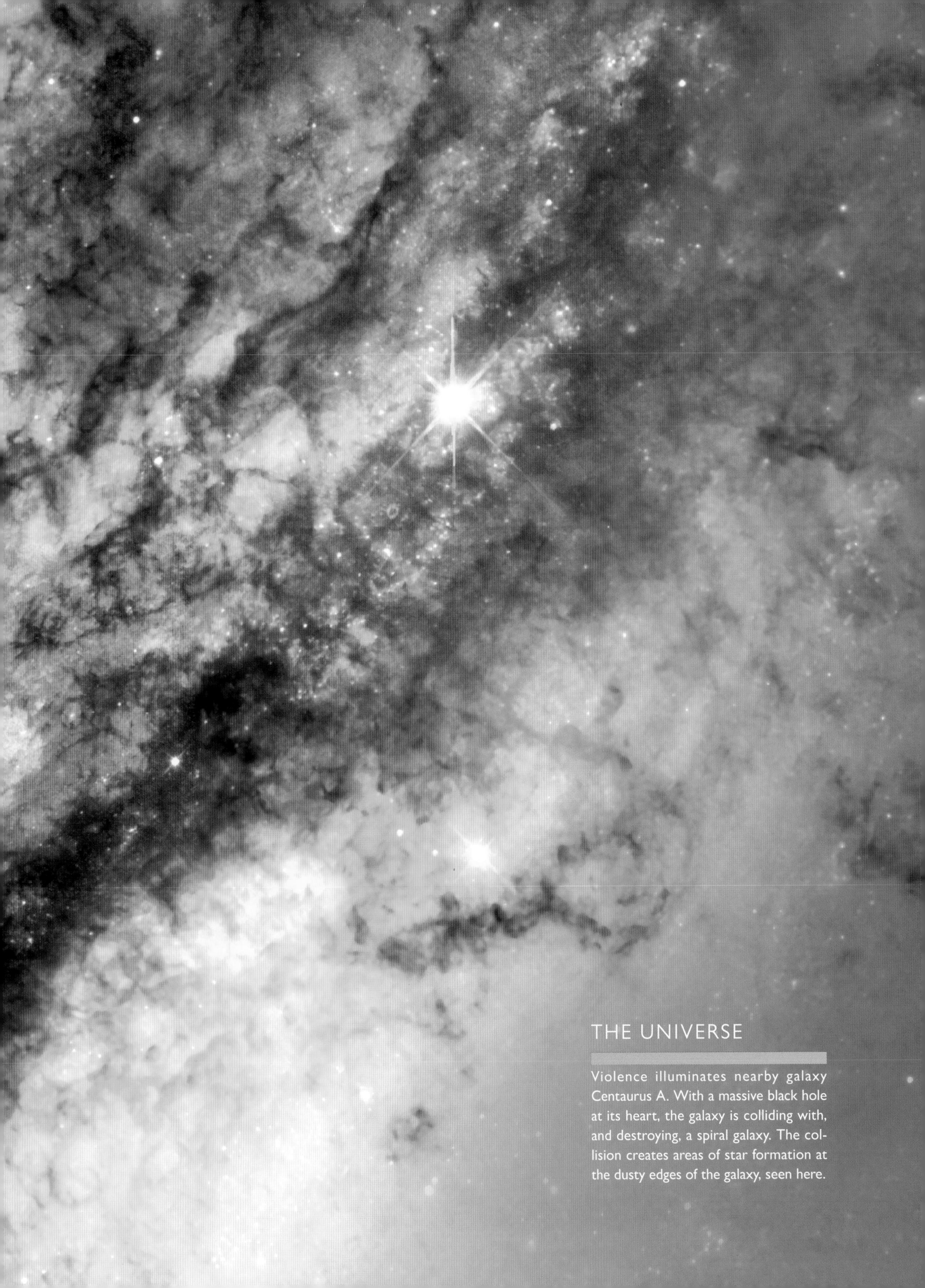

THE UNIVERSE

Violence illuminates nearby galaxy Centaurus A. With a massive black hole at its heart, the galaxy is colliding with, and destroying, a spiral galaxy. The collision creates areas of star formation at the dusty edges of the galaxy, seen here.

FOREWORD

Year by year, generation by generation, the way we look at our universe continues to evolve, thanks both to new technologies and new ways of thinking, spurred on by our ability to view stars and galaxies that are distant in space and time. We are sharing new ways of seeing as well, as space telescopes and interplanetary probes transmit information across millions of miles, information that we capture and transform into remarkable visual displays. From that information, ever new maps can be created—maps such as you have never seen before; maps like the ones in this beautiful volume.

This National Geographic *Space Atlas* has special meaning for me. It is an enduring honor to have been one of the few humans to have stood on the moon. Just 12 years after the launch of the Soviet Union's first Earth-orbiting artificial satellite, Sputnik 1, Neil Armstrong and I set foot on the moon on July 20, 1969. The moon to me is not a distant object in space but a real place where I spent time, and a real landscape that I remember in my mind's eye. Looking at the maps of Earth's moon on these pages is for me a little like retracing a vacation on the map that was carried along.

All those who remember—and all those born since, who don't—can recognize that Apollo 11 capped an extraordinary decade of technological development and space exploration. An unwavering focus on long-term goals guided the United States in developing new materials, rockets, orbits, space capsules, and space suits, and in making the scientific breakthroughs necessary to get humans to the moon and back again.

Our successful moon missions both required and informed maps of different kinds. Terminal phase maps of the trajectory approach and descent maneuver, especially the target area for our landing, were critically important. In the more than four decades since that first moon walk, more than 500 astronauts have lived and worked in space, all in orbit near Earth. Human spaceflight, although still complex, dangerous, and expensive, has become common enough that plans for space tourism are under way, and customers are lining up.

During this same period, brilliantly designed unmanned systems—from space-based telescopes to planetary rovers to deep-space probes—have made spectacular discoveries, including evidence of the current or past presence of water on our moon, on the moons of Jupiter, and on Mars. And so far we have located more than 400 planets orbiting other stars like our sun. All this information gets added to the maps—to use the term very broadly—of our solar system, galaxy, and universe.

Now the time has come for another long-term road map to allow humans back into the space exploration business. Destination—Mars. To reach that tantalizing

An iconic image of the space age shows astronaut Buzz Aldrin walking on the surface of the moon during the 1969 Apollo 11 mission.

goal will require intelligence, resourcefulness, and the ability to adapt to changing circumstances. The effort will undoubtedly foster several awe-inspiring achievements along the way. To get to Mars, we will travel aboard new, reusable spacecraft. These successors to the magnificent space shuttle fleet will be special-purpose ships capable of many missions, including long-duration trips cycling between the orbits of Earth and Mars, and we will need new charts of the heavens, just as European explorers had to generate charts of the New World as they began to explore it.

Just as the names of the lunar seas became part of America's vocabulary in the days of Apollo, the names of comets and other near-Earth objects will become

The dry, rusty, windswept terrain of Mars was captured by the Mars exploration rover Opportunity in the winter of 2012. The rover rested atop an outcropping known as Greeley Haven on the rim of the crater Endeavour, just south of the Martian equator.

familiar as we rendezvous with them on our initial journeys—names like Wirtanen, Hartley 3, and Apophis. We will intercept asteroids and sift their rocky soil, developing tools and equipment as we uncover the building blocks of the universe. We will fly to the golden tail of an ancient comet, gathering and sampling material from the birth of the galaxy. Our picture of the world will change, maps changing with it.

Step by step we will continue outward, eventually landing on the Martian moon Phobos. From there we

will directly control robotic operations on the surface of Mars. This incremental exploration will allow us to develop the techniques required for long-duration human space travel in preparation for our historic mission to homestead the red planet itself. If we put the plan in motion and don't waver, we can set foot on Mars by 2035—66 years after Tranquility Base, which itself was 66 years after Kitty Hawk. But we need to get started now.

The success of this ambitious plan will require a unified commitment to stay the course through economic upturns and downturns and several presidential administrations. Indeed, it will take a concerted effort to make it to Mars. Perhaps most crucial, it will require tapping the limits of human ingenuity to overcome the fundamentals of time, space, mass, and energy that rule our universe. But from where I stand—and from where I stood 43 years ago—I know we can do it.

The human race would not be where it is today without maps—and we won't get very far beyond the horizon of our present reckoning without them either. A book like this National Geographic *Space Atlas* by James Trefil is so exciting because it refines our sense of the frontiers of space by taking the surge of information coming to us from space explorations, both manned and unmanned, and translating all of that new data into graceful text, remarkable imagery, and elegant maps.

—Buzz Aldrin

ABOUT THIS BOOK

Space Atlas is a visual guide to the universe, starting with our own solar system and reaching out to the farthest galaxies and the mysteries of multiple universes. The book combines clear, nontechnical text that explains the nature of planets, stars, galaxies, and exotic objects, such as black holes, with photos and art that illustrate the strange beauty of outer space. Over 90 pages of detailed maps show us the surfaces of distant planets and explain how the planets, stars, and galaxies fit into the big picture.

Space is organized from near to far in three major chapters. The first, "The Solar System," encompasses the formation of the system, the planets from tiny Mercury to icy Neptune, the major moons, and smaller but significant objects such as dwarf planets, asteroids, and comets. The second, "The Galaxy," describes the stars and structures of our own Milky Way and the nature of galaxies in general. The third chapter, "The Universe," digs into the big issues of cosmology: the structure of the universe and its birth and death. An epilogue, "Mysteries," explores some of the hottest topics in modern physics: string theory and the possibility of a multiverse.

Introducing the chapters are four star charts that depict the stars and constellations of the night sky as seen from Earth; wrapping up the book are appendices that fill in statistics about the planets, satellites, and interesting deep-sky objects.

Each of the three main chapters is marked by colored tabs in the margin; a header along the edge of the page identifies the subject. Lists of useful statistics (some slightly rounded off) and dates appear on the opening pages of every topic.

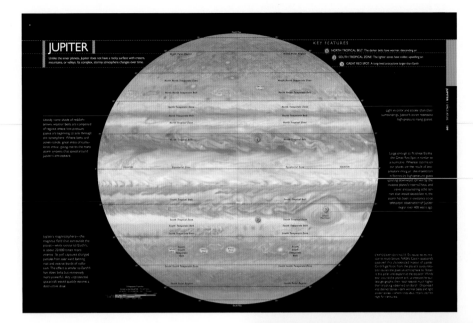

Maps in this atlas include planets, moons, galaxies, and orbits. Features such as craters or atmospheric structures are labeled. Particularly interesting features are marked by colored, numbered bullets and identified in a key. At the bottom of the page, a cartographer's note gives background information about the map.

Many entries contain explanatory cutaway art or diagrams that give insight into planetary anatomies or help to explain scientific concepts. Boxes on full-page photos contain relevant facts.

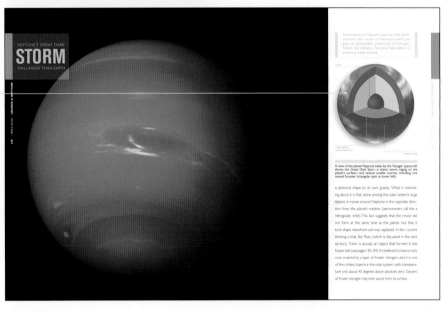

Sidebars—boxed text on special subjects—throughout the book highlight interesting stories related to the main entry. These include biographical sketches and the author's interviews with famous contemporary astronomers.

INTRODUCTION

The universe used to be such a simple place. • After all, for most of human history the people who thought about such things pictured Earth as sitting still in the center of creation, while the heavenly bodies such as stars and planets moved around it. In ancient mythologies, Earth was usually flat, and the motion of the sun across the sky was due to the actions of a god or goddess. Starting in about the fifth century B.C., though, a new way of thinking about the universe began to develop around the eastern Mediterranean—a way that didn't depend on the whims of the gods, a way that began to take humanity out of what Carl Sagan called the "demon-haunted world." Greek philosophers began to construct models of the universe that may seem primitive to us, but had the strange new feature of operating solely according to natural laws, without supernatural intervention. Many scholars consider this development to be the beginning of science.

All of these models shared two basic, unquestioned assumptions. One was that Earth sat motionless at the center of the universe, while everything else—sun, moon, and planets—swung around it. The second assumption was that in the heavens, which were pure and eternal, everything moved in a circular path. (This assumption is based on the notion that a circle is the most perfect geometrical shape, and therefore the appropriate figure for the realm of the pure.) In these models, the stars and planets were embedded in solid crystalline spheres whose rotations governed their motion across the sky. (This, incidentally, explains why comets were such a problem for early astronomers, since their orbits would shatter those spheres. It's one reason why Aristotle argued that comets had to be burning vapors in Earth's atmosphere.) Eventually, these models became quite complicated, with the planets embedded in small spheres that rolled within larger spheres.

Given the cozy nature of that model, it's a little surprising that there was a spirited debate in the ancient world centered on the question of whether the universe was finite or infinite in extent. The philosopher Archytas of Tarentum (428–327 B.C.) made an interesting argument that the universe must be boundless. Suppose, he argued, that the universe really has a boundary, an edge. Then a spearman could walk up to that edge and throw a spear outward. The spear would have to land somewhere, and that landing place would be outside of the boundary. No matter how far out you make the boundary, the spearman can always find something outside of it. Therefore, he argued, the universe must have no boundary, but must be infinite.

Following Archytas's argument, we can identify events that expanded the human view of the universe, each event like another flight of a spear. In what follows,

in fact, we will encounter three such spearmen, each of whom widened the universe we live in.

THE FIRST SPEARMAN

The first was the Polish cleric Nicolaus Copernicus (1473–1543). He produced the first serious model of the solar system in which the sun was at the center and Earth moved in orbit, like the other planets. In his book *On the Revolutions of the Celestial Spheres,* he wrote, "[I]t will be realized that the sun occupies the middle of the universe. All these facts are disclosed to us by the principle governing the order in which the planets follow one another, and by the harmony of the entire universe, if only we look at the matter, as the saying goes, with both

An illustration of an armillary sphere—a model of the heavens, with Earth at the center and the constellations of the zodiac around the edge—shows the standard depiction of the universe until the age of Copernicus. Flanking the sphere are diagrams of the theories of astronomers Ptolemy and Tycho Brahe.

Polish astronomer Nicolaus Copernicus overturned the astronomical world with a new, sun-centered vision of the universe. His landmark book, *On the Revolutions of the Celestial Spheres,* was published in 1543.

I Cælarum firarum Iphæra immobilis.
II Saturnus anno. XXX. reuoluitur
III Iouis XII annorum reuolutio
IIII Martis bima reuolutio
V Telluris cum orbe lunari annua reuolutio
VI Veneris nouem mensium reuolutio
VII Mercurius octuaginta dierum

CL PTOLEMAEI

eyes." With Copernicus, the human universe suddenly changed. No longer were Earth and humanity at the very center of existence. Human beings were now inhabitants of just one of many bodies circling the sun. Throughout this book the so-called Copernican principle will come into play—the notion that there is nothing particularly special about humanity and the planet that gave it birth.

With Copernicus, the universe also became much larger. No longer did human beings live in a cosmos bounded by the sky hanging a few miles over their heads and Earth under their feet. Astronomers after Copernicus came to realize that the solar system is huge compared with Earth. Imagine Earth as a sphere the size of a city block located in New York City. The sun would then be a large sphere somewhere around the Mississippi River, and the outer planets would be in Asia. Quite a shift in perspective for people who had spent their entire existence on that city block.

THE SECOND SPEARMAN

The second spear was thrown by a German astronomer named Friedrich Bessel in the early years of the 19th century. Using state-of-the-art telescopes, he was the first to determine the distance to nearby stars. Once again, the universe expanded enormously. If we imagined the solar system crammed into a space the size of a football field, then a typical star would be in a city several hundred miles away.

As the century wore on, astronomers realized that our own sun was just one star—a pretty ordinary one at that—in a mighty city of stars they called the Milky Way.

The five galaxies known as Stephan's Quintet are actually four interacting galaxies and a fifth (the bluish spiral at upper left) that is younger and much closer to our own. The central two galaxies are colliding, and in the process giving birth to new stars, seen in the blue star clusters around their edges. The gravitational influence of the nearby galaxies has also distorted the arms of the barred spiral galaxy at top right.

Our own sun and solar system, which had loomed so large in our minds, were now relegated to just one system among billions. Astronomers began to realize that not all stars are the same, and they started cataloging what they saw. They also noticed faint patches of light in the sky that they called nebulae, although their telescopes lacked the ability see what those clouds were made of. The world was getting ready for the third spearman.

THE THIRD SPEARMAN

His name was Edwin Hubble, and he was an American working at a brand-new telescope on California's Mount Wilson in the 1920s. With that telescope Hubble was able to examine nebulae in detail, picking out individual stars embedded in them. From this he was able to determine how far away those nebulae are. Once again the spear flew outward. He found that many nebulae were, in fact, their own gigantic cities of stars, like the Milky Way. Hubble established that the universe was composed of what we now call galaxies. In a sense, this was simply an extension of what Copernicus taught us so long ago. Earth is just one planet among many in the solar system, the sun just one star among many in the Milky Way, the Milky Way just one galaxy among billions in the universe.

But that wasn't the only thing that Hubble discovered. He found that those other galaxies are moving away from us, that the universe is expanding. This discovery led, in turn, to our current best picture of the origins of the universe, the scenario we call the big bang. In this account the universe began in an unimaginably hot, dense state about 14 billion years ago and has been expanding and cooling ever since. The amazing thing is that scientists have produced models that can reliably trace backward through this process, back to a fraction of a second after the initial event.

Today it may be that another spearman is approaching the boundaries of our universe—his or her identity is still unclear. If, however, some modern theories prove to be successful, it may turn out that our own universe is just one among a multitude of universes in what scientists are starting to call the multiverse. Such a development would, of course, be the ultimate vindication of Nicolaus Copernicus, though he could scarcely have imagined such an outcome.

FOUR UNIVERSES

This book is organized to follow the progress of Archytas's spearmen. Think of our world as being composed of a series of nested "universes," each seeming to encompass all of creation until a spearman comes along and takes us into the next one.

The first, most intimate universe is, of course, our own solar system. Before the invention of the telescope it consisted of the six innermost planets, and

most scientific attention was focused on understanding how the planets moved—essentially, astronomers wanted to know where the planets were. In the 19th century this began to change, until today we focus on how the planets are built—on what (as opposed to where) they are. Furthermore, we have discovered that the solar system is a lot more complex than those early scientists had imagined. Starting with Galileo's discovery of Jupiter's moons, we began to realize that there was a lot more to the solar system than just a few planets. Each moon was, in effect, a new world with its own history, its own characteristics, its own mysteries. Even the cold, dark region beyond Pluto revealed structures and complexities of which we had never dreamed. This new solar system is the subject of the first section of this book.

THE GALAXY

The second universe is our own Milky Way galaxy. As was the case with the solar system, scientists began with the simple tasks—Where are stars located? how bright are they? and so forth. And as with the solar system, the 19th century slowly brought a new set of questions: What are stars made of and how do they work? It wasn't until the 1930s, though, that the new science of nuclear physics revealed the source of a star's energy in nuclear reactions and it began to sink in that the stars are not eternal: They have a life cycle—a beginning, a middle, and an end—like everything else. In fact, we have come to see the Milky Way (and other galaxies as well) as a gigantic factory that

Studies of the collection of galaxies known as Pandora's Cluster yield clues to the presence of mysterious dark matter. The colliding galaxies in the cluster make up only about 5 percent of its mass, gases (colored here in red) another 20 percent. The rest of the cluster's mass (colored in blue) is dark matter. Dark matter is not visible, but it can be detected by its gravitational influence.

converts the primordial hydrogen of the universe into the heavier elements from which planets and human beings (among other things) are made. In the process, we have discovered all sorts of new and interesting objects out there, from black holes to planetary systems around other stars. We also discovered that most of the matter in the Milky Way and other galaxies is not the familiar stuff from which we are made, but something new called dark matter. The exploration of this particular universe is the subject of the second section of this book.

THE UNIVERSE

The third universe is the massive collection of galaxies to which we usually apply that term. The study of the beginning and end of our universe has engaged scientists over the last few decades. We have traced our way back to the beginning using the tools of elementary particle physics, and to the end using the tools of observational astronomy. Against all expectations, astronomers in the 1990s discovered that the expansion of the universe is not slowing down but speeding up. This knowledge, in turn, has led to the realization that most of the matter in the universe is in an unknown form that has been named dark energy. The fate of the universe depends on what dark energy is and what its properties are. In the third section of this book, we will explore the universe as best we know it today.

Finally, for our fourth and last universe, we leave the realm of hard data and enter the speculative world of the theoretical physicist. Some modern theories suggest that our universe is just one among a huge multitude of universes—what some theorists call a multiverse. With this excursion, we will have completed our tour of the universes that make up the environment in which we live.

THE NIGHT SKY

Have you ever had the experience of seeing the night sky in its full glory, far away from city lights? If so, you remember the brilliance of the stars against the velvet blackness of the sky.

For most of the history of *Homo sapiens,* this is what the sky looked like every night. No wonder early people grouped the stars together into constellations and incorporated them into their mythology. No wonder that the earliest astronomers were motivated by the (false) idea that the stars and planets had a profound effect on the affairs of humans.

This interest grew into an endeavor that modern astronomers would recognize. On bamboo books in China and clay tablets in Babylon we find records of naked-eye observations of the night sky. The Greeks took these observations, added their own, and produced a picture of the night sky in which stars and planets circled Earth on crystal spheres.

Fifteen hundred years later the Polish cleric Nicolaus Copernicus changed all that, introducing the notion that Earth circled the sun, rather than vice versa. Once Galileo turned his telescope on the heavens, our vision of the night sky changed forever. No longer was Earth the center of the universe. Our planet was just one among many. Our moon was a world of craters and mountains, and other planets had their moons as well, circling them like miniature solar systems.

Today we see the few thousand stars in the night sky as a small sample of the billions of stars in our own Milky Way galaxy and our own galaxy as one of billions in our universe. Even so, today's sky-watchers still use sky maps that resemble those of long ago. Stars are plotted out against a dome-like sky and contained in 88 official constellation regions. The next eight pages should allow you to find and appreciate the brightest stars and objects in the sky just as our ancestors did, thousands of years ago.

An 18th-century illustration of the constellations of the Northern Hemisphere shows the figures that have been assigned to the night skies since antiquity. Although it's hard actually to detect these shapes in the stars, they form a useful guide to the sky, and today's astronomers still use constellations to map the stars.

HOW TO READ THE SKY MAPS

Unlike terrestrial maps, which show land from above, star charts represent the sky from below. These four maps depict the skies over Earth's Northern and Southern Hemispheres, with the north or south celestial poles at top center. Objects around the edge of the maps might be seen from either hemisphere.

To pinpoint locations, astronomers use celestial coordinates: Right ascension is similar to longitude, and measured in hours, minutes, and seconds denoted by Roman numerals around the rims of the star charts. Declination corresponds to latitude, measuring position north or south of the celestial equator in degrees and minutes. Parallels of declination are concentric blue circles on the charts.

Astronomers organize the sky into 88 constellation regions, marked here by yellow boundary lines. Main stars in a constellation have Greek letter designations: alpha for the brightest, beta for second brightest, proceeding more or less in order of magnitude, or apparent brightness (see keys).

Declination

Constellation boundaries

Constellation figure

Right ascension

Zodiac

Ecliptic (path of the sun across the sky)

New General Catalogue object

Messier object

NORTHERN SKY | SUMMER–FALL

Stars and constellations on this map are highest in the Northern Hemisphere sky from July to December.

Star magnitude

-1 0 1 2 3 4 5

Variable star

- ○ Open star cluster
- ⊕ Globular star cluster
- ⬭ Galaxy
- ▢ Diffuse nebula
- ○ Planetary nebula
- ✦ Supernova remnant
- --- Constellation boundary
- — Ecliptic

SEPTEMBER

XXIII

AUGUST

XXII

XXI

AQUARIUS

γ

θ

ι

ζ

Baham
θ

Homam
ξ

α
Markab

51 Pegasi

PEGASUS

Sadalbari
μ

Scheat
β

Enif
ε

Kitalpha α
β

EQUULEUS

⊕ M15

γ
δ

η Matar

CAPRICORNUS

XX

DELPHINUS

ε
β δ
γ

α

ANDROMEDA

LACERTA

η

Alshain β

Altair α

Tarazed

γ

M

AQUILA

δ

I

γ

Dumbbell ○
M27

β

α

SAGITTA

θ Alya

SERPENS

Veil Nebula

L

ε

61
Cygni

CYGNUS

γ Deneb
Sadr
α

χ
φ

η

β

Albireo

δ

LYRA

K

North
America
NGC 7000

②

Y

α

δ

Garnet
Star
μ

ζ

α Alderamin
CEPHEUS

Alfirk
β

Sulafat γ

Ring ○ M57 δ
Sheliak β ζ
ε

α Vega

OPHIUCHUS

ε
δ σ
Altais

DRACO

Err

XIX

JULY

KEY FEATURES

(1) POLARIS: The North Star

(2) MILKY WAY

(3) ANDROMEDA: The only northern galaxy visible to the naked eye

Celestial Sphere

PISCES

ECLIPTIC

OCTOBER

δ

ζ

μ

α
Alrescha

ARIES

γ
Algenib

PISCES

η

o

CETUS

Kaffaljidhma
ξ₂   ν   γ

μ

α
Menkar

λ

γ
β   Mesarthim
Sheratan

Alpheratz

δ
π

α
Hamal

ARIES

α Mothallah
M33
TRIANGULUM

ρ

β Mirach

γ

β

μ

M31
Andromeda
Galaxy (3)

Almach
γ

NOVEMBER

Pleiades
M45
η

λ

CASSIOPEIA

M34

ρ

Algol
β

TAURUS

TAURUS

hedar
α
η

Caph

W

NGC 869   NGC 884

κ

o

Hyades

γ

δ
ε   α
Aldebaran

Ruchbah

δ

Double
Cluster

ε

τ
η   ι
γ   Mirfak

PERSEUS

α

ν

ξ
ε

Menkib
California
NGC 1499

π₃ π₄
π₂   π₅

o₁

π₆

A

λ

μ

o₂

ι

Bellatrix
γ

DECEMBER

Y
ε   ζ
η

M38

El Nath
β

Meissa
λ

The Kids

Capella
α

M36
Crab
M1

ζ

ORION

β

AURIGA

Polaris

CAMELOPARDALIS

Menkalinan   β   M37

χ₁

Betelgeuse
α

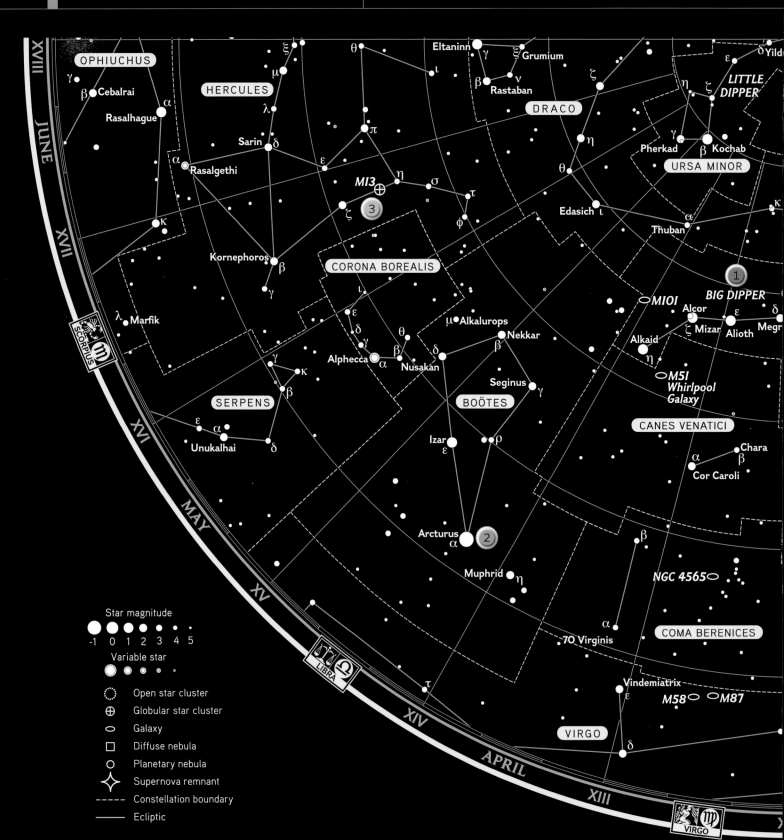

NORTHERN SKY | WINTER–SPRING

Stars and constellations on this map are highest in the
Northern Hemisphere sky from January to June.

XVIII

JUNE

XVII

SCORPIUS

XVI

MAY

XV

LIBRA

XIV

APRIL

XIII

VIRGO

OPHIUCHUS

γ

β Cebalrai

Rasalhague

HERCULES

ξ

μ

λ

α

Sarin

δ

π

α

Rasalgethi

ε

η

MI3 ⊕

σ

τ

3

ζ

φ

κ

Kornephoros

β

γ

λ Marfik

CORONA BOREALIS

ι

ε

δ

γ

θ

β

Alphecca α

Nusakan

μ Alkalurops

Nekkar

δ

β

Seginus

γ

BOÖTES

SERPENS

γ

κ

β

ε α

Unukalhai

δ

Izar

ε

ρ

Arcturus

α

2

Muphrid η

Eltaninn

γ

ξ Grumium

ι

β

ν

Rastaban

ζ

DRACO

η

θ

Edasich ι

Thuban

δ Yild

ε

LITTLE
DIPPER

η

ζ

γ

Pherkad

β Kochab

URSA MINOR

κ

α

BIG DIPPER

MIOI

Alcor

ζ Mizar

Alioth

ε

δ

Megr

Alkaid

η

M5I
Whirlpool
Galaxy

CANES VENATICI

Chara

α

β

Cor Caroli

β

NGC 4565

α

70 Virginis

COMA BERENICES

Vindemiatrix

ε

M58

M87

VIRGO

δ

τ

KEY FEATURES

(1) BIG DIPPER: This star pattern, called an asterism, is part of Ursa Major.

(2) ARCTURUS: At the tip of Boötes, this is the fourth brightest star in the sky.

(3) M13: This globular cluster of stars is the brightest in the northern skies.

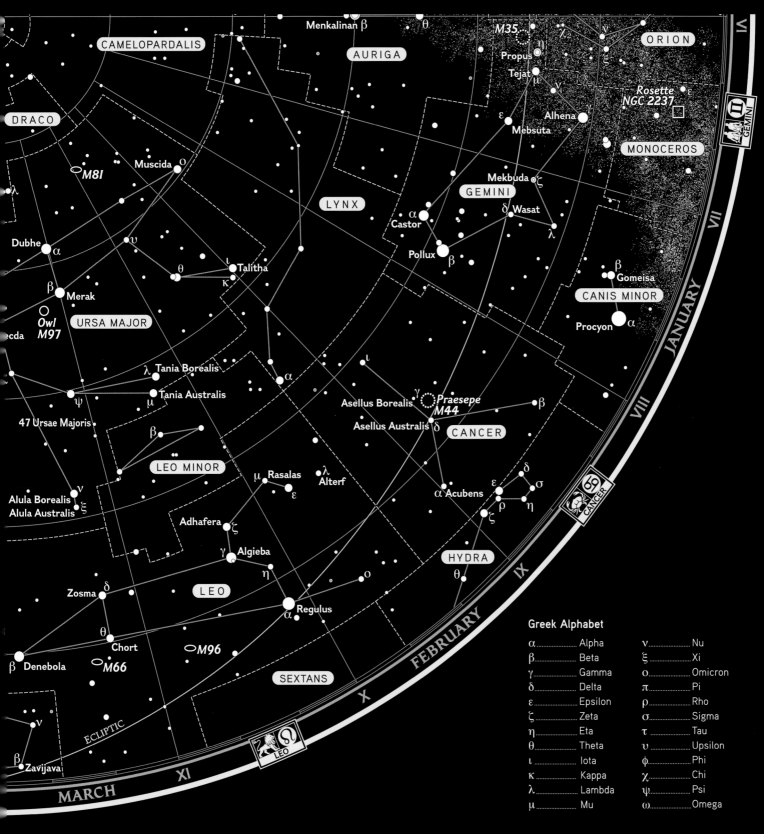

Greek Alphabet

α	Alpha	ν	Nu
β	Beta	ξ	Xi
γ	Gamma	ο	Omicron
δ	Delta	π	Pi
ε	Epsilon	ρ	Rho
ζ	Zeta	σ	Sigma
η	Eta	τ	Tau
θ	Theta	υ	Upsilon
ι	Iota	φ	Phi
κ	Kappa	χ	Chi
λ	Lambda	ψ	Psi
μ	Mu	ω	Omega

CARTOGRAPHER'S NOTE: The easiest way to find an object is to start with a familiar shape, such as the Big Dipper, and orient yourself from there. Note that stars, like the moon, appear to travel from east to west through the sky during the night as Earth rotates.

SOUTHERN SKY | SUMMER–FALL

Stars and constellations on this map are highest in the Southern Hemisphere sky from July to December.

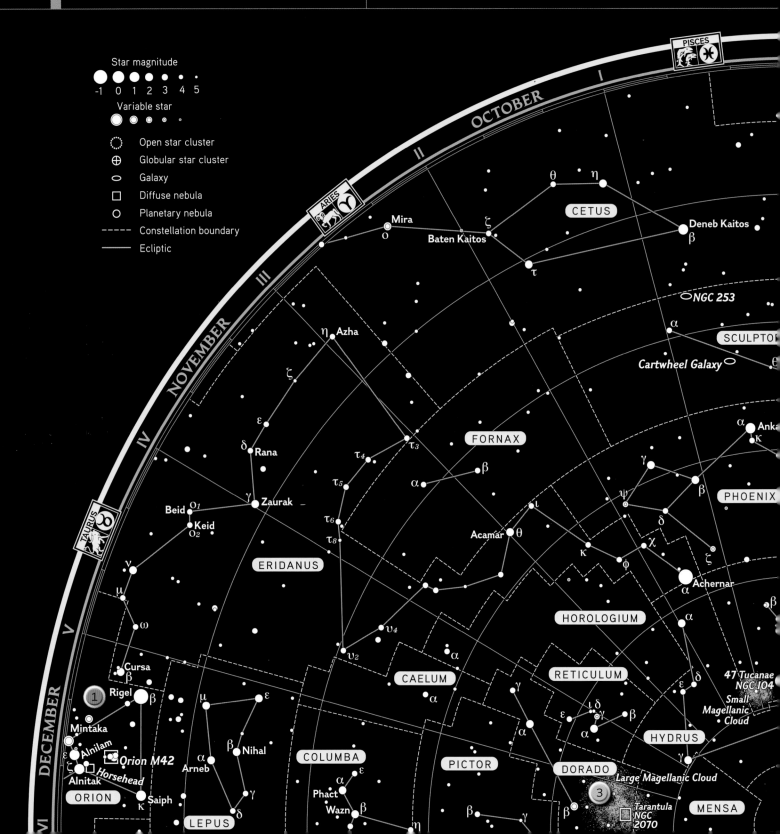

Star magnitude

-1 0 1 2 3 4 5

Variable star

◌ Open star cluster
⊕ Globular star cluster
○ Galaxy
□ Diffuse nebula
○ Planetary nebula
---- Constellation boundary
— Ecliptic

PISCES

OCTOBER

NOVEMBER

DECEMBER

ARIES

TAURUS

CETUS

θ η
ζ

Mira
ο
Baten Kaitos

Deneb Kaitos
β
τ

NGC 253
α
SCULPTOR

η Azha

ζ

Cartwheel Galaxy ○

ε

FORNAX

α Anka
κ

δ Rana
τ₄
τ₃
β
γ

τ₅
α

ψ
β
PHOENIX

Beid ο₁
Keid
ο₂
γ Zaurak

τ₆

τ₈

Acamar θ
κ
φ
χ
ζ

ERIDANUS

δ

γ
μ
υ₄
κ
α Achernar

HOROLOGIUM
α

ω
υ₂

CAELUM
α
γ

RETICULUM

47 Tucanae
NGC 104
Small
Magellanic
Cloud

Cursa
β
Rigel
β

Mintaka
μ
ε
β Nihal
COLUMBA

ε
δ γ
β

HYDRUS
γ

Alnilam
Orion M42
α Arneb
Horsehead
Alnitak
ORION
κ Saiph
γ
δ
LEPUS

PICTOR
α

Phact
Wazn β

DORADO
Large Magellanic Cloud
③
Tarantula
NGC
2070

MENSA

KEY FEATURES

1. ORION: One of the most prominent constellations, visible north and south
2. HELIX NEBULA: A large and spectacular planetary nebula
3. LARGE MAGELLANIC CLOUD: One of two galaxies visible from the Southern Hemisphere

SEPTEMBER

XXIII

PISCES

ECLIPTIC

AQUARIUS

η ζ
κ Sadachbia
Sadalmelik
α

λ

XXII

AUGUST

Ancha
θ

τ₂
δ Skat

2 Helix
NGC 7293
υ

Sadalsuud
β

XXI

α ε
Fomalhaut

PISCIS AUSTRINUS
δ γ
β

β

δ Deneb Algedi
Nashira
γ

M30 ⊕ ε

ζ

ν Saturn
NGC 7009
Albali
ε

CAPRICORNUS

θ

XX

μ
ι

θ
ι

δ₁ μ₁ λ
δ₂ γ
β
GRUS

α
Al Na'ir
ε

MICROSCOPIUM

α

ω ψ

β
Dabih
α
Algedi

AQUILA

ι

JULY

γ
α ε δ

TUCANA

INDUS

α

η

θ₁ θ₂

ι

λ

SAGITTARIUS

β

XIX

β
Peacock
α

Rukbat
β₁
Arkab β α
δ γ

ζ τ
Ascella
σ
Nunki
φ

SCUTUM
α

SAGITTARIUS

γ
β

TELESCOPIUM

δ

ν

PAVO

ε

CORONA
AUSTRALIS

M22 ⊕

λ Kaus Borealis

Omega □ □ Eagle M16
M17

η

SERPENS

OCTANS

λ

ζ
ξ

π

θ

Kaus
Australis
ε
ζ
α

δ

γ
Alnasl

Kaus Media
μ
Lagoon
M8 ▫
Trifid
M20

XVIII

CARTOGRAPHER'S NOTE: Stars and constellations around the edges of this map will be visible
from much of the Northern Hemisphere. There is no prominent southern polestar, but the
southern sky features the brightest stars and some of the most dramatic celestial bodies.

SOUTHERN SKY | WINTER–SPRING

Stars and constellations on this map are highest in the
Southern Hemisphere sky from January to June.

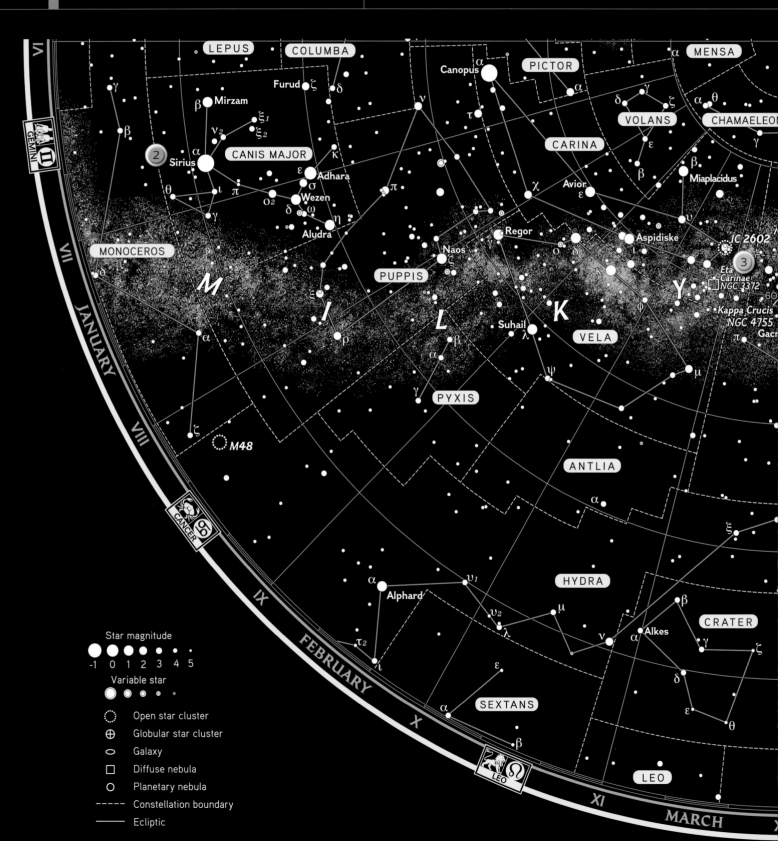

LEPUS   COLUMBA   PICTOR   MENSA

Canopus   α

Furud   ζ   δ   ν   τ   α   VOLANS   γ   α   θ   CHAMAELEON

Mirzam   β   CARINA   δ   ζ   γ

ν₂   ξ₁   CANIS MAJOR   ξ₂   β   ε

α   κ   χ   Avior   β   Miaplacidus

Sirius   ε   Adhara   ε

θ   ι   π   σ   Regor   Aspidiske   IC 2602   ν

γ   o₂   Wezen   π   Naos   γ   o   ι   Eta Carinae   3

δ   ω   η   ζ   κ   NGC 3372   60

MONOCEROS   Aludra   PUPPIS   VELA   Kappa Crucis

M   Suhail   φ   NGC 4755   Gacr

I   L   λ   K   π

α   ρ   β   ψ   μ

α

γ   PYXIS

Star magnitude   M48   ANTLIA   ξ

-1 0 1 2 3 4 5   α

Variable star

⊙ ● ● ● ● ·   α   υ₁   HYDRA   β

○ Open star cluster   Alphard   υ₂   CRATER

⊕ Globular star cluster   λ   μ   Alkes

○ Galaxy   τ₂   ν   α   γ   ζ

□ Diffuse nebula   ι   δ

○ Planetary nebula   α   SEXTANS   ε

----- Constellation boundary   ε   θ

—— Ecliptic   β   LEO

KEY FEATURES

1. SOUTHERN CROSS (CRUX): A prominent constellation visible only from the Southern Hemisphere

2. SIRIUS: Brightest star in the sky (after the sun)

3. ETA CARINAE: An unstable massive star, likely to end in a supernova

Greek Alphabet

α	Alpha	ν	Nu
β	Beta	ξ	Xi
γ	Gamma	ο	Omicron
δ	Delta	π	Pi
ε	Epsilon	ρ	Rho
ζ	Zeta	σ	Sigma
η	Eta	τ	Tau
θ	Theta	υ	Upsilon
ι	Iota	φ	Phi
κ	Kappa	χ	Chi
λ	Lambda	ψ	Psi
μ	Mu	ω	Omega

CARTOGRAPHER'S NOTE: Many star names are from the Greek, including Sirius, which means "searing." The large number of Arabic star names, such as Algorab, which means "the crow," reflects the prominent role of Arab astronomers in charting the sky.

SPACE

Glowing in infrared light, the nebula RCW 120 is illuminated by the radiation from two supermassive stars in its center (not visible in this image). The stars' light and solar winds heat dust inside the nebula (colored red) as well as tiny dust grains at its edges (colored green).

The Chinese say that a journey of a thousand miles begins with a single step. It is fitting, then, that we begin our exploration of the universe in our own astronomical backyard—our own solar system. In this section you will find amazing maps of our nearest neighbors, with a level of detail that would have been unimaginable a generation ago. They are the result of a new method of exploration: the space probe. Every planet in our system has been visited by one or more spacecraft, some of them actually landing on the planet (as on Mars and Venus) and some simply sending back images (as for Jupiter and Saturn). We have explored not only the planets but also their moons. We have come to realize that every world in our system has its own unique

THE SOLA

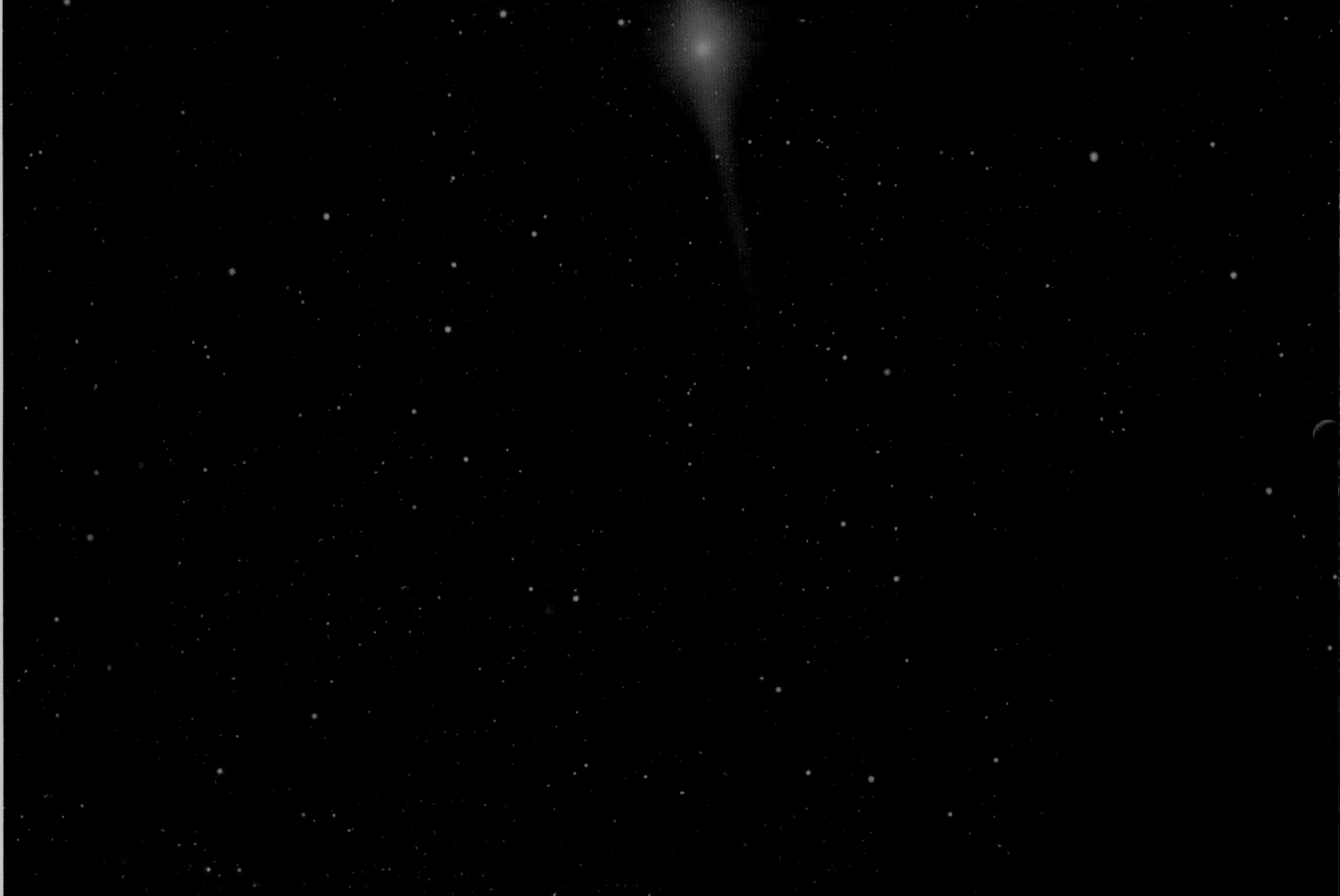

story to tell. The old notion that Mars and Venus are places where life could have developed (and might possibly still exist) has been replaced by a focus on the frigid but possibly life-sustaining moons of Jupiter and Saturn, and this change is reflected in the amount of attention this section gives to these moons.

Finally, modern research has extended our idea of the limits of the solar system beyond the orbit of Pluto to what are called the Kuiper belt and Oort cloud. Planet-size worlds have been discovered out there, and we are now seeing the inner planets as just a small part of the entire system. The well-publicized "demotion" of Pluto is a result of this new way of looking at our home system.

R SYSTEM

SOLAR SYSTEM

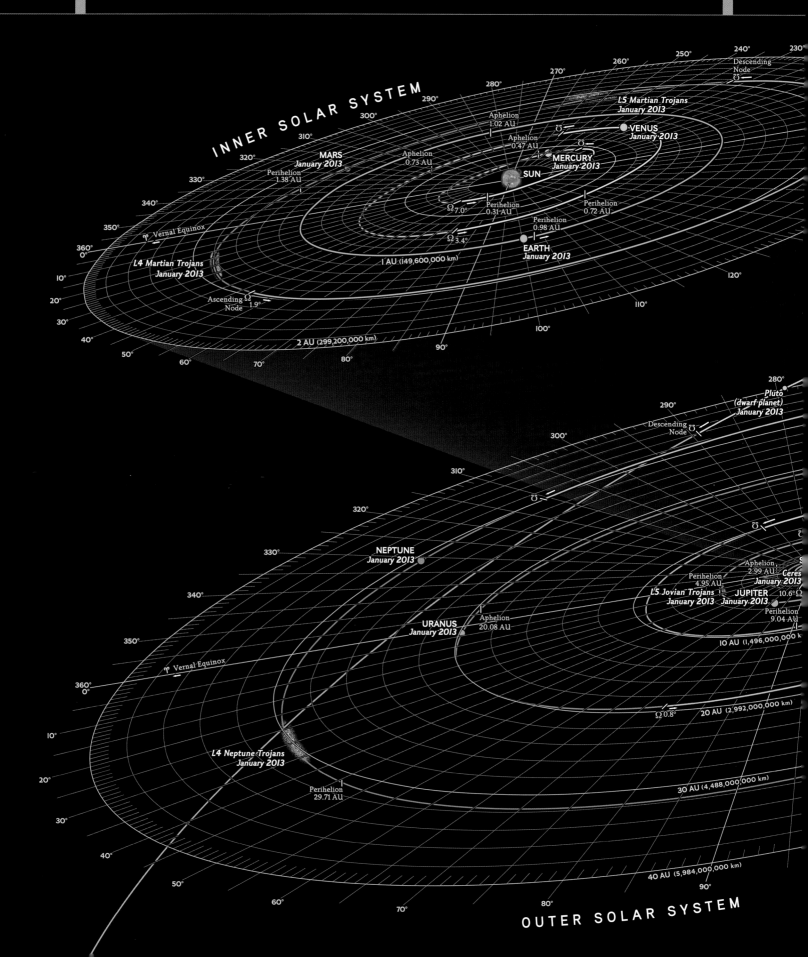

INNER SOLAR SYSTEM

360°
0°
350°
340°
330°
320°
310°
300°
290°
280°
270°
260°
250°
240°
230°

Descending Node

L5 Martian Trojans
January 2013

VENUS
January 2013

MARS
January 2013

Aphelion
1.02 AU

Aphelion
0.47 AU

Aphelion
0.73 AU

MERCURY
January 2013

Perihelion
1.38 AU

SUN

Perihelion
0.31 AU

Perihelion
0.72 AU

Ω 7.0°

Perihelion
0.98 AU

Vernal Equinox

Ω 3.4°

EARTH
January 2013

1 AU (149,600,000 km)

L4 Martian Trojans
January 2013

Ascending Ω
Node 1.9°

2 AU (299,200,000 km)

10°
20°
30°
40°
50°
60°
70°
80°
90°
100°
110°
120°

280°
290°
300°
310°
320°
330°
340°
350°
360°
0°
10°
20°
30°
40°
50°
60°
70°
80°
90°

Pluto
(dwarf planet)
January 2013

Descending
Node

NEPTUNE
January 2013

Aphelion
2.99 AU
Ceres
January 2013

Perihelion
4.95 AU

L5 Jovian Trojans
January 2013

JUPITER
January 2013

10.6° Ω

Perihelion
9.04 AU

URANUS
January 2013

Aphelion
20.08 AU

10 AU (1,496,000,000 k

Vernal Equinox

Ω 0.8°

20 AU (2,992,000,000 km)

L4 Neptune Trojans
January 2013

Perihelion
29.71 AU

30 AU (4,488,000,000 km)

40 AU (5,984,000,000 km)

OUTER SOLAR SYSTEM

Eight planets, five dwarf planets, well over a hundred moons, and countless asteroids and comets orbit our massive sun. The four terrestrial planets form a relatively compact family in the inner solar system. Across the asteroid belt, the large, gaseous outer planets grow increasingly remote from the sun's warmth. All the planets orbit on roughly the same plane as Earth, known as the ecliptic. Pluto, now designated a dwarf planet, is an exception.

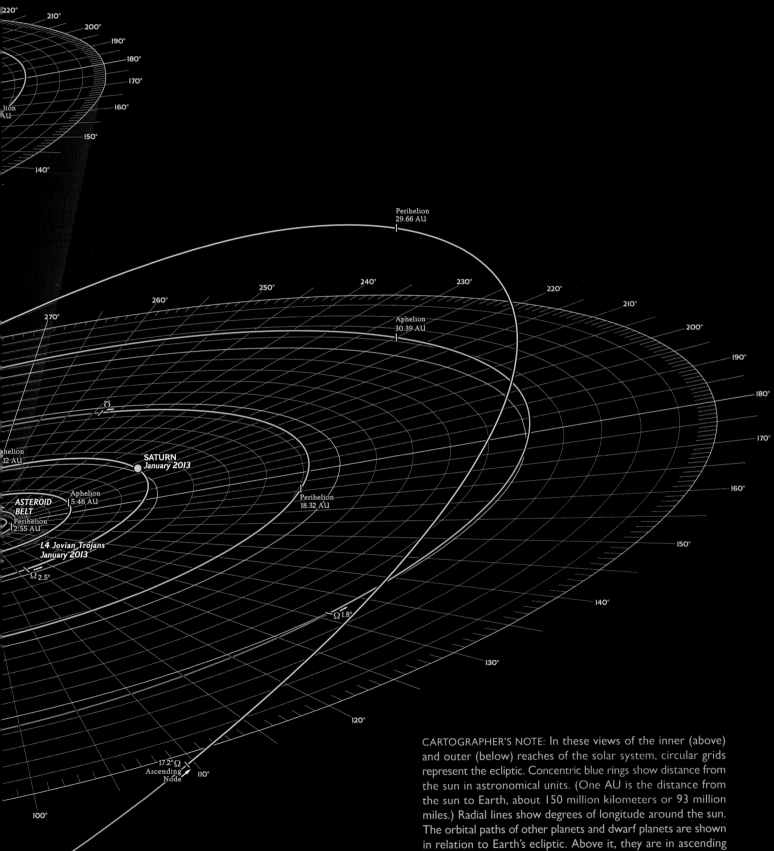

220°
210°
200°
190°
180°
170°
160°
150°
140°

lion
AU

Perihelion
29.66 AU

250°   240°   230°   220°
260°   210°
270°   200°

Aphelion
30.39 AU   190°

180°

170°

helion
12 AU

SATURN
January 2013

Aphelion
5.46 AU

ASTEROID
BELT
Perihelion
2.55 AU

Perihelion
18.32 AU

160°

L4 Jovian Trojans
January 2013

Ω 2.5°

Ω 1.8°

150°

140°

130°

120°

17.2° Ω
Ascending
Node   110°

100°

CARTOGRAPHER'S NOTE: In these views of the inner (above) and outer (below) reaches of the solar system, circular grids represent the ecliptic. Concentric blue rings show distance from the sun in astronomical units. (One AU is the distance from the sun to Earth, about 150 million kilometers or 93 million miles.) Radial lines show degrees of longitude around the sun. The orbital paths of other planets and dwarf planets are shown in relation to Earth's ecliptic. Above it, they are in ascending mode, with their paths shown as a solid line; below it, they are in descending mode, with broken lines.

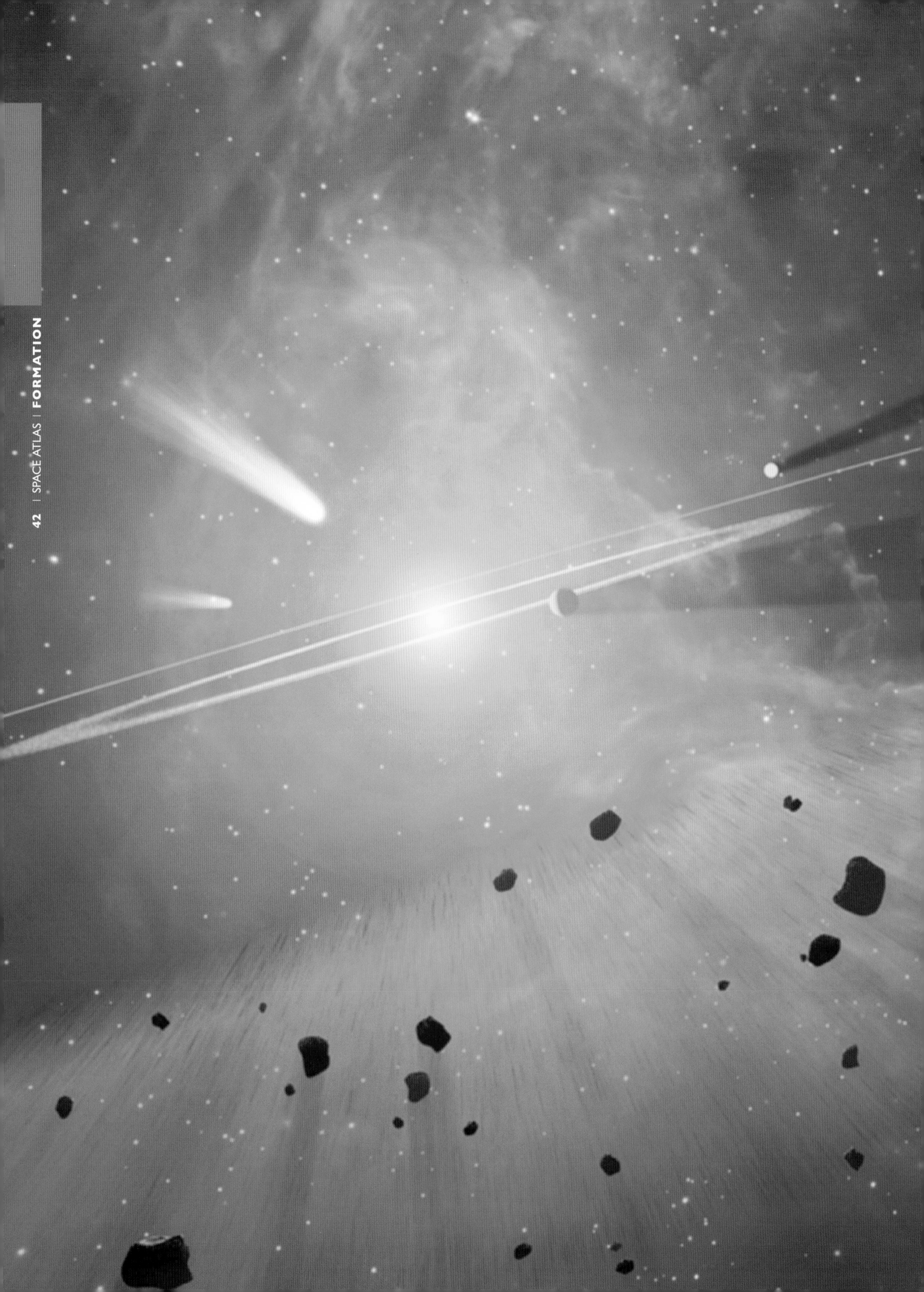

t all started about 4.5 billion years ago with a huge interstellar cloud floating in space. Today we see an orderly collection of planets circling a rather ordinary star. The question: How did we get here from there? • Looking at our familiar solar system, we can see regularities that give us hints about the answer to this question: First, all the planets orbit in the same plane; second, all the planets orbit in the same direction; and third, the planets closest to the sun are small and rocky, while those farther out are gas giants. The explanation for these regularities—and many others—begins with work by scientists in the 18th century, most notably the French physicist Pierre-Simon Laplace (1749–1827).

FORMATION

[BIRTH OF THE SOLAR SYSTEM]

AGE: 4.5–4.6 BILLION YEARS

DISTANCE FROM CENTER OF MILKY WAY: 28,000 LIGHT-YEARS

TYPE OF STAR: MAIN SEQUENCE, G2-V

MAIN ELEMENTS: HYDROGEN, HELIUM, OXYGEN, CARBON, NITROGEN

NUMBER OF PLANETS: 8

TERRESTRIAL PLANETS: MERCURY, VENUS, EARTH, MARS

GAS AND ICE GIANTS: JUPITER, SATURN, URANUS, NEPTUNE

NUMBER OF DWARF PLANETS: 5

NUMBER OF MOONS: 169

DISTANCE FROM SUN TO NEPTUNE'S ORBIT: 30 AU

DISTANCE FROM SUN TO EDGE OF OORT CLOUD: ~100,000 AU

Artist's conception of a planetary system forming around the star Epsilon Eridani. (Inset) Art of young solar system

Laplace reasoned that the ordinary laws of gravity would produce something like our solar system if they operated on diffuse interstellar clouds—the objects in the sky that scientists at the time christened nebulae (singular: nebula), Latin for "clouds." Easily observed with telescopes, these cloudy patches of light can be seen everywhere in the night sky, and Laplace's theory of how they could evolve into a solar system was duly named the nebular hypothesis. With many details filled in, it is essentially our modern theory.

Our understanding of how the solar system formed begins with an examination of its interstellar cloud. Like the others of its kind, it was made up primarily of the primordial gases created in the big bang—hydrogen and helium—with a small admixture of heavier elements produced in stars (see pages 277–283). Modern work indicates that one or more large stars exploded in our interstellar cloud, producing regions where the mass was more concentrated. These mass concentrations exerted a strong gravitational force, pulling in surrounding material. Eventually, the cloud (originally some tens of light-years across) began to break up and collapse around the places where these mass concentrations started, and one of those accretions of matter, which we now call the pre-solar nebula, eventually became our own solar system. As the gases collapsed, the nebula began to rotate. Laplace's picture of how this nebula became the solar system depicts an orderly, rather placid process in which the force of gravity and the effects of heat radiated by the sun produced the system we see today, more or less in its modern form. As we shall see, notions about how orderly the process actually was have changed drastically in the last few years.

SUN AND FROST

Of course, gravity never quits, and it continued its work once the pre-solar nebula formed. Two important things happened as the inward collapse progressed: First, most of the mass of the pre-solar nebula became concentrated at the center, where it eventually became the star we call the sun. Second, as the cloud contracted, its spin increased, much as a skater's spin increases when she pulls in her arms. The various forces acting on the spinning, contracting cloud—gravity, pressure, centrifugal force, and even magnetism—caused the small amount

CLUES TO EARTH'S FORMATION

People often ask how scientists can know about events like the formation of Earth, which took place billions of years ago. Let's look at one part of that story—the differentiation of the planet into core, mantle, and crust described on page 76. Unraveling this process is a fascinating scientific detective story.

The story starts in the dust cloud from which the solar system formed. In that cloud were a certain number of hafnium-182 atoms. (Hafnium is a relatively rare material, usually seen as a silvery gray metal.) The nuclei of these atoms are unstable, decaying with a half-life of about nine million years into atoms of tungsten-182, which is stable. (Tungsten is the metal customarily used as the filament in incandescent lightbulbs.) What makes these atoms interesting is that hafnium is chemically attracted to the kinds of materials found in Earth's mantle, while tungsten is attracted to the iron and nickel found in the core. This means that if the iron materials sank into the core quickly, before the hafnium had time to decay, most of the tungsten-182 would be found in the mantle. If, on the other hand, the differentiation happened after the hafnium had all decayed, the tungsten-182 would be in the core.

By comparing the amount of tungsten-182 in mantle rocks to the amount found in meteorites (which never underwent differentiation), scientists have concluded that Earth's core formed about 30 million years after the solar system started to condense from its gas cloud.

One small piece of the puzzle in place.

of material that hadn't been taken into the nascent sun to flatten out into a rotating disk surrounding the central sphere. With the formation of this disk, the solar system was starting to take shape.

As the disk was forming, the sun was also firing up, heating nearby material. Out to a place between the present orbits of Mars and Jupiter, the temperature was hot enough that volatile materials like water and methane couldn't exist in solid form. Beyond this boundary, which astronomers refer to loosely as the "frost line," these materials remained as solid ices. Thus, the building blocks available for the inner planets were different from those available for planets beyond the frost line. It's no wonder, then, that the so-called terrestrial (Earthlike) planets near the sun are different from the Jovian (Jupiter-like) planets farther out.

Our main tools for understanding what happened from that point on are massive computer simulations of the early solar system—simulations that take into account the kinds of forces mentioned above as well as the effects of the sun's energy output and radiation. The descriptions that follow are largely a summary of these calculations.

TERRESTRIAL PLANETS

Let's start with the terrestrial planets. Because the light-weight, volatile materials had mostly been removed from the inner disk, these planets formed primarily from materials with high melting points (think iron, nickel, and the rocky compounds of silicon). As they circled the sun, grains of these materials collided and stuck together, eventually forming boulder-size objects that then aggregated into mountain-size bodies called planetesimals. It is these bodies that eventually came together to form the planets.

Until the last decade of the 20th century, it was assumed that the planets formed in pretty much their present orbits and states. This scenario is not, however, how the computers tell us things happened. In fact, the picture they give us is truly staggering. The end result of the process described above was an inner solar system with dozens of moon-size planetary embryos zipping around. What followed was an impossible game of planetary billiards, with the embryos colliding, melding, breaking apart, and, occasionally, getting kicked out of the solar system completely. As we shall see later, we believe

PLANET-BUILDING DUST AROUND A YOUNG STAR

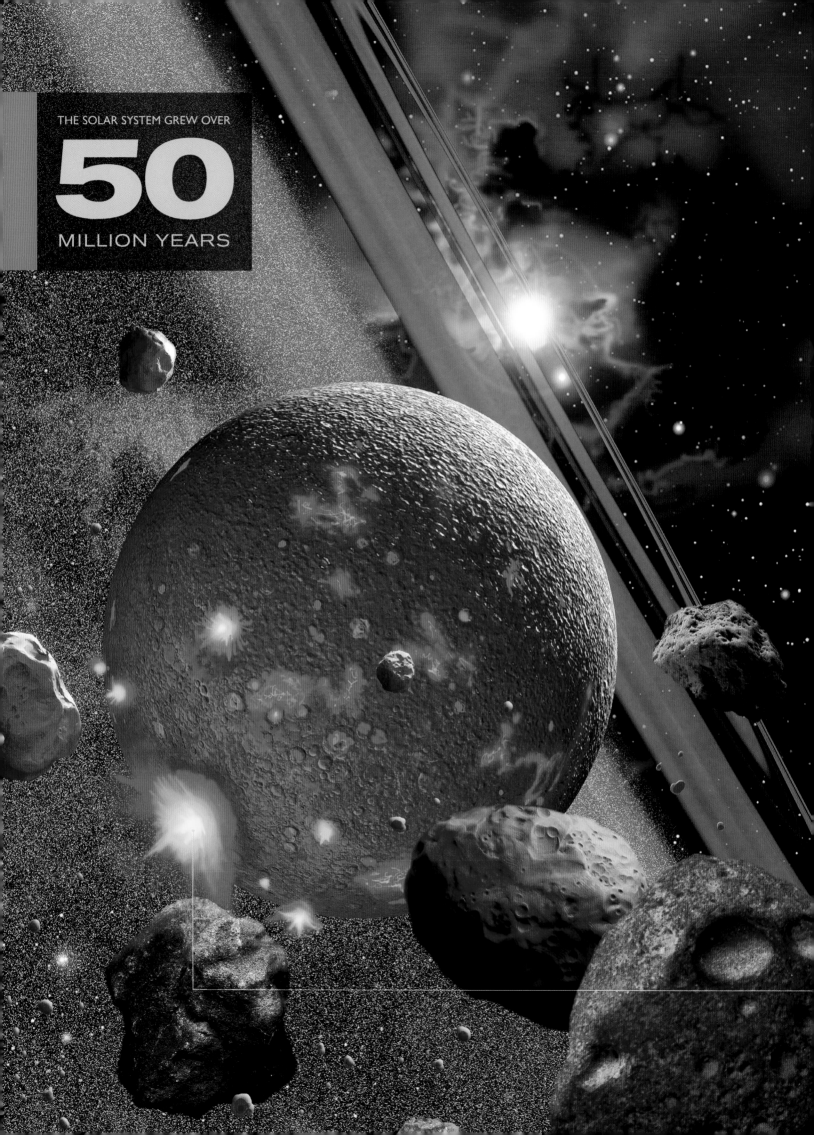

THE SOLAR SYSTEM GREW OVER

50

MILLION YEARS

that Earth's moon was created in just such a collision, and scientists have suggested that another late collision stripped Mercury of its lightweight outer coating, leaving the small, dense planet we see today.

In any case, regardless of the details, by the time the game of planetary billiards was over, the inner solar system was left with the four planets—Mercury, Venus, Earth, and Mars—that we see today.

GIANT PLANETS

Meanwhile, a different scenario was playing out beyond the frost line. Because there was so much undisturbed material out there, the planetesimals grew more quickly and to a larger size than was the case for the terrestrial planets. The great mass of these bodies allowed them to start capturing the abundant hydrogen and helium that surrounded them. These planets are known as gas giants, particularly Jupiter and Saturn, and they are the largest planets in our system.

The subsequent story of the outer planets is somewhat more complicated than it is for the terrestrials. The gas giants Jupiter and Saturn formed quickly, as described above. Apparently the next two planets—Uranus and Neptune—formed later and much closer to the sun than they are now. They also formed at a time when the sun was emitting huge streams of particles into space, streams that blew much of the primordial hydrogen and helium out of the solar system. Consequently, these two planets wound up being smaller and different in composition from Jupiter and Saturn. In fact, they are often referred to as ice giants rather than gas giants to emphasize this difference.

An illustration depicts the violent formation of Earth during the solar system's early years, as the inner planets were bombarded by countless planetesimals and heated by the collisions.

The four giant planets as well as all of the leftover planetesimals and other materials continued in their orbits, interacting with each other through the force of gravity. Then, about 500 million years after the whole scenario started (that is, about 4 billion years ago), Jupiter and Saturn fell into what scientists call a 2:1 resonance: Jupiter started to orbit the sun twice for every time Saturn orbited once. The gravitational effects of this kind of situation are enormous. Neptune's orbit was pushed outward, sending it careering into the remnants of the protoplanetary disk like a bowling ball among tenpins. At that time, the disk extended to about the present orbit of Uranus, but by the time the planetary migrations were over, the system was clear out to well past the present orbit of Pluto.

Another effect of the migrations was to move Jupiter closer to the sun, a shift that had the effect of disturbing the orbits of objects in the asteroid belt between that planet and Mars. Much of that material was ejected from the solar system, leaving the asteroid belt in more or less its present condition. As a result of all these moves, there was a period of a couple of hundred million years when every body in the inner solar system suffered intense collisions. Scientists have taken to calling this period the Late Heavy Bombardment. Its scars can be seen in the craters that still survive on airless worlds like Mercury and the moon, although any craters that might have formed on Earth would have long since weathered away.

In any case, it has become clear to astronomers over the last few decades that the early life of the solar system was far from the placid, orderly collapse that Laplace had in mind in the 18th century. But once the initial fireworks died down, the solar system became a much more orderly and predictable kind of place—just the sort of place we need in order to begin our tour of the first of our "universes."

INNER PLANETS

M A R S

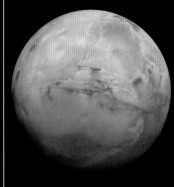

Average distance from the sun:	227,900,000 km
Perihelion:	206,620,000 km
Aphelion:	249,230,000 km
Revolution period:	687 days
Average orbital speed:	24.1 km/s
Average temperature:	-65°C
Rotation period:	24.6 hours
Equatorial diameter:	6,792 km
Mass (Earth=1):	0.107
Density:	3.93 g/cm^3
Surface gravity (Earth=1):	0.38
Known satellites:	2
Largest satellites:	Phobos, Deimos

Image by: Mars Global Surveyor

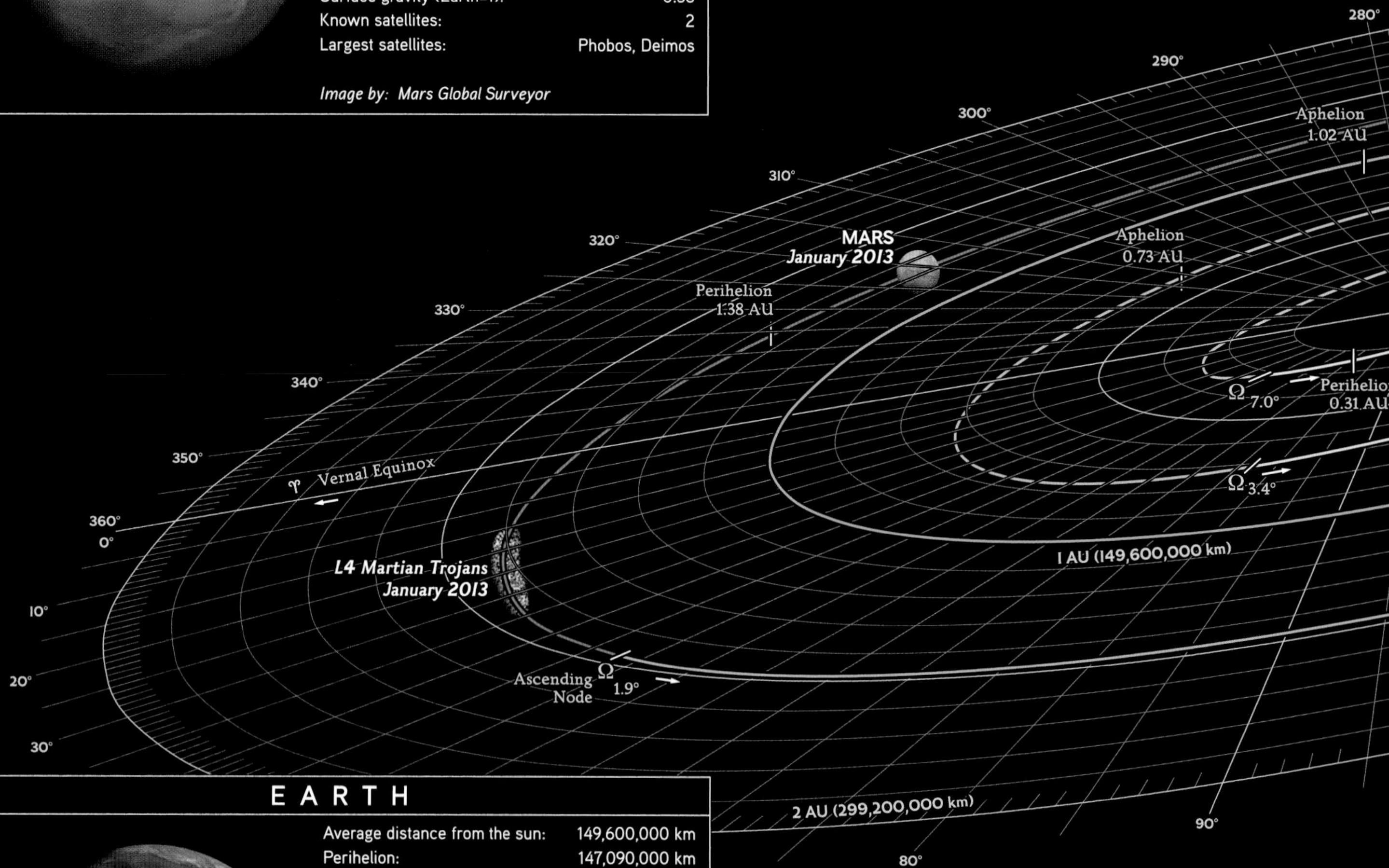

280°
290°
300°
Aphelion
1.02 AU
310°

320°
MARS
January 2013
Aphelion
0.73 AU

330°
Perihelion
1.38 AU

340°
Ω 7.0°
Perihelion
0.31 AU

350°
Ω 3.4°

Υ Vernal Equinox

360°
0°
1 AU (149,600,000 km)

10°
L4 Martian Trojans
January 2013

20°
Ascending Ω 1.9°
Node

30°

E A R T H

Average distance from the sun:	149,600,000 km
Perihelion:	147,090,000 km
Aphelion:	152,100,000 km
Revolution period:	365.2 days
Average orbital speed:	29.8 km/s
Average temperature:	15°C
Rotation period:	23.9 hours
Equatorial diameter:	12,756 km
Mass:	5,973,600,000,000,000,000,000 metric tons
Density:	5.52 g/cm^3
Surface gravity:	9.78 m/s^2
Known satellites:	1
Largest satellite:	Earth's moon

Image by: Galileo Orbiter

2 AU (299,200,000 km)
90°
80°

The terrestrial planets, small and rocky, form the inner core of the solar system. All possess a secondary atmosphere (appearing after their formation), though Mercury's is barely detectable. Craters from the era of early bombardment mark Mercury and Earth's moon; atmospheric weathering and the effects of volcanism or plate tectonics have erased many of these early traces from Venus and Earth.

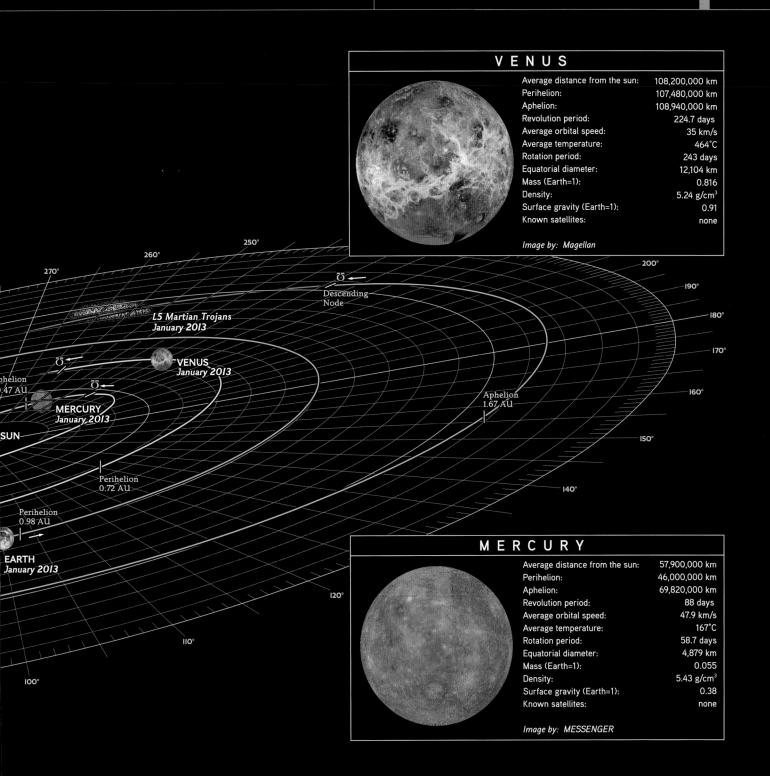

VENUS

Average distance from the sun:	108,200,000 km
Perihelion:	107,480,000 km
Aphelion:	108,940,000 km
Revolution period:	224.7 days
Average orbital speed:	35 km/s
Average temperature:	464°C
Rotation period:	243 days
Equatorial diameter:	12,104 km
Mass (Earth=1):	0.816
Density:	5.24 g/cm³
Surface gravity (Earth=1):	0.91
Known satellites:	none

Image by: Magellan

MERCURY

Average distance from the sun:	57,900,000 km
Perihelion:	46,000,000 km
Aphelion:	69,820,000 km
Revolution period:	88 days
Average orbital speed:	47.9 km/s
Average temperature:	167°C
Rotation period:	58.7 days
Equatorial diameter:	4,879 km
Mass (Earth=1):	0.055
Density:	5.43 g/cm³
Surface gravity (Earth=1):	0.38
Known satellites:	none

Image by: MESSENGER

CARTOGRAPHER'S NOTE: The orbits of the inner planets take them from the scorching temperatures of Mercury to the deep winter chill of Mars. Swift Mercury races around the sun, its hemispheres burning and freezing. Torrid Venus bakes under an atmosphere that holds in most of the sun's energy, the greenhouse effect writ large. Uniquely sited, Earth is in the habitable zone, where water can exist as a liquid. Frigid Mars offers tantalizing evidence of a warmer, wetter past. Well studied in the space age, each terrestrial planet has been visited and mapped in detail by spacecraft.

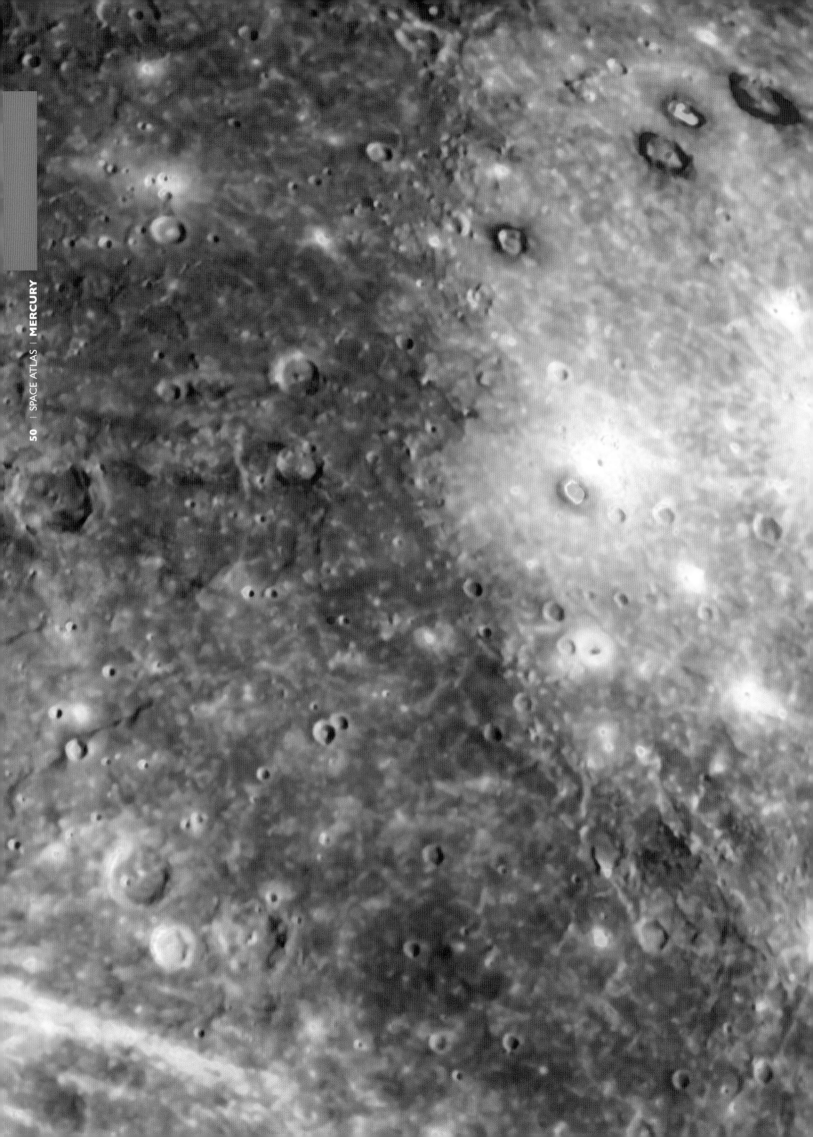

Mercury is the closest planet to the sun, which means that observers on Earth will always see it near the sun in the sky. During the day, light from the planet is overwhelmed by light from the sun, so we can see it only at dawn or sunset when the sun is safely below the horizon. The planet can never appear high in the night sky, for the simple reason that in that situation it will always be on the other side of Earth from the observer. Like Venus, then, Mercury always appears as a morning or evening "star" to observers. Records of naked-eye sightings of Mercury by Assyrian astronomers go back to the 14th century B.C., and by the 4th century B.C. the Greeks had realized that their morning and evening stars were actually a single body.

MERCURY

[SWIFT, SUN-BLASTED WORLD]

DISCOVERER: UNKNOWN
DISCOVERY DATE: PREHISTORIC
NAMED FOR: ROMAN MESSENGER GOD

MASS: 0.06 × EARTH'S
VOLUME: 0.06 × EARTH'S
MEAN RADIUS: 2,440 KM (1,516 MI)
MIN./MAX. TEMPERATURE: -173/427°C (-279/801°F)
LENGTH OF DAY: 58.65 EARTH DAYS
LENGTH OF YEAR: 87.97 EARTH DAYS
NUMBER OF MOONS: 0
PLANETARY RING SYSTEM: NO

False-color image of Caloris Basin on
Mercury's surface. (Inset) Mercury

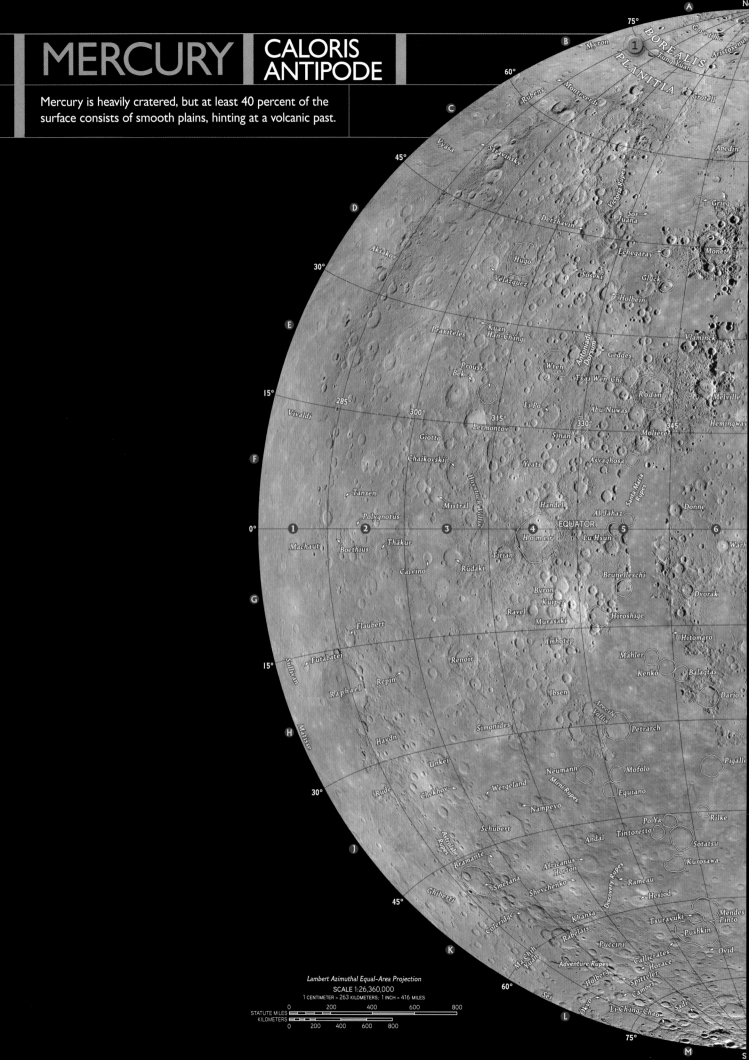

MERCURY | CALORIS ANTIPODE

Mercury is heavily cratered, but at least 40 percent of the surface consists of smooth plains, hinting at a volcanic past.

Lambert Azimuthal Equal-Area Projection
SCALE 1:26,360,000
1 CENTIMETER = 263 KILOMETERS; 1 INCH = 416 MILES

STATUTE MILES
0   200   400   600   800

KILOMETERS
0   200   400   600   800

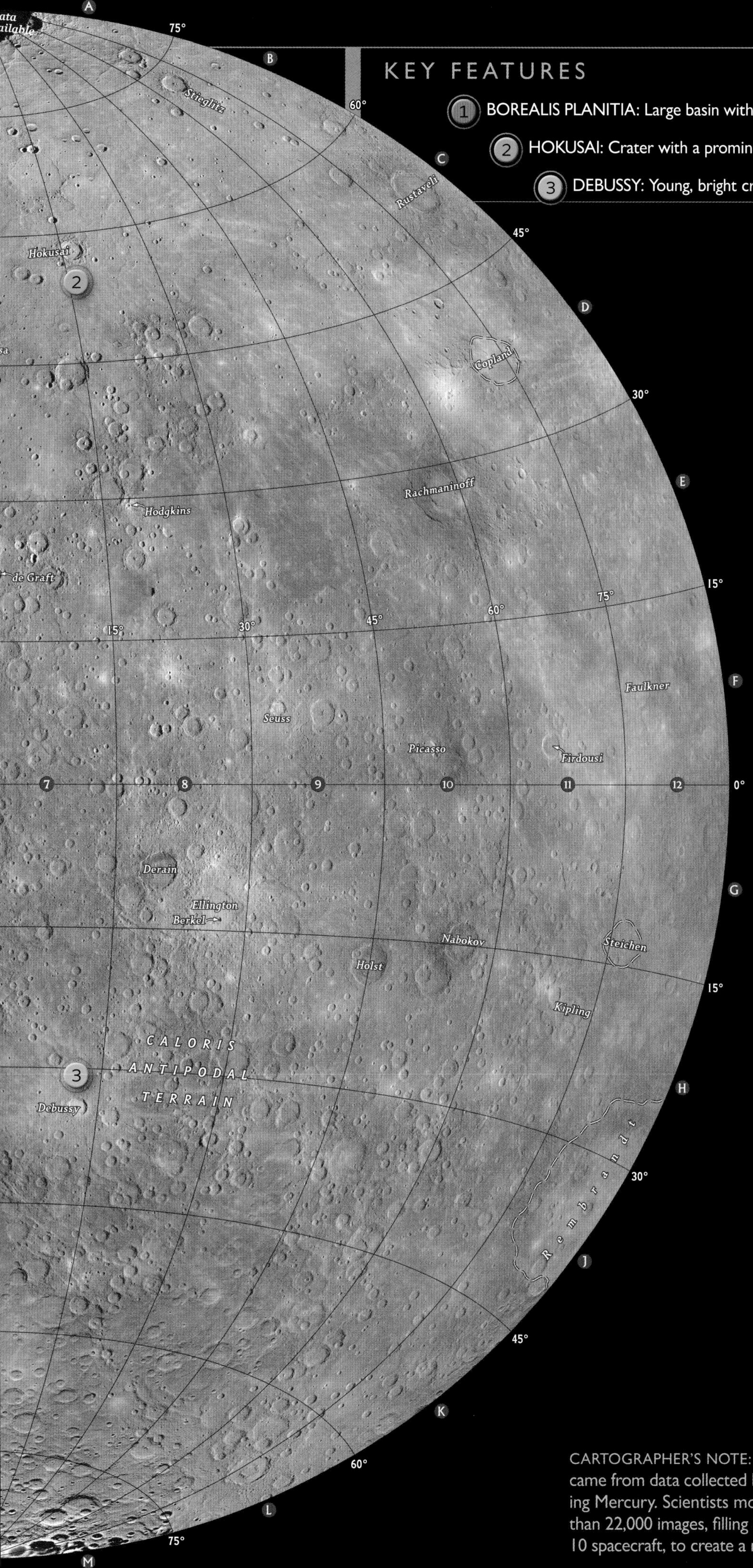

KEY FEATURES

1 BOREALIS PLANITIA: Large basin with a smooth floor

2 HOKUSAI: Crater with a prominent ray system

3 DEBUSSY: Young, bright crater with a prominent ray system

CARTOGRAPHER'S NOTE: The global mosaic that underpins these maps came from data collected by NASA's MESSENGER probe, currently studying Mercury. Scientists modified and combined an amalgamation of more than 22,000 images, filling in any gaps with data from the previous Mariner 10 spacecraft, to create a nearly complete view of the planet's topography.

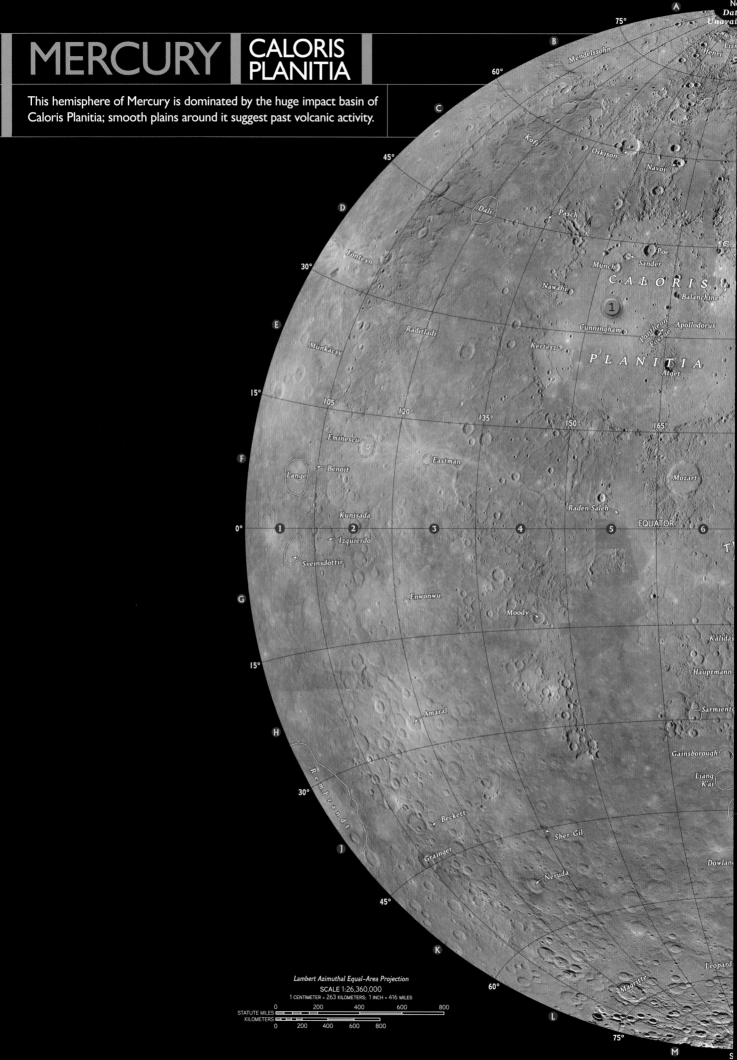

MERCURY | CALORIS PLANITIA

This hemisphere of Mercury is dominated by the huge impact basin of
Caloris Planitia; smooth plains around it suggest past volcanic activity.

CALORIS

PLANITIA

Lambert Azimuthal Equal-Area Projection
SCALE 1:26,360,000
1 CENTIMETER = 263 KILOMETERS; 1 INCH = 416 MILES

STATUTE MILES
0   200   400   600   800
KILOMETERS
0   200   400   600   800

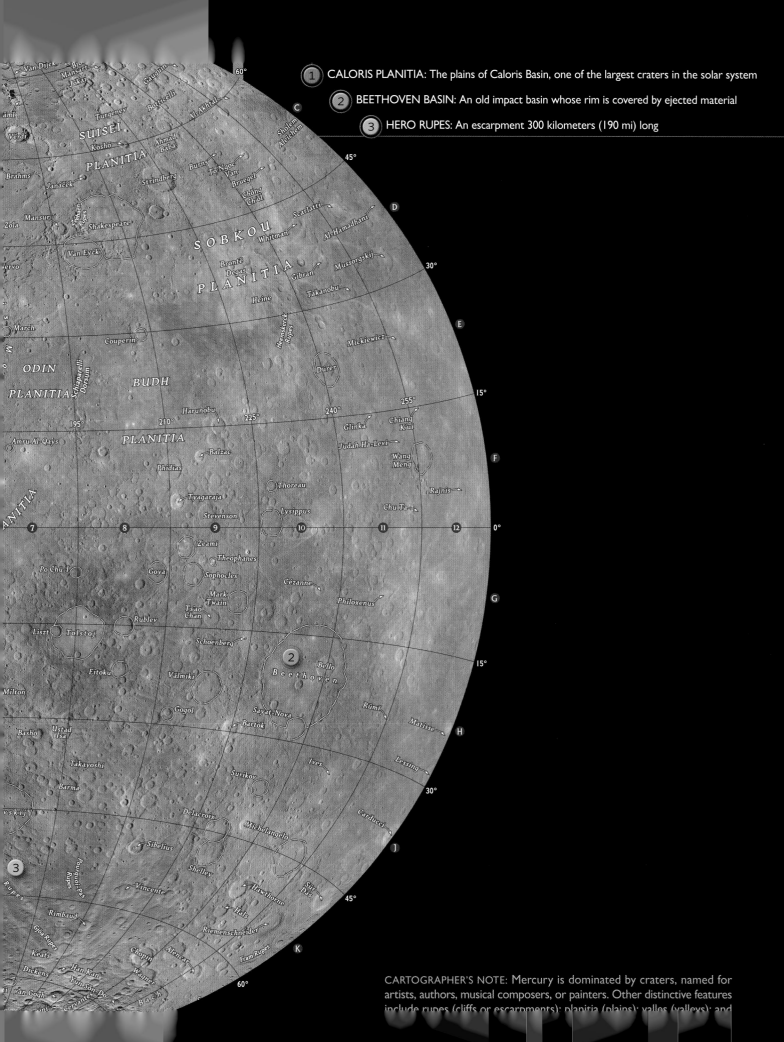

1 CALORIS PLANITIA: The plains of Caloris Basin, one of the largest craters in the solar system

2 BEETHOVEN BASIN: An old impact basin whose rim is covered by ejected material

3 HERO RUPES: An escarpment 300 kilometers (190 mi) long

CARTOGRAPHER'S NOTE: Mercury is dominated by craters, named for artists, authors, musical composers, or painters. Other distinctive features include rupes (cliffs or escarpments); planitia (plains); valles (valleys); and

The Romans gave Mercury its current title, naming it after the swift-footed messenger of the gods (presumably because its rapid motion in the sky imitated the god's swift flight). It is interesting that the Babylonians named the planet Nabu, after the messenger of the gods in their pantheon, probably for the same reason.

HOT AND COLD

Mercury is small—it has only about 5 percent the mass of Earth—and it has long since lost whatever atmosphere it ever had to space and the blistering radiation from the sun. Like our own moon, it is a dead world, with no geological activity to raise mountains or atmosphere to erode its surface. The main features of the planet, like those of the moon, are craters that bear mute testimony to long-ago impacts. The planet rotates on its axis every 176 Earth days and circles the sun every 88 days or so. Thus, every part of Mercury's surface is exposed to both the direct rays of the sun (during its day) and the cold of space (at night).

As you might expect for a planet so close to the sun, Mercury's surface temperatures can get quite hot—427°C (801°F) at the equator at high noon. This is hotter than the melting point of lead. What you might not expect is that it can also get quite cold: minus 173°C (-279°F) at midnight. The reason for this is that once a part of the planet moves into the night side, there is no atmosphere to act as a blanket and keep it warm. The energy accumulated during the day is quickly radiated into space and the temperature falls quickly. (To be exact, there is a thin mist of atoms above Mercury's surface—scientists call it an exosphere. It's made up of atoms boiled off the surface and eventually lost to space.)

CRATERED WORLD

Because the planet has no atmosphere, the craters on Mercury, like those on the moon, last for a long time. The largest of these, the Caloris Basin, is almost 1,600 kilometers (1,000 mi) across, and must be the scar of a very large impact. In fact, on the exact opposite side of the planet from Caloris is a region of jumbled hills that goes by the descriptive name Weird Terrain. Some scientists think that this area was created by a shock wave from the impact that created Caloris. Many of the larger craters on Mercury appear to have a smooth surface similar to the maria on the moon. It is thought that these smooth regions are caused by lava outflows, probably as a result of the impact itself. Between the craters are regions of rolling

A cutaway view shows Mercury's structure. The planet has a large, molten core surrounded by a solid mantle 500 to 700 kilometers (300 to 400 mi) thick. This, in turn, is covered by 100 to 300 kilometers (60 to 180 mi) of crust.

CORE

MANTLE

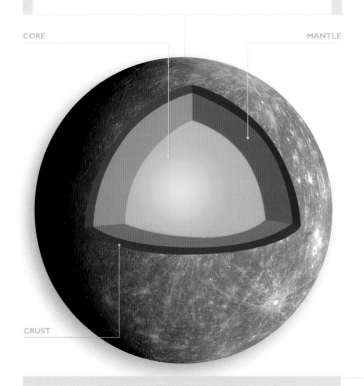

CRUST

Half in sunlight, Mercury's south pole displays the planet's characteristic craters. Like the moon, the planet has almost no atmosphere, so its craters do not erode; thus, some of these structures are billions of years old.

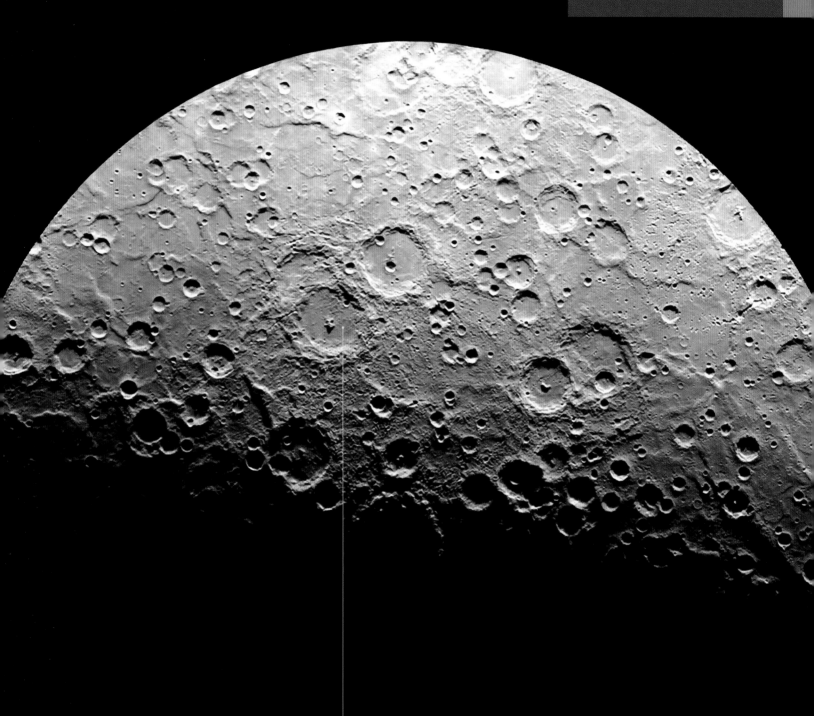

hills, representing Mercury's oldest surviving surface. These plains are crisscrossed with long ridges that may have been created by surface wrinkling as Mercury cooled (think of an apple shriveling as it dries out).

Like the other terrestrial planets, Mercury is a rocky world. It has a small magnetic field, about one percent as strong as the one on Earth, which supports the idea that, like Earth, it has a core composed mainly of iron—you can think of the planet as a giant permanent magnet.

In fact, based largely on data gathered from the space probes described below, scientists believe that the planet has an unusually large iron core, comprising over 42 percent of its volume. Several theories have been advanced to explain this unusual composition. The most popular explanation is that after it had gone through the process of differentiation (see page 76), Mercury was struck by a large planetesimal during the Late Heavy Bombardment about four billion years ago. The collision blew off a lot of the lighter outer layer of the planet, leaving behind the iron in

MERCURY'S WEIRD TERRAIN

Although Mercury is visible to the naked eye, only since the launch of the MESSENGER spacecraft have we been able to explore it in detail. Mercury is an airless world, so geological features, once formed, do not erode. This means that the history of the planet can still be read on its unchanging surface.

By far the most striking feature on Mercury's surface is the Caloris Basin, a crater almost 1,600 kilometers (1,000 mi) across. The collision that formed it is thought to have occurred about 3.8 billion years ago, at about the same time as the impacts that created the maria on Earth's moon. It is one of the largest craters in the solar system.

The power of that impact can be seen in the fact that the walls thrown up around the edges of the crater are over 1.6 kilometers (1 mi) high. More interesting, though, is a jumbled region of irregular hills on the opposite side of Mercury that scientists have given the name Weird Terrain. This region was clearly created by the impact that formed the Caloris Basin. There are two theories to explain how this happened. In one, seismic waves created by the impact traveled around the planet to converge at the antipode, breaking up the previously smooth surface. The other theory is that material thrown up by the impact traveled around the planet and fell back to the surface, creating the irregular overlay we see today.

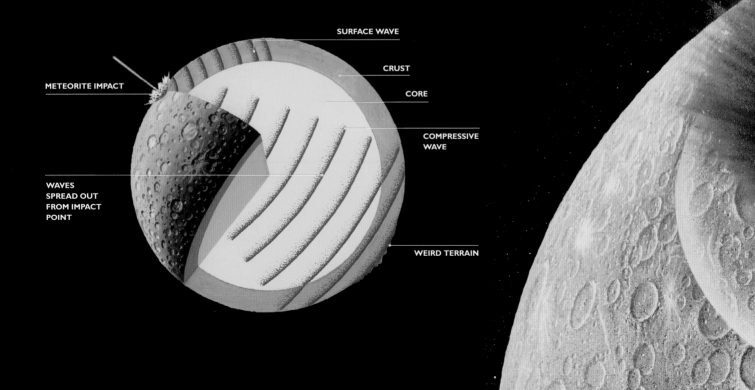

METEORITE IMPACT

WAVES SPREAD OUT FROM IMPACT POINT

SURFACE WAVE

CRUST

CORE

COMPRESSIVE WAVE

WEIRD TERRAIN

the core. Alternatively, the planet may have formed before the sun had settled down, in which case solar activity could have blown lighter materials away from the region where the planet was forming. In this scenario, Mercury would have formed initially in an iron-rich environment.

SPACECRAFT TO MERCURY

Because it is so close to the sun, observing Mercury with ground-based telescopes is difficult. In fact, most of our information about the details of the planet's structure comes from two space probes. The first of these, Mariner 10, arrived at the planet in 1974 and made three close approaches before it ran out of fuel. It gave us photos of about half of Mercury's surface. Most likely the craft is still in orbit around the sun, making undocumented close approaches to the planet as you read this.

The MESSENGER (MErcury Surface, Space ENvironment, GEochemistry, and Ranging) spacecraft was launched from Cape Canaveral in 2004, and, after visiting Venus and Earth, it made its first flyby of Mercury on January 14, 2008. On March 18, 2011, it went into orbit around the planet, where it began to return detailed data about the planet's surface, topography, and magnetic field, producing more than 70,000 images of the small planet's craters and ridges. Mission scientists hope the craft will provide some answers to questions about Mercury's unusual density and the structure of its core. A European Space Agency probe called Bepi Columbo is expected to arrive at Mercury in 2019, and will represent the next wave of exploration of the planet.

We can't leave the planet without mentioning what may be its greatest contribution to the advance of science. Like the other planets, Mercury orbits the sun in an elliptical path, with the point of closest approach being called the perihelion. Because of the gravitational pull of the other planets, scientists expected that each time the planet came around, the perihelion would shift a little—think of the ellipse rotating a little in space each time around the sun. In the late 19th century, calculations showed that there was more shift than could be explained by simple gravitational effects—about 43 seconds of arc per century, in fact. When Albert Einstein published the theory of general relativity in 1915, it turned out that he was able to explain exactly this much more shift in Mercury's perihelion. Thus, the planet provided one of the initial tests of our best current theory of gravity.

A MASSIVE IMPACT CREATES CALORIS BASIN.

he second world from the sun, Venus has often been called Earth's sister planet. Indeed, it is the planet closest in mass to our own, weighing in at 85 percent of Earth mass. Like Mercury, Venus is seen only as a morning or evening "star," and it also is named for a god in the Roman pantheon. Venus was the goddess of love, and a number of ancient civilizations saw it in the same way—the Babylonians, for example, named it Ishtar, after their own goddess of desire. • Aside from the moon, Venus is the brightest object in the night sky, and it can be seen even through the glare of modern cities. Perhaps because of its prominence, Venus also has the dubious distinction of being the most frequently reported UFO.

VENUS

[A BEAUTIFUL INFERNO]

DISCOVERER: UNKNOWN
DISCOVERY DATE: PREHISTORIC
NAMED FOR: ROMAN GODDESS OF LOVE

MASS: 0.82 × EARTH'S
VOLUME: 0.86 × EARTH'S
MEAN RADIUS: 6,052 KM (3,760 MI)
SURFACE TEMPERATURE: 462°C (864°F)
LENGTH OF DAY: 243 EARTH DAYS (RETROGRADE)
LENGTH OF YEAR: 224.7 EARTH DAYS
NUMBER OF MOONS: 0
PLANETARY RING SYSTEM: NO

A computer-generated view of volcano Gula Mons, Venus.
(Inset) Eastern hemisphere of Venus

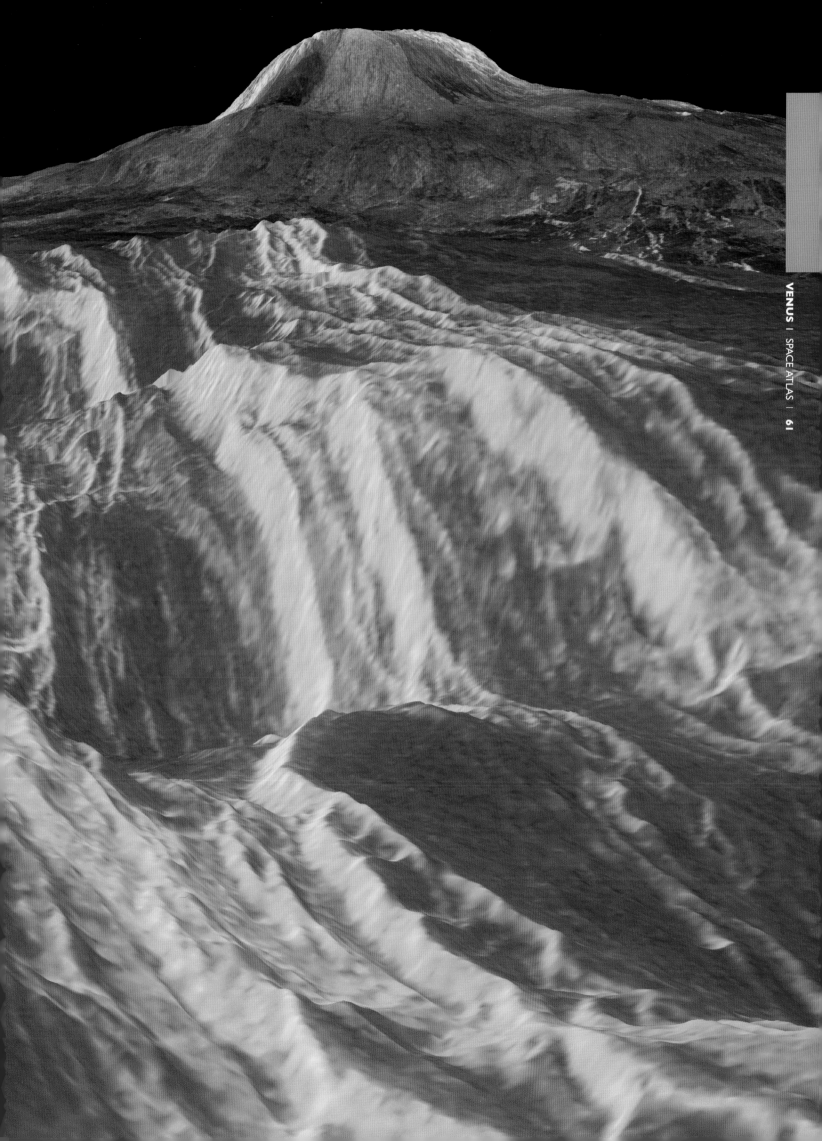

VENUS | WEST

Under a smothering atmosphere,
Venus has a varied, volcanic landscape.

75°

A
N
B
60°
C
45°

Mamapacha Fluctus
Laūma Dorsa
Semele Tholi
Bathkol Tessera
Pandrosos Dorsa
Dorsa
Barsova
Okipeta Dorsa
Bachue Corona
Metis Mons
Upunusa Tholus
Ranane Coron
Tikoiwuti Dorsa
Vinmara Planitia
Virilis Tesserae
Mokosha Mons
Bau Corona
Yablochkina
Ahsonnutli Dorsa

D
GANIKI PLANITIA
Ashtart Tholus
Sekmet Mons

30°
Lahevhev Tesserae
Kokyanwuti Mons
Bellona Fossae
KAWELU PLANITIA
Yuki-Onne Tessera
Fornax Rupes
Sakwap-mana Mons
Mumtaz-Mahal
Venilia Mons
Sudenitsa Te

E
Tubman
Boleyn
O'Keeffe
Polik-mana Mons
AS
Nökomis Montes
Pani Corona
Nazit Mons
Kono Mon
Wheatle

15°
Yolkai-Estsan Mons
Nahas-tsan Mons
210° Batten
225°
Perchta Corona
240°
Taranga Corona 255°
R

F
Tkashi-mapa Chasma
Zisa Corona
ULFRUN REGIO
Hanwi Chasma
Nipa Tholus
Aruru Corona
Paor Tho
Mem Loimis Mons
Lama Tholus
HINEMOA PI

Ozza Mons
REGIO
Lengdin Corona

2
Maat Mons
A
3
4
5
EQUATOR
6
Rusalka
0°
1
2
Javine Corona

Ongwuti Mons
Kiehela Chasma
Itoki Fluctus
Grechukha Tholi
Planitia
Ningyo Fluctus
Maram Corona
Chimon

G
Veledar Linea
Dziwica Chasma
Atete Corona
Uretsete Mons
Žemina Corona
3
Mbokomu Mons
Chondi Chasma
Spandarmat Mons
Gu Plan

15°
Jokwa Linea
Darline
Lalohonua Corona
S
M
A
Thaukhud Linea
T
Stanton
Wawalag Planitia
Achek Dorsa

H
Isabella
Chuginadak Mons
Aditi Dorsa
Wollstonecraft

30°
Stowe
Tsovinar Dorsa
IMDR REGIO
Idunn Mons

J
Etain Dorsa
Tinianavyt Dorsa
HELEN PL
Nsomeka
Rokapi Dorsa
ISHKUS REGIO
Gende
45°
Citlalpul Vallis
Barrymore
Planitia
Durant
Va
Saule Dorsa
Vejas-mate Dorsa
sadi

K
Evelyn
Nup
Leonard
Planiti
60°
Sayers
Nambi Dorsum

L
75°

Lambert Azimuthal Equal-Area Projection
SCALE 1:58,994,000 at the Equator
1 CENTIMETER = 590 KILOMETERS; 1 INCH = 931 MILES

STATUTE MILES
0   500   1000   1500   2000
KILOMETERS
0   500   1000   1500   2000

✳ Spacecraft landing or impact site
◌ Crater

KEY FEATURES

1. ATLA REGIO: A region of old volcanic flows
2. MAAT MONS: Highest volcano on Venus
3. ŽEMINA CORONA: A domed feature in the steepest terrain on Venus

75° 60° 45° 30° 15° 345° 330° 315° 300° 285° 0° 15° 30° 45° 60° 75°

Lagerölf
Uorsar Rupes
Freyja Mts.
ISHTAR
Lakshmi
Planum
TERRA
Obukhova
Duncan
Omosi-Mama Corona
Sacajawea Patera
Danu Montes
Bagbartu Mons
Lampedo Linea
Vesta Rupes
Lind
tessori
Libuše Planitia
Demeter Corona
Beiwe Corona
Senectus Tesserae
Galina
SEDNA PLANITIA
Manzan-Gurme Tesserae
Urash Corona
Evaki Tholus
Vassi
de Staël
Agrona Linea
Lenore
Breksta Linea
karra-mahte Fossae
Toci Tholus
Sanger
Venera 9 (U.S.S.R)
Landed October 22, 1975
Barton
Gula Mons
HYNDLA
GUINEVERE PLANITIA
Sif Mons
Theia Mons
Aurelia
REGIO
Centlivre
Venera 10
(U.S.S.R) Landed October 25, 1975
Shih Mai-Yu
Seymour
Tuli Mons
Rosa Bonheur
Undine Planitia
Atanua Mons
Tuulikki Mons
LAUFEY REGIO
Ngone
Hellman
Heng-o
Nedolya Tesserae
Pioneer Venus 2 Large Probe (U.S.) Landed December 9, 1978
Var Mons
Mortim-Ekva Fluctus
Comnena
Corona
Rhpisunt Mons
7   8   9   10   11   12   0°
Pólóznitsa Corona
Perynya Tholus
Venera 12 (U.S.S.R) Landed December 21, 1978
Venera 13 (U.S.S.R) Landed March 1, 1982
PHOEBE
Dolya Tessera
Navka
Aleksota Mons
Perunitsa Fossae
(U.S.S.R) Venera 7 Landed Dec. 15, 1970
REGIO
Planitia
Venera 8 (U.S.S.R) Landed Kuly 22, 1972
Kanykey
Khosedem Fossae
Gulam Fossae
Dzerassa Planitia
Darago Fluctus
Venera 14 (U.S.S.R) Landed March 5, 1982
VASILISA REGIO
Planitia
Yunya-mana Mons
Liv
Venera 11 (U.S.S.R) Landed December 25, 1978
Avviyar
Atai Mons
Ushas Mons
Ts'an Nu Mons
Mielikki Mons
Kwannon Tholus
Aglaonice
ALPHA
tseger
Angerona Tholus
Pioneer Venus 2 Day Probe (U.S.) Landed December 9, 1978
DIONE
Saskia
REGIO
THEMIS REGIO
Tamiyo Corona
Nevelson
REGIO
Shiwanokia Corona
Nepthys Mons
Innini Mons
beona Mons
Dix
Faravari Mons
Hathor Mons
Bibi-Patma Corona
Hippolyta Linea
ITIA
Ponselle
NERINGA
REGIO
Grey
Meitner
Kaiwan Fluctus
Morrigan Linea
Nike Fossae
Penardun Linea
LAVINIA PLANITIA
Mylitta Fluctus
Kalaipahoa Linea
Alcott
Cavillaca Fluctus
LADA TERRA
Quetzalpetlatl Corona
Juturna Fluctus

CARTOGRAPHER'S NOTE: Hidden by clouds of acrid sulfur dioxide, the surface of Venus can't be seen from orbit. NASA's Magellan mission surveyed the surface of the planet using synthetic aperture radar to pierce the thick atmosphere. Once processed and analyzed back on Earth, that data produced this high-resolution relief model, colorized to approximate how the landscape might look if we could view it from above.

VENUS | EAST

The highlands of Aphrodite Terra, about half the size of Africa, dominate the equatorial region.

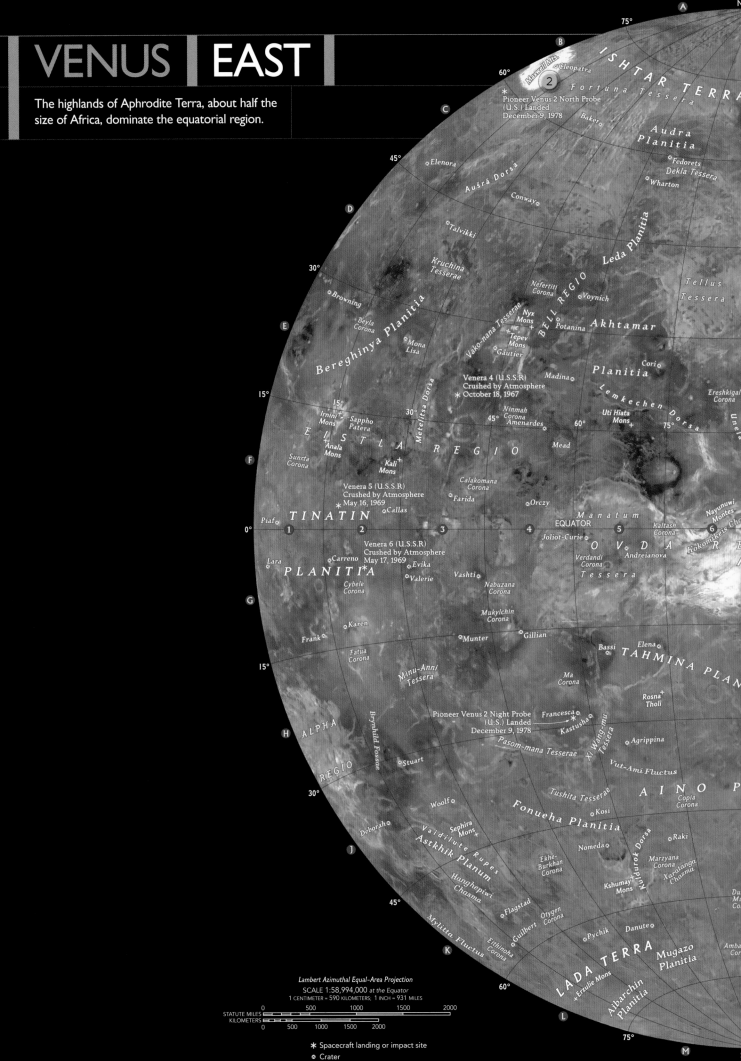

Lambert Azimuthal Equal-Area Projection
SCALE 1:58,994,000 *at the Equator*
1 CENTIMETER = 590 KILOMETERS; 1 INCH = 931 MILES

STATUTE MILES
0   500   1000   1500   2000

KILOMETERS
0   500   1000   1500   2000

✳ Spacecraft landing or impact site
⊙ Crater

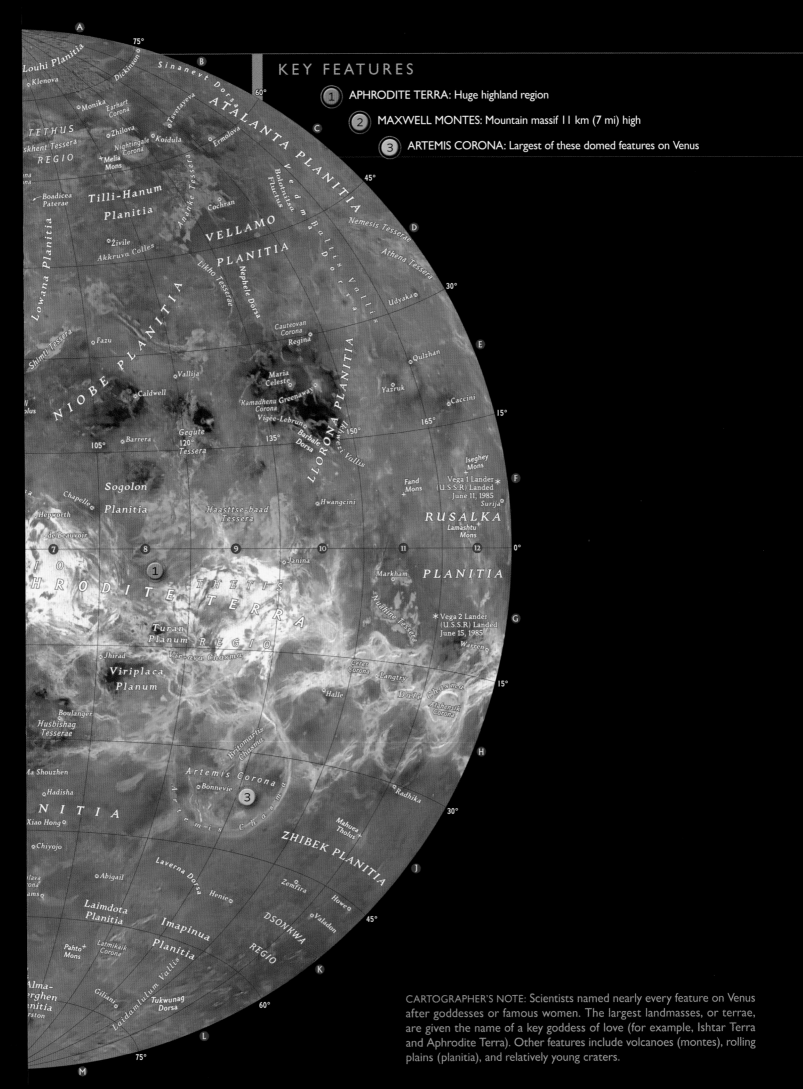

KEY FEATURES

(1) **APHRODITE TERRA:** Huge highland region

(2) **MAXWELL MONTES:** Mountain massif 11 km (7 mi) high

(3) **ARTEMIS CORONA:** Largest of these domed features on Venus

Louhi Planitia
Klenova
Dickinson
Sinanevt Dorsa
TETHUS
Monika
Earhart
Corona
Zhilova
skhent Tessera
REGIO
Nightingale
Corona
Koidula
Tsvetayeva
Ermolova
ATALANTA PLANITIA
Melia
Mons
Anahke Tessera
Boadicea
Paterae
Tilli-Hanum
Planitia
Cochran
Bolatutsa
Fluctus
Veden
Ma
Baltis
Dorsa
Nemesis Tesserae
Athena Tessera
Živile
VELLAMO
Akkruva Colles
Lowana Planitia
Likho Tessera
PLANITIA
Udyaka
Shimti Tessera
Nephele Dorsa
Fazu
Cauteovan
Corona
Regina
Quizhan
NIOBE PLANITIA
Vallija
Maria
Celeste
Yazruk
Caldwell
Kamadhenu
Corona
Greenaway
Caccini
Vigée-Lebrun
Barrera
Gegute
Tessera
Barbale
Dorsa
LLORONA PLANITIA
Iseghey
Mons
Sogolon
Chapelle
Haasttse-baad
Tessera
Hwangcini
Vega 1 Lander
(U.S.S.R) Landed
June 11, 1985
Hepworth
Planitia
Fand
Mons
Surija
de Beauvoir
RUSALKA
Lamashtu
Mons
(7) (8) (9) (10) (11) (12) 0°
Janina
PLANITIA
(1)
Markham
HRODITE TERRA
THETIS
Turan
Planum
REGIO
Nuatine Tessera
Vega 2 Lander
(U.S.S.R) Landed
June 15, 1985
Jhirad
Vir-ava Chasma
Warren
Viriplaca
Planum
Ceres
Corona
Langtry
Dali Chasma
Halle
Atahensik
Corona
Boulanger
Husbishag
Tesserae
Britomartis
Chasma
Ma Shouzhen
Artemis Corona
Hadisha
Bonnevie
(3)
Radhika
NITIA
Mahuea
Tholus
Xiao Hong
Chiyojo
ZHIBEK PLANITIA
Abigail
Laverna Dorsa
Henie
Zemfira
Howe
Laimdota
Planitia
Valadon
Imapinua
DSONKWA
Pahto
Mons
Latmikaik
Corona
Planitia
REGIO
Alma-
erghen
Giliani
Tukwunag
Dorsa
nitia
Laidamlulum Vallis

CARTOGRAPHER'S NOTE: Scientists named nearly every feature on Venus
after goddesses or famous women. The largest landmasses, or terrae,
are given the name of a key goddess of love (for example, Ishtar Terra
and Aphrodite Terra). Other features include volcanoes (montes), rolling
plains (planitia), and relatively young craters.

Venus completes an orbit around the sun in about 225 days, but its rotation is somewhat unusual. Viewed from above, the planets in the solar system all move around the sun in a counterclockwise direction, and most of them rotate on their axes in the same direction, counterclockwise as seen from above. This situation is a holdover from the spinning disk of matter from which the planets formed. Venus, however, rotates backward and completes a "day" in 243 Earth days, the slowest rate of rotation of all the planets. Scientists suggest that these anomalies are the result of a collision during the violent early days of planetary formation (see pages 44–46).

Despite the planet's proximity to Earth, astronomers learned little about Venus until the latter part of the 20th century. This is because a thick layer of white clouds perpetually obscures the surface of the planet. Then, in the 1960s, the United States and the Soviet Union began systematic programs of sending space probes to the planet, first to observe it from orbit or in a flyby and then to land probes on the surface itself. In 1962 Mariner 2 flew by the planet and probed it with microwave and infrared sensors. This was when we learned that the surface of Venus is extremely hot—around 462°C (864°F), which is hotter than the surface of Mercury, despite Venus's greater distance from the sun.

The high temperature of the Venusian surface came as something of a surprise. In the science fiction of the mid-20th century, Venus was often described as a hot, swampy place, but fit for human colonization. However, the rudimentary data revealed by these first space probes were enough to remove Venus from the list of hospitable worlds in the solar system.

MISSIONS TO VENUS

In 1966 the Soviet probe Venera 3 crash-landed on the Venusian surface. The mission's goal was to land a probe and return data, but the intense pressure of the atmosphere crushed the capsule on the way down. In 1967 the strengthened Venera 4 spacecraft entered the atmosphere and sent back data, but the parachute that slowed its velocity was too effective—the craft took so long to descend that its batteries gave out before it reached its destination. Finally, in 1970, Venera 7, bolstered against the pressure and equipped with a smaller parachute, actually made it to the surface and sent back pictures. This was followed by several more spacecraft that landed successfully, typically sending back data for less than an hour before succumbing to the extreme conditions on the planetary surface.

In 1978 the American Pioneer Venus mission produced the first detailed maps of Venus, using radar

A cutaway view of the interior of Venus. The core is mainly solid iron, the mantle is thick, and the crust is thin—about half as thick as Earth's. Venus's thick atmosphere supports a greenhouse effect that makes it the hottest planet in the solar system.

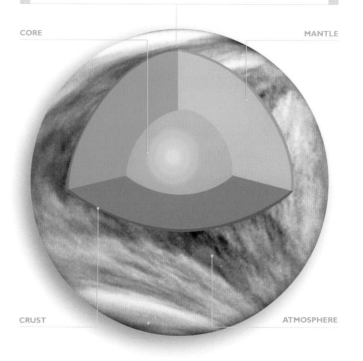

CORE   MANTLE

CRUST   ATMOSPHERE

These images, taken 24 hours apart by the European Space Agency's probe Venus Express, show a rapidly evolving storm, or vortex, over Venus's south pole (marked by a yellow dot).

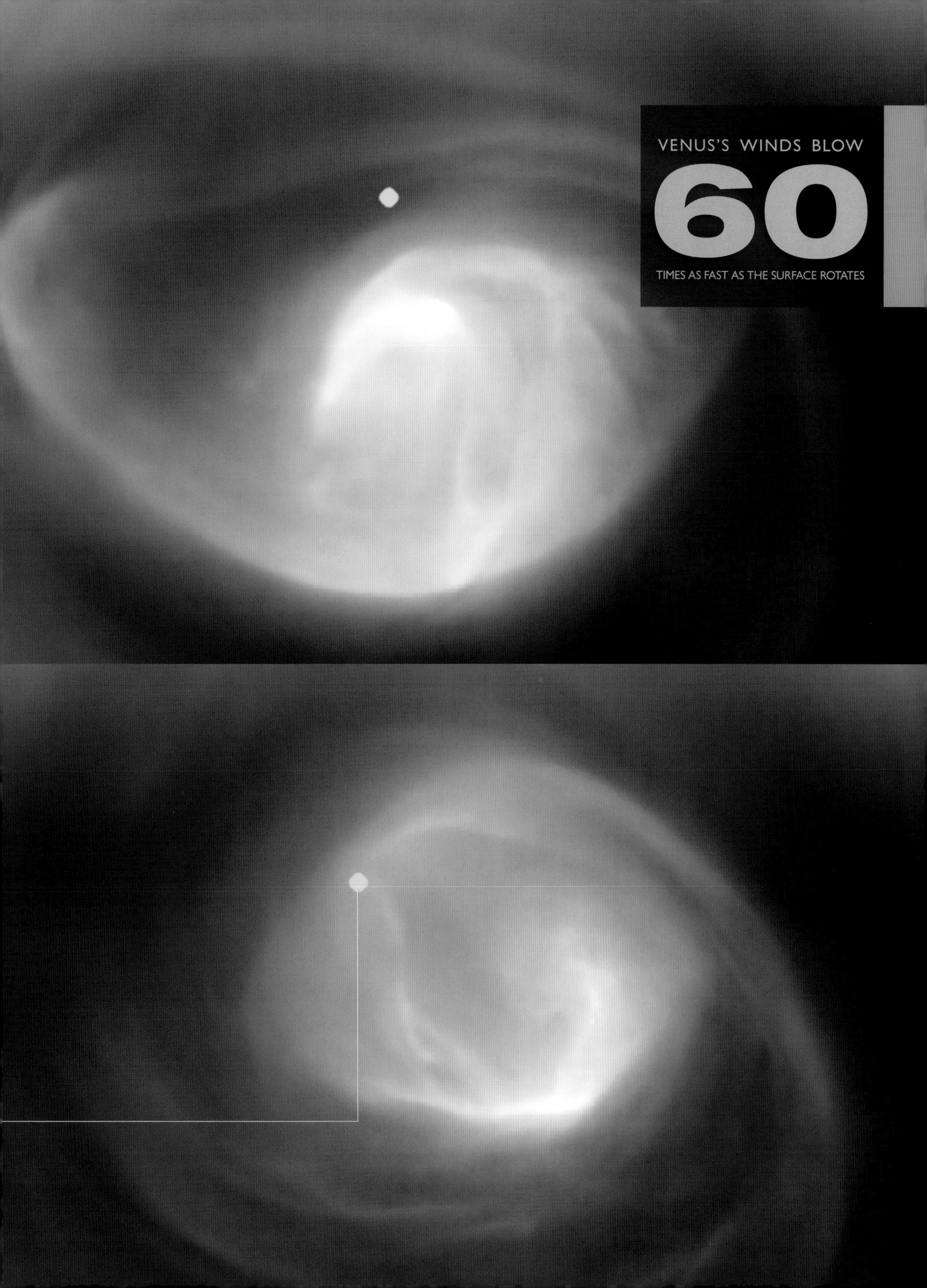

VENUS'S WINDS BLOW

60

TIMES AS FAST AS THE SURFACE ROTATES

AT VENUS'S SURFACE, AIR WEIGHS

90

TIMES MORE THAN ON EARTH

signals to penetrate the clouds. Subsequent missions from the United States and the Soviet Union continued the exploration, with radar maps from the Magellan probe in 1989 revealing the surface in unprecedented, three-dimensional form. In 2005 the European Space Agency launched Venus Express, which entered a polar orbit around the planet in 2006 and has since been sending back data about Venus's dramatic atmosphere.

HOT, TOXIC, AND VOLCANIC

Venus's atmosphere is almost pure carbon dioxide—over 95 percent—with most of the rest being nitrogen. Atmospheric pressure at the surface is 92 times that at sea level on Earth. This is about the same pressure that you would experience at a depth of one kilometer (about 3,000 ft) below the surface of the ocean on Earth. No wonder those first space probes were crushed! Scientists suggest that early on in its history, Venus may have had oceans, but that they were lost because the sun's intense radiation evaporated them. Without oceans to pull carbon dioxide out of the atmosphere, the concentration of this gas grew as volcanoes spewed it out. The planet experienced a runaway greenhouse effect that produced the intense temperatures we see today.

Because of Venus's dense atmosphere, the surface temperature is essentially constant everywhere. The surface winds are slow (only a few miles an hour), but it would be difficult to stand up to them on the surface because of the high density of the atmosphere—think of the slow winds as being more like a tidal wave than a gentle breeze.

An artist's conception of a lightning bolt on Venus. With an atmosphere made mainly of carbon dioxide, a crushing surface pressure, and sulfuric acid clouds, Venus is not the abode of life so often imagined in early science fiction.

Venus's clouds are composed mainly of sulfur dioxide and sulfuric acid. Intense winds at these higher altitudes have speeds clocking at several hundred miles an hour; we do not yet understand what causes them. It can actually rain sulfuric acid on Venus, but the drops evaporate as they fall through the thick atmosphere, never reaching the surface. The clouds also generate the planet's small magnetic field by means of a complex interaction with particles streaming from the sun. Venus's slow rotation precludes it from generating a magnetic field like Earth's, which depends on the rotation of its liquid iron core.

Radar mapping reveals a Venusian surface shaped primarily by volcanic activity. About 80 percent of the surface consists of smooth plains, with two highland "continents" making up the rest. One hundred sixty-seven volcanoes on the Venusian surface are bigger than the one forming the Big Island in Hawaii, the largest volcano on Earth. There is evidence that many Venusian volcanoes are still active.

The plains are dotted with impact craters, most of which show no evidence of weathering. Scientists argue that this fact implies that about 500 million years ago Venus underwent a "resurfacing" event, during which lava flows covered the old surface (with its craters), creating the smooth plains we see now and presenting a new surface for incoming meteorites. Models suggest that over time the temperature of the mantle rises, the crust weakens, and, every 100 million years or so, a new surface is created. This sort of sporadic repaving of the planetary surface is something we don't see on Mercury, but it illustrates an important point about our solar system. Every world we will encounter will be different in some way from all the others—new geology, new atmospheric physics, new phenomena. This is certainly true if we compare Venus to the next planet we'll visit—our own, Earth.

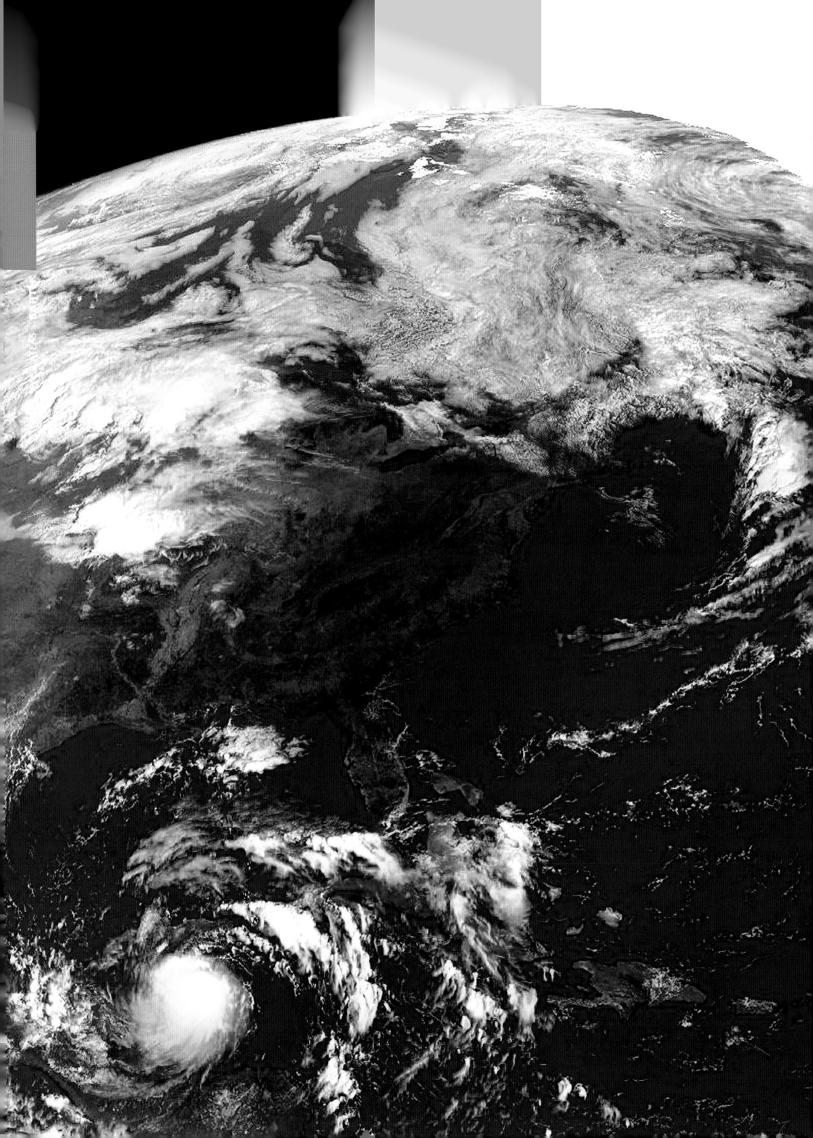

For obvious reasons, we know a lot more about Earth than we do about any other object in the solar system. But there is a lot to be learned by looking at Earth as just one more object in a solar system full of planets and moons. • So what distinguishes Earth from other worlds? There are two important distinctions: First, it is the largest terrestrial planet. As we shall see below, Earth's size is related to the fact that its surface is constantly changing, constantly moving. And second, Earth's orbit lies in a narrow band around the sun known as the habitable zone, which is defined as the region in which liquid water can stay on the planetary surface for long periods of time. Because of this, Earth is the only object in the solar system where we know life exists. • Let's look at these distinctions separately.

EARTH

[OCEAN PLANET]

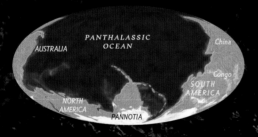 

DISCOVERER: UNKNOWN
DISCOVERY DATE: PREHISTORIC
NAMED FOR: OLD ENGLISH WORD *ERTHA*, MEANING "GROUND"

....

MASS: 5,972,190,000,000,000,000,000,000 KG
VOLUME: 1,083,206,916,846 KM³ (259,875,159,532 MI³)
MEAN RADIUS: 6,371 KM (3,959 MI)
MIN./MAX. TEMPERATURE: -88/58°C (-126/136°F)
LENGTH OF DAY: 23.93 HOURS
LENGTH OF YEAR: 365.26 DAYS
NUMBER OF MOONS: 1
PLANETARY RING SYSTEM: NO

Cyclones cross the Atlantic. (Inset) Earth's continents
600 million (left) and 90 million years ago (right)

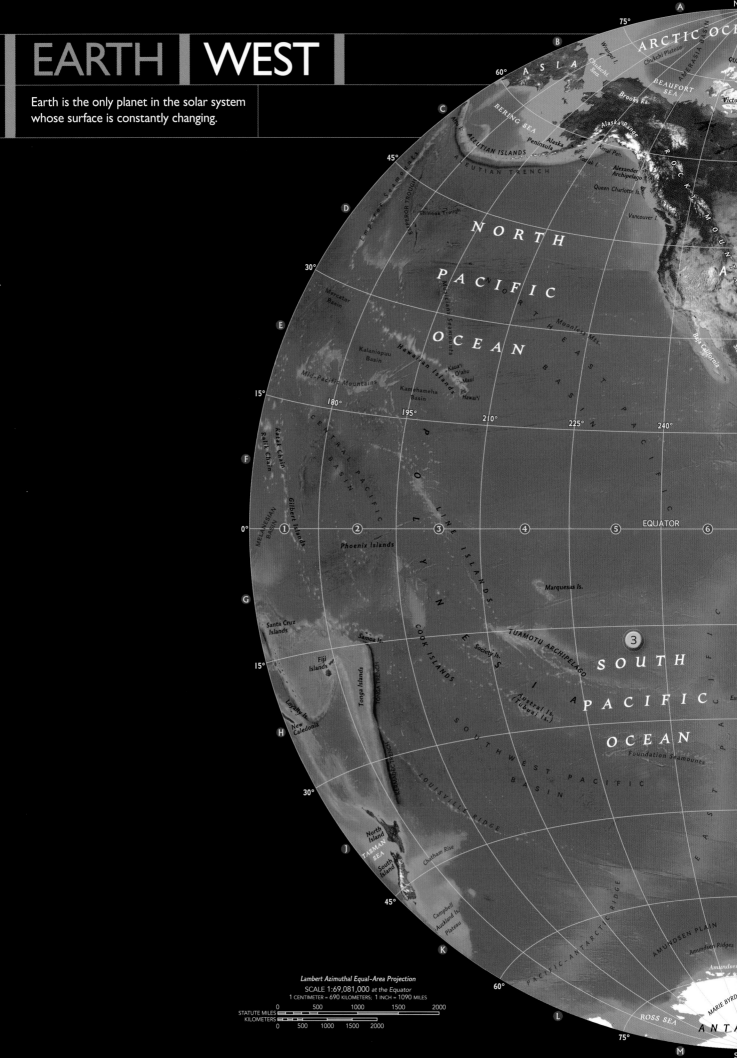

EARTH | WEST

Earth is the only planet in the solar system whose surface is constantly changing.

ARCTIC OCEAN

ASIA

Wrangel I.
Chukchi Plateau
Chukchi Sea
BEAUFORT SEA
Victoria I.
Brooks Ra.
Amerasia Basin
QU

BERING SEA
Alaska Range
Alaska Peninsula
Kenai Pen.
Kodiak I.
Alexander Archipelago
ROCKY MOUNTAINS

ALEUTIAN ISLANDS
ALEUTIAN TRENCH
Queen Charlotte Is.
Vancouver I.

NORTH AMERICA

Emperor Seamounts
Chinook Trough
EMPEROR TROUGH
Mendocino

NORTH

PACIFIC

OCEAN

Baja California

Mercator Basin

NORTHEAST PACIFIC BASIN

Murray Seamounts
Moonless Mts.

Kalaniopuu Basin

Hawaiian Islands
Kaua'i
O'ahu
Maui
Hawai'i

Mid-Pacific Mountains

Kamehameha Basin

180°   195°   210°   225°   240°

CENTRAL PACIFIC BASIN

P O L Y N E S I A

LINE ISLANDS

Ratak Chain
Ralik Chain

MELANESIAN BASIN

Gilbert Islands

EQUATOR
① ② ③ ④ ⑤ ⑥

Phoenix Islands

Marquesas Is.

TUAMOTU ARCHIPELAGO

③

Santa Cruz Islands

Samoa Is.

Society Is.

COOK ISLANDS

SOUTH

Fiji Islands

Tonga Islands

Austral Is. (Tubuai Is.)

PACIFIC

Loyalty Is.
New Caledonia

TONGA TRENCH

SOUTHWEST PACIFIC BASIN

OCEAN

Foundation Seamounts

EAST PACIFIC

KERMADEC TRENCH

LOUISVILLE RIDGE

30°

North Island

TASMAN SEA

South Island

Chatham Rise

Campbell
Auckland Is.
Plateau

45°

PACIFIC-ANTARCTIC RIDGE

AMUNDSEN PLAIN

Amundsen Ridges

60°

ROSS SEA

MARIE BYRD

ANTA

75°

Lambert Azimuthal Equal-Area Projection

SCALE 1:69,081,000 *at the Equator*

1 CENTIMETER = 690 KILOMETERS; 1 INCH = 1090 MILES

STATUTE MILES
0   500   1000   1500   2000

KILOMETERS
0   500   1000   1500   2000

KEY FEATURES

① ANDES MOUNTAINS: Created by the subduction of tectonic plates

② ICELAND: The northernmost reach of the Mid-Atlantic Ridge

③ PACIFIC OCEAN: Part of a liquid, surface ocean, a feature unique to Earth

CARTOGRAPHER'S NOTE: Satellite imagery provides the detail of lake and mountain, desert and forest, showing every corner of our living planet. Peering beneath the watery ocean depths, global seafloor topographic data were used to create a representation of the ocean crust invisible from the surface.

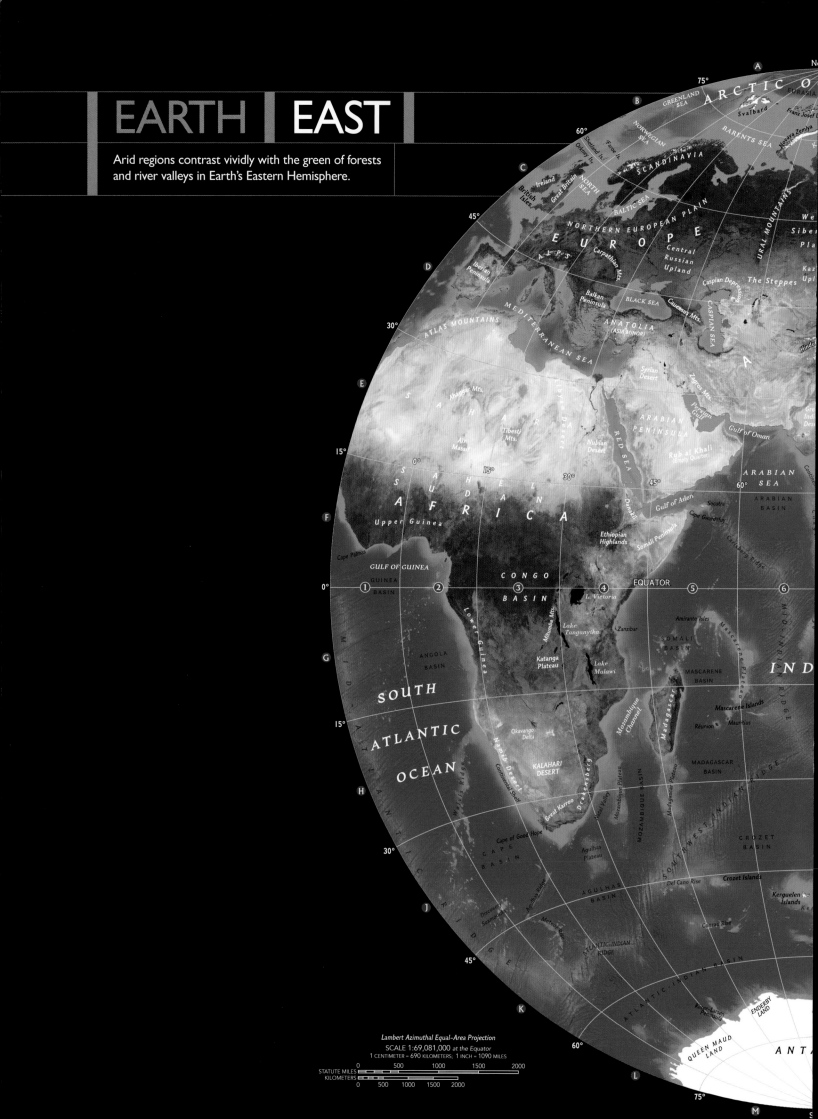

EARTH | EAST

Arid regions contrast vividly with the green of forests
and river valleys in Earth's Eastern Hemisphere.

Lambert Azimuthal Equal-Area Projection
SCALE 1:69,081,000 at the Equator
1 CENTIMETER = 690 KILOMETERS; 1 INCH = 1090 MILES

STATUTE MILES
0   500   1000   1500   2000

KILOMETERS
0   500   1000   1500   2000

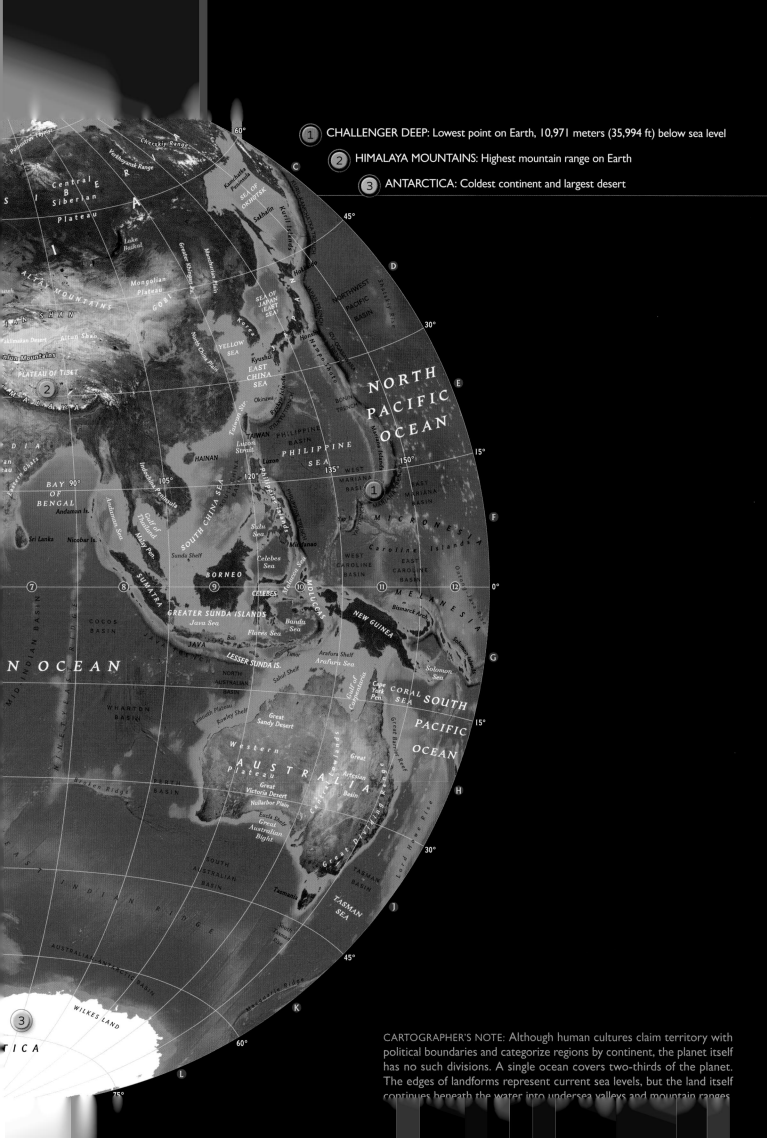

1 CHALLENGER DEEP: Lowest point on Earth, 10,971 meters (35,994 ft) below sea level

2 HIMALAYA MOUNTAINS: Highest mountain range on Earth

3 ANTARCTICA: Coldest continent and largest desert

CARTOGRAPHER'S NOTE: Although human cultures claim territory with political boundaries and categorize regions by continent, the planet itself has no such divisions. A single ocean covers two-thirds of the planet. The edges of landforms represent current sea levels, but the land itself continues beneath the water into undersea valleys and mountain ranges.

For the first half billion years of its existence, give or take a few hundred million years, Earth swept around its orbit collecting debris from the process of planetary formation. Had anyone been standing on the surface, he or she would have seen the fiery impact of large meteorites all around. Scientists refer to this period as the Great Bombardment. (Please note that the Great Bombardment happened early in the formation of the solar system, and is different from the Late Heavy Bombardment, which occurred when the solar system was half a billion years old.) Each impact added a certain amount of energy to the newly forming planet, energy that was converted to heat. Eventually, Earth either melted all the way through or heated up to the point that it became soft enough for materials to flow easily—scientists are still debating the details. What is not under debate, however, is that as a result of this early heating the planet went through a process known as differentiation. The heaviest materials—mainly iron—sank to the center to form Earth's core, while lighter materials formed the mantle and crust. Like a salad dressing left in one place too long, the materials within Earth separated themselves out under the influence of gravity.

Earth's differentiation created the magnetic field in which we all live. The heavy iron and nickel sank down, and at the very center of the planet the pressure was high enough to force those atoms into a solid. Above this solid core, however, is a region where the temperatures and pressure are only high enough to form a liquid layer. It is the rotation of this fluid metal core that ultimately produces the planet's magnetic field.

THE BOILING EARTH

The interior of the nascent planet had two sources of heat—the leftover heat from the Great Bombardment and the heat generated by the decay of radioactive elements in the rock. Like a pot of water on a stove, the interior of the planet had to find a way of getting this heat to the surface, where it could radiate away. And like a pot on the stove, Earth "boiled." Over hundreds of millions of years, the solid rocks in the mantle circulated, with hot material rising in one place, cooling, and sinking somewhere else. As this boiling went on, the lighter material that had risen to just under Earth's surface during differentiation was carried along, like leaves on a stream. And on top of this, a collection of the lightest material of all—the stuff we call the "solid ground" of continents—rode along like passengers on a raft.

The best way to picture the working of the planet, then, is to imagine a thin layer of oil on a pot of boiling water. The action of the water will break the layer of oil

A cutaway view of Earth shows a solid iron-nickel core at the center, surrounded by a liquid layer of the same material. This two-layer core is overlain by a thick mantle and an outer crust.

CORE

CRUST

MANTLE

A volcanic eruption on the Big Island of Hawaii brings many different kinds of materials to Earth's surface—particles lofted into the atmosphere, gases (not visible here), and red-hot lava flowing into the sea.

EARTH HAS
12
MAJOR TECTONIC PLATES

into pieces, creating a kind of jigsaw puzzle. In just the same way, the boiling of the mantle breaks the planet's surface into pieces called plates.

The theory of plate tectonics is based on the idea that the surface of the planet is made from plates that respond to the movement of rocks in the mantle. ("Tectonics" comes from the same Greek root as "architect," and refers to the process of building.) Some plates carry continents, some do not, but because of the constant boiling, the plates and everything on them are always in motion, reshaping the face of the planet. There have been times, for example, when all of the continents were strung around the Equator, like some titanic necklace, and times when they were stuck together in a single land mass called Pangaea ("all Earth"). Other planets don't operate this way. Mercury and Mars (as well as our moon) are small enough that they got rid of their heat early on and are now "frozen." The volcanic activity on Venus is a sign of geological activity, but the planet is apparently just a little too small to foster anything like plate tectonics.

HABITABLE ZONE

Earth is what astronomers call a Goldilocks planet—not too cold, not too hot, but juuuust right. For the last four billion years, while the luminosity of the sun increased by a third (see page 203), processes on our planet adjusted so that the atmospheric temperature stayed between the freezing and boiling points of water. This meant that liquid water could always be found at the surface. As we saw on page 70, the presence of liquid water is thought to be a necessary prerequisite for the development of life. Had Earth been closer to the sun than it is, it might have followed the same path as our sister planet Venus, with a runaway greenhouse effect extinguishing whatever life-forms had developed. Had it been farther from the sun, it might have frozen solid. In either case, there would be no life on our planet.

There is a narrow band around any star in which the surface temperature of a planet will remain below the boiling point and above the freezing point of water.

EARTH'S FIRST MEASUREMENT

Despite what used to be taught in elementary schools, the idea that Earth is round has been with us since antiquity; by the time Columbus sailed, it had been known for millennia. In fact, the first recorded measurement of Earth's diameter was made about 240 B.C. by the Greek geographer Eratosthenes of Cyrene (276–194 B.C.), who became chief librarian of the celebrated Library of Alexandria. His method worked this way: He knew that at noon on the summer solstice, light from the sun penetrated all the way to the bottom of a well in the city of Syene (now Aswan), indicating that the sun was directly overhead. At the same time on the same day he measured the length of a shadow cast by a pole of known height in Alexandria. From this measurement, using geometry he concluded that the distance between Alexandria

and Aswan was one-fiftieth of the circumference of Earth. How he determined the distance between these two cities remains one of those amiable scholarly mysteries that will never be resolved, but he reported that Earth had a circumference of 252,000 stadia (singular: stade).

The problem is that there were several definitions of the stade in the ancient world, just as we have the statute mile (5,280 ft) and the nautical mile (6,076 ft) today. All of the stadia were around 180 meters (200 yd) long—twice the length of a football field, and the most probable choice has Eratosthenes' measurement of Earth's circumference at about 47,000 kilometers (29,000 mi), compared to its current value, which is just under 40,000 kilometers (25,000 mi). He may also have used this baseline to estimate the distances to the moon and sun

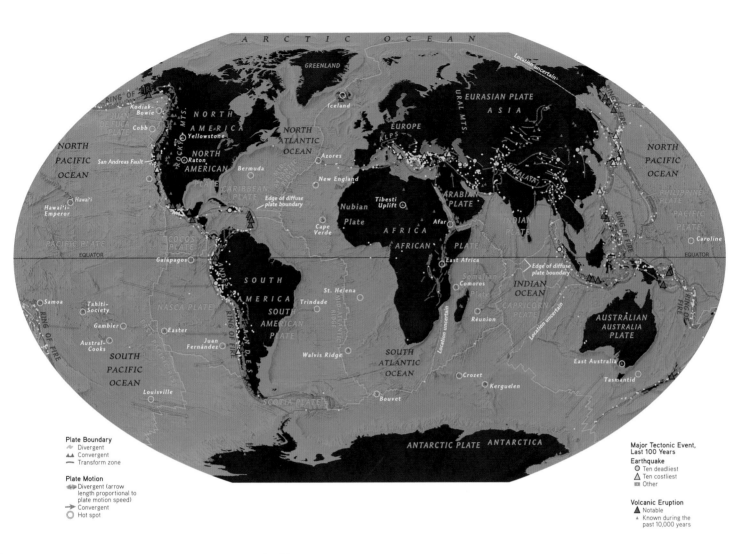

Plate Boundary
↗ Divergent
▲▲ Convergent
— Transform zone

Plate Motion
⬅▶ Divergent (arrow length proportional to plate motion speed)
➡ Convergent
◯ Hot spot

Major Tectonic Event, Last 100 Years
Earthquake
◯ Ten deadliest
△ Ten costliest
▧ Other

Volcanic Eruption
▲ Notable
▴ Known during the past 10,000 years

A map shows Earth's tectonic plates. The plates move in response to the slow churning of Earth's mantle, a churning driven in part by heat from radioactive decay. Alone among the planets in the solar system, Earth features a surface that is constantly changing because of this motion.

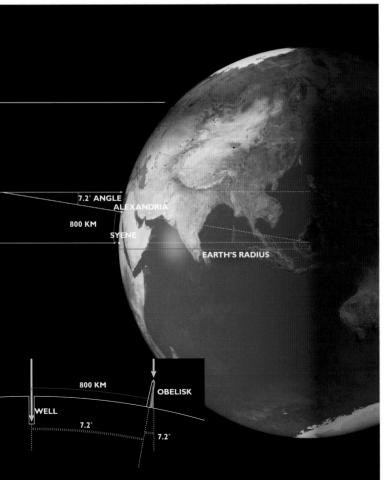

This narrow band is called the continuously habitable zone (CHZ) of the star. Earth is in the CHZ of the sun, which is why life exists here.

The presence of life changes a planet dramatically. On Earth, for example, the inclusion of the corrosive element oxygen in our atmosphere is a result of the metabolic activity of life, as are many of the organic processes that break up rocks and create soil. Astronomers look for these chemical signs of life as they search for an Earth-type planet in the CHZ of another star.

The moon, familiar to all of us from childhood, is the brightest object in the night sky. It completes an orbit around Earth in a little over 27 days and turns on its axis in exactly the same amount of time. This means that it always keeps the same face toward us. In the jargon of astronomers, we say that the moon has been de-spun because of a complex series of gravitational interactions between its mass and that of Earth. We shall see that this is a common phenomenon for moons in the solar system. These gravitational interplays are also pulling the moon away from Earth at the rate of about four centimeters (1.5 in) per century.

EARTH'S MOON

[OUR STEADY NEIGHBOR]

DISCOVERER: UNKNOWN
DISCOVERY DATE: PREHISTORIC
NAMED FOR: OLD ENGLISH WORDS FOR "MOON" AND "MONTH"
DISTANCE FROM EARTH: 384,400 KM (238,855 MI)

MASS: 0.012 × EARTH'S
VOLUME: 0.02 × EARTH'S
MEAN RADIUS: 1,738 KM (1,080 MI)
SURFACE GRAVITY: 0.17 × EARTH'S
MIN./MAX. TEMPERATURE: -233/123°C (-387/253°F)
LENGTH OF DAY: 27.32 EARTH DAYS
LENGTH OF YEAR: 27.32 EARTH DAYS

Moonrise above rock formations, Djibouti.
(Inset) The phases of the moon.

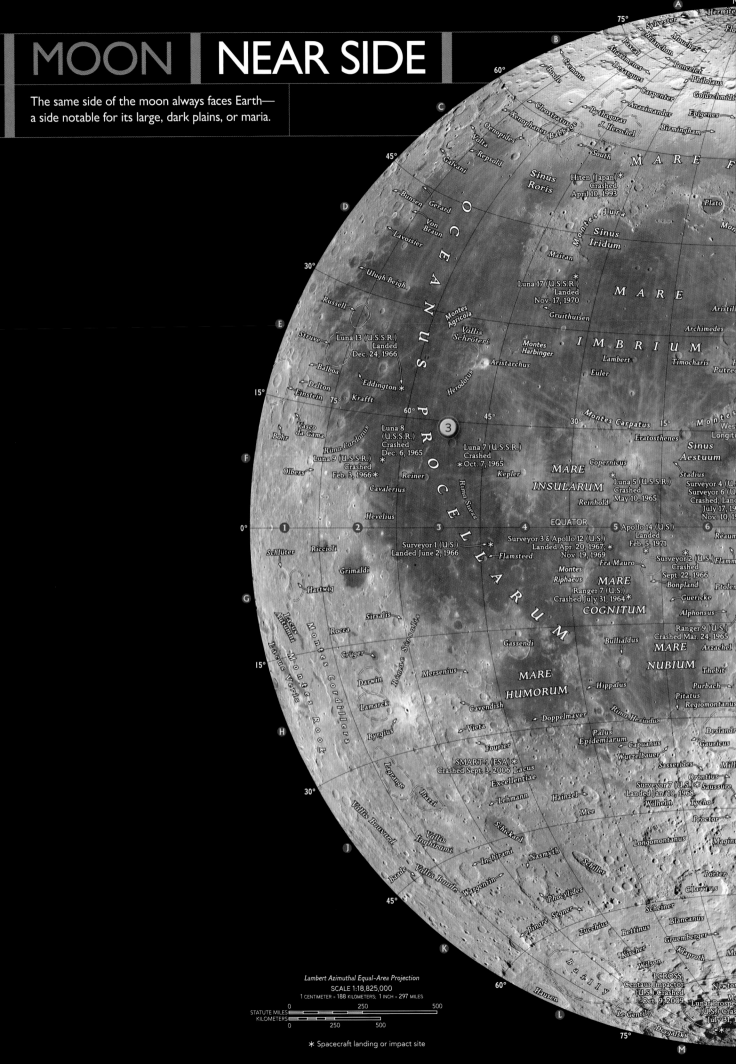

MOON | NEAR SIDE

The same side of the moon always faces Earth—
a side notable for its large, dark plains, or maria.

Lambert Azimuthal Equal-Area Projection
SCALE 1:18,825,000
1 CENTIMETER = 188 KILOMETERS; 1 INCH = 297 MILES

STATUTE MILES
KILOMETERS
0   250   500

∗ Spacecraft landing or impact site

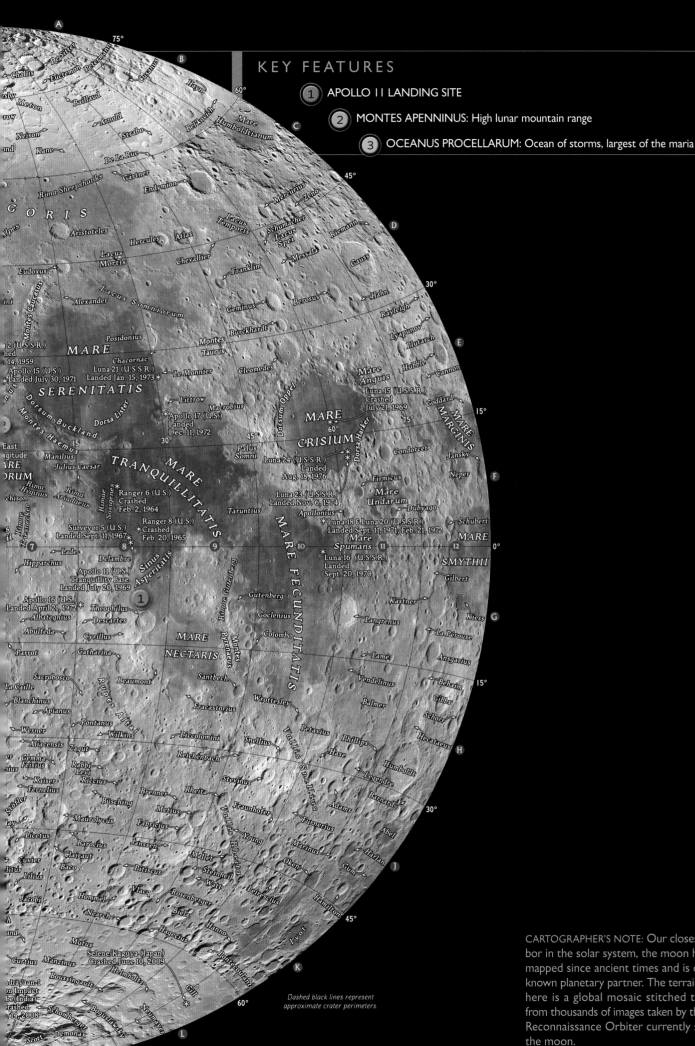

KEY FEATURES

1. APOLLO 11 LANDING SITE

2. MONTES APENNINUS: High lunar mountain range

3. OCEANUS PROCELLARUM: Ocean of storms, largest of the maria

Dashed black lines represent approximate crater perimeters.

CARTOGRAPHER'S NOTE: Our closest neighbor in the solar system, the moon has been mapped since ancient times and is our best known planetary partner. The terrain shown here is a global mosaic stitched together from thousands of images taken by the Lunar Reconnaissance Orbiter currently studying the moon.

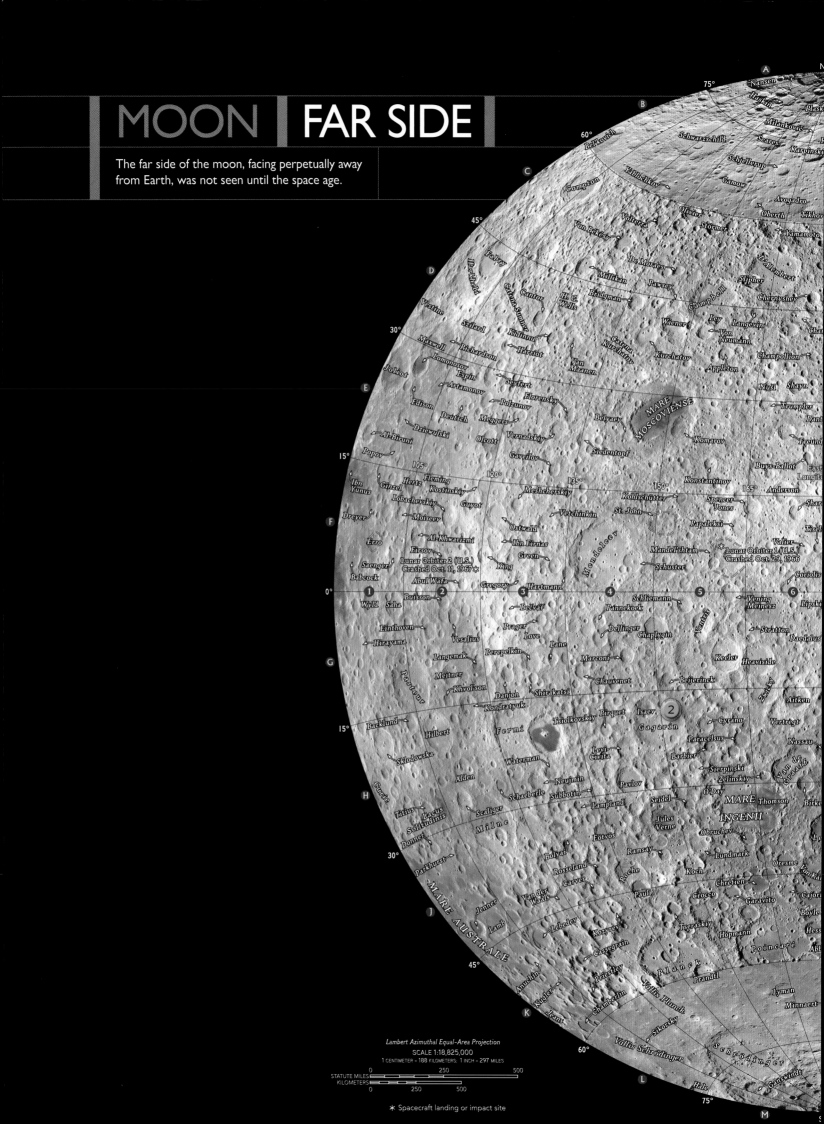

MOON | FAR SIDE

The far side of the moon, facing perpetually away
from Earth, was not seen until the space age.

Lambert Azimuthal Equal-Area Projection
SCALE 1:18,825,000
1 CENTIMETER = 188 KILOMETERS; 1 INCH = 297 MILES

STATUTE MILES 0 250 500
KILOMETERS 0 250 500

✱ Spacecraft landing or impact site

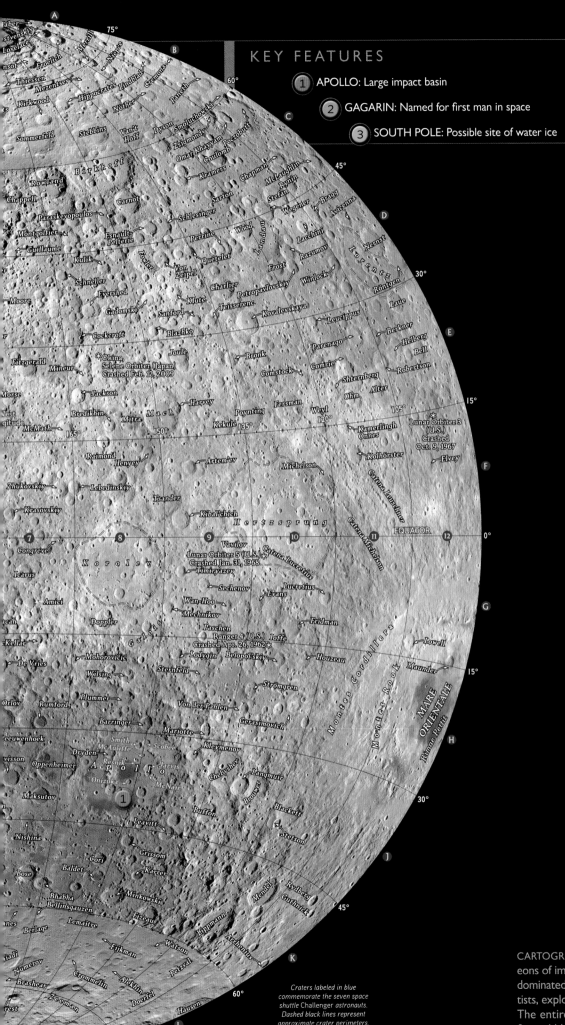

KEY FEATURES

(1) **APOLLO:** Large impact basin

(2) **GAGARIN:** Named for first man in space

(3) **SOUTH POLE:** Possible site of water ice

Craters labeled in blue commemorate the seven space shuttle Challenger *astronauts. Dashed black lines represent approximate crater perimeters.*

CARTOGRAPHER'S NOTE: Pockmarked from eons of impacts, the surface of the far side is dominated by craters, most named for scientists, explorers, astronauts, and cosmonauts. The entire surface is covered by very dry, fine rubble known as regolith.

There is a long history of scientific studies of the moon. Greek, Chinese, and Indian astronomers all realized that the moon shone because it reflected light from the sun, and Aristotle taught that the moon marked the boundary between the earthly and heavenly spheres. The astronomer Claudius Ptolemy (ca 100 A.D.), expanding upon earlier work by Greek astronomers, estimated both the distance to the moon and the satellite's size to within a few percent of the currently accepted values.

In 1609 Galileo used his new telescope to produce drawings of the lunar surface, showing mountains, plains, and craters. The realization that the moon had geological features similar to those on Earth played a role in discrediting the old, sun-centered theories of the universe, since those theories had held that the moon was a smooth, featureless sphere. The far side of the moon—never visible from Earth—was first photographed by the Soviet probe Luna 3 in 1959, and both American and Soviet unmanned spacecraft landed on the lunar surface in the same year.

ONE SMALL STEP

The space race, fueled by the Cold War, resulted in the first human beings setting foot on the moon in 1969, when astronaut Neil Armstrong uttered his famous line "That's one small step for a man, one giant leap for mankind" as he stepped off of the ladder of the Apollo 11 spacecraft. More important from a scientific viewpoint, astronauts on the six lunar landing missions brought back 380 kilograms (840 lb) of lunar rock samples for scientific study, some of which date back to the solar system's earliest years.

Since the end of the Apollo program in 1972, only unmanned probes have gone to the moon. In addition to the United States, the European Space Agency, India,

Japan, and China have engaged in lunar exploration over the past few decades.

THE MOON'S ANATOMY

We now know that the moon, like Earth, was formed about 4.5 billion years ago. Scientists have long debated how this happened. The basic problem is that the moon is significantly less dense than Earth, primarily because the moon has such a small iron core. How could Earth and the moon, both of which apparently formed in the same part of the planetary cloud, end up looking so different?

The currently favored theory is that early in the formation of Earth (but after differentiation had occurred) Earth and a Mars-size object (on which differentiation

A cutaway view of the moon. The moon is a dead world with a small, solid iron core, a thick mantle, and a crust full of craters. Because the moon has no atmosphere, the craters, once formed, never disappear.

INNER CORE

OUTER CORE

CRUST

MANTLE

Apollo 17 astronaut plants the American flag on the moon. The lunar vehicle was driven to places near the landing site to collect geological samples. Much of what we know about the formation of the moon resulted from the study of those rocks. The footprints you see are still there.

had also occurred) collided. This collision blew off a large chunk of Earth's lower-density mantle, and some of this ejected material, along with material from the other object, went into orbit around Earth. At this point, the same process of accretion that built the terrestrial planets came into play, and the moon formed from that orbiting material. Although the moon is only the fifth largest satellite in the solar system, it is the largest in relation to its planet. Its radius is a quarter that of Earth's, and it weighs in at one eighty-oneth of Earth's mass.

Earth's densest material (its iron core) did not contribute to the makeup of the moon, and this explains the density difference between the two. The young moon went through the same kind of differentiation process as Earth (see page 76), so its interior has a layered structure although, for the reasons we discussed above, it has a much smaller core. The main features on the near side (the side visible from Earth) are the large dark plains that cover about a third of its surface (the far side does not have many of these features). These plains are called maria (singular: mare), Latin for "oceans," because early astronomers thought they were seas. They are actually massive outflows of lava, with the biggest ones dating back to between 3 and 3.5 billion years ago. The lighter areas on the moon, usually referred to as highlands, are older—perhaps 4.4 billion years old—and represent the first materials to crystallize from molten material as the moon cooled. The maria and highlands together produce the familiar "man in the moon" image that you can see on any clear night.

THE SURFACE

Craters, the result of meteorite impacts over the eons, dot the lunar surface. Since the moon has no atmosphere to speak of, and since it is now a frozen world with no geological activity, there is nothing to remove

TIME AND TIDES

SUN

SPRING TIDE

Everybody knows that tides are caused by the moon's gravitational pull on the oceans, but two things about tides make the story a little complicated: First, there are two tides a day rather than one; and second, high tide occurs when the moon is on the horizon, not when it is directly overhead. Thus, the tides are not simply a matter of the moon pulling the ocean water toward it. In that case, there would be one high tide every day, and it would occur when the moon is directly overhead.

We usually think of the moon as going around Earth, but in fact the center of mass, its motion creates a centrifugal force that raises a second tidal bulge in the oceans directly opposite to the one raised by the moon's gravity. This is why there are two tides every day.

The fact that these high tides occur when the moon is on the horizon rather than overhead is related to the fact that Earth's oceans are relatively shallow—their average depth is only 5 kilometers (3 mi). This means that the tidal bulges can't keep up with the spot underneath the moon as Earth rotates; they lag behind. If the oceans

these craters once they form. Consequently, there are literally hundreds of thousands of them visible on the moon today.

In fact, the only change agent on the moon's surface is the arrival of new meteorites. Small impacts break up the surface rock, creating small, glassy pieces that weld themselves together (think of damp Rice Krispies). This material, called the lunar regolith, is found everywhere on the moon except on steep surfaces. It is about 10 to 20 meters (30 to 60 ft) deep on the older highlands and about 3 to 5 meters (9 to 15 ft) deep in the maria.

Because people often talk about building permanent bases or colonies on the moon, the question of whether or not there is water at the surface has garnered a lot of scientific attention over the years. The best place to look for water is in deep craters at the moon's poles, since these areas are never exposed to direct sunlight. In 2009 the first Indian moon mission, named Chandrayaan-1, found evidence for water in light reflected from the lunar surface. A few weeks later the American spacecraft LCROSS dropped an impactor the size of a pickup truck into one of the shadowed craters and saw in the resulting debris indications of enough water to fill a small wading pool.

We can't leave the subject of the moon without dealing with some popular misconceptions:

- There is no statistically significant evidence that more people are admitted to psychiatric wards during a full moon than at any other time.
- There is no evidence of an alien UFO base on the far side of the moon.
- The fact that the moon looks bigger on the horizon than when it is overhead is an optical illusion. You can verify this yourself by marking the apparent size of the moon on a stick, first when the moon is on the horizon, then a few hours later when it is overhead. You'll find the same result in both cases.

LOW TIDE, BAY OF FUNDY

Mars, the fourth planet from the sun, is the most thoroughly explored of all the planets except Earth, and has probably figured in more science fiction than any other astronomical object. Named for the Roman god of war, it often appears to have a reddish cast due to the iron oxide (rust) on its surface. Mars is smaller than Earth, having about half the radius and only about 11 percent the mass of our home planet. Because of its small mass, Mars lost most of its atmosphere to space long ago, and today has only a thin atmosphere composed mostly of carbon dioxide. The average air pressure on Mars is roughly equivalent to the pressure 35 kilometers (24 mi) above sea level on Earth.

MARS

[THE RED DESERT]

DISCOVERER: UNKNOWN
DISCOVERY DATE: PREHISTORIC
NAMED FOR: ROMAN GOD OF WAR

MASS: 0.11 × EARTH'S
VOLUME: 0.15 × EARTH'S
MEAN RADIUS: 3,390 KM (2,106 MI)
MIN./MAX. TEMPERATURE: -87/-5°C (-125/23°F)
LENGTH OF DAY: 1.03 EARTH DAYS
LENGTH OF YEAR: 1.88 EARTH YEARS
NUMBER OF MOONS: 2
PLANETARY RING SYSTEM: NO

The walls of Victoria crater, Mars.
(Inset) Icy clouds drifting over planet Mars

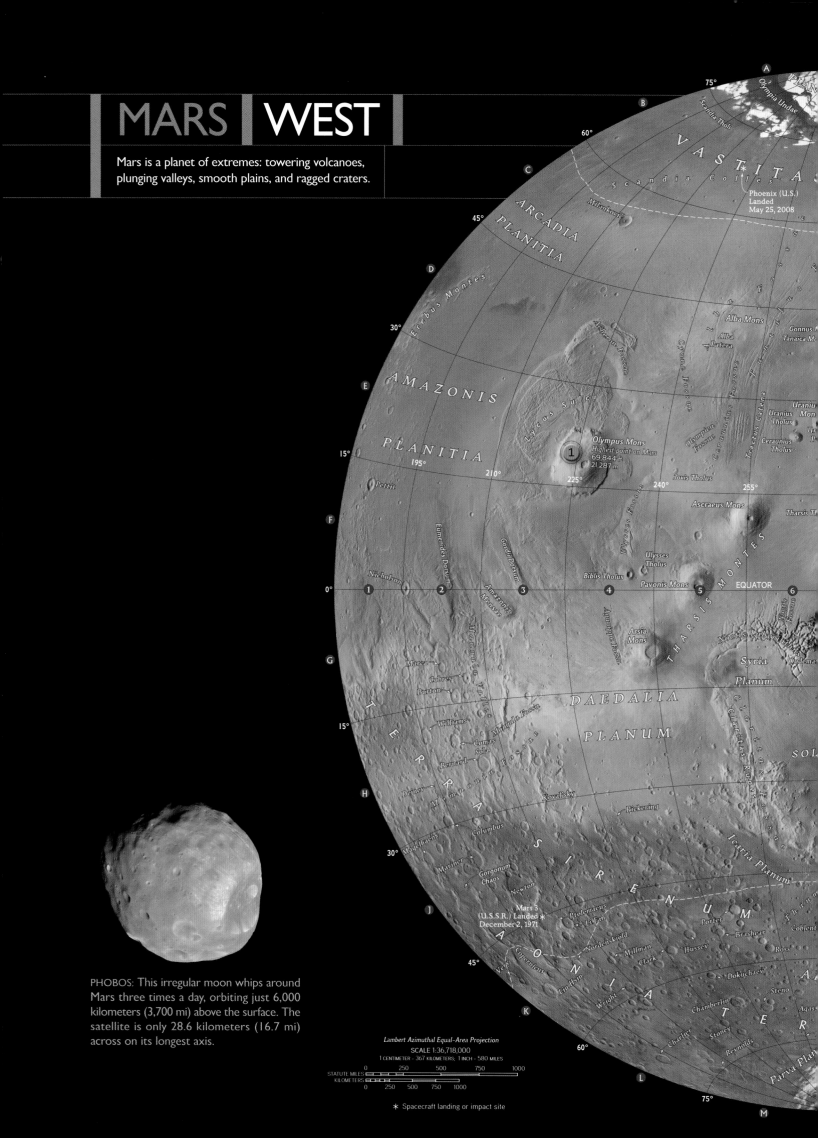

MARS | WEST

Mars is a planet of extremes: towering volcanoes, plunging valleys, smooth plains, and ragged craters.

PHOBOS: This irregular moon whips around Mars three times a day, orbiting just 6,000 kilometers (3,700 mi) above the surface. The satellite is only 28.6 kilometers (16.7 mi) across on its longest axis.

Lambert Azimuthal Equal-Area Projection
SCALE 1:36,718,000
1 CENTIMETER = 367 KILOMETERS; 1 INCH = 580 MILES

STATUTE MILES
0   250   500   750   1000

KILOMETERS
0   250   500   750   1000

✳ Spacecraft landing or impact site

VASTITAS

Olympia Undae

Scandia Tholi

Scandia Colles

ARCADIA PLANITIA

Milankovic

Phoenix (U.S.)
Landed
May 25, 2008

Erebus Montes

Alba Mons

Alba Patera

Gonnus

Tanaica Mo

AMAZONIS

Acheron Fossae

PLANITIA
195°

Lycus Sulci

Olympus Mons
Highest point on Mars
69,844 ft
21,287 m

Uraniu Mon

Uranius Tholus

Ceraunius Tholus

Olympica Fossae

Ceraunius Fossae

Tractus Catena

Petrit

210°

225°

240°

255°

Jovis Tholus

Ascraeus Mons

Tharsis T

Eumenides Dorsum

Gordii Dorsum

Ulysses Fossae

Ulysses Tholus

Biblis Tholus

Pavonis Mons

EQUATOR

Nicholson

Amazonis Mensa

THARSIS MONTES

Noctis

Mareotis Fossae

Syria

Ondema

Agnitha Fossae

Arsia Mons

Planum

Marca

Cobres

Burton

Memnonia Fossae

Mangala Fossa

DAEDALIA

Claritas Fossae

Claritas Rupes

TERRA

Williams

Comas Sola

PLANUM

Bernard

Gigas Sulci

Kovalsky

Icaria Planum

Depira

Mangala Valles

Pickering

SOL

Columbus

Midaethaius

Mariner

Gorgonum Chaos

SIRENUM

Newton

Mars 3
(U.S.S.R.) Landed ✳
December 2, 1971

Ptolemaeus

Li Fan

Porter

Brashear

Coblent

Nordenskiold

Millman

Hussey

Ross

Clark

Dokuchaev

Indus

Wright

Chamberlin

Agass

Copernicus

Charlier

Stoney

Reynolds

AONIA

TER

Steno

Parva Plan

KEY FEATURES

① **OLYMPUS MONS:** An enormous shield volcano

② **VALLES MARINERIS:** A canyon system 4,000 km (2,500 mi) long and over 8 km (5 mi) deep

③ **CHRYSE PLANITIA:** A smooth plain that shows evidence of water erosion

Viking 1 (U.S.)
Landed
July 20, 1976 ✳

Mars Pathfinder
(U.S.) Landed
July 4, 1997
✳

Opportunity
(U.S.) Landed
Jan. 25, 2004

Mars 6
(U.S.S.R.)
Crashed
March 12, 1974

*With the absence of sea level,
elevations are referenced to a
3,390 km radius sphere.*

CARTOGRAPHER'S NOTE: This color mosaic
of the red planet shows how Mars would
look to human eyes observing from orbit.
Constructed from many thousands of images
returned from NASA's Mars Global Surveyor,
the rocky, desolate terrain of the surface
and the distinctive red hue of the planet's
regolith are detailed. At the poles frigid ice
caps coat the surface, advancing and receding
with the Martian seasons.

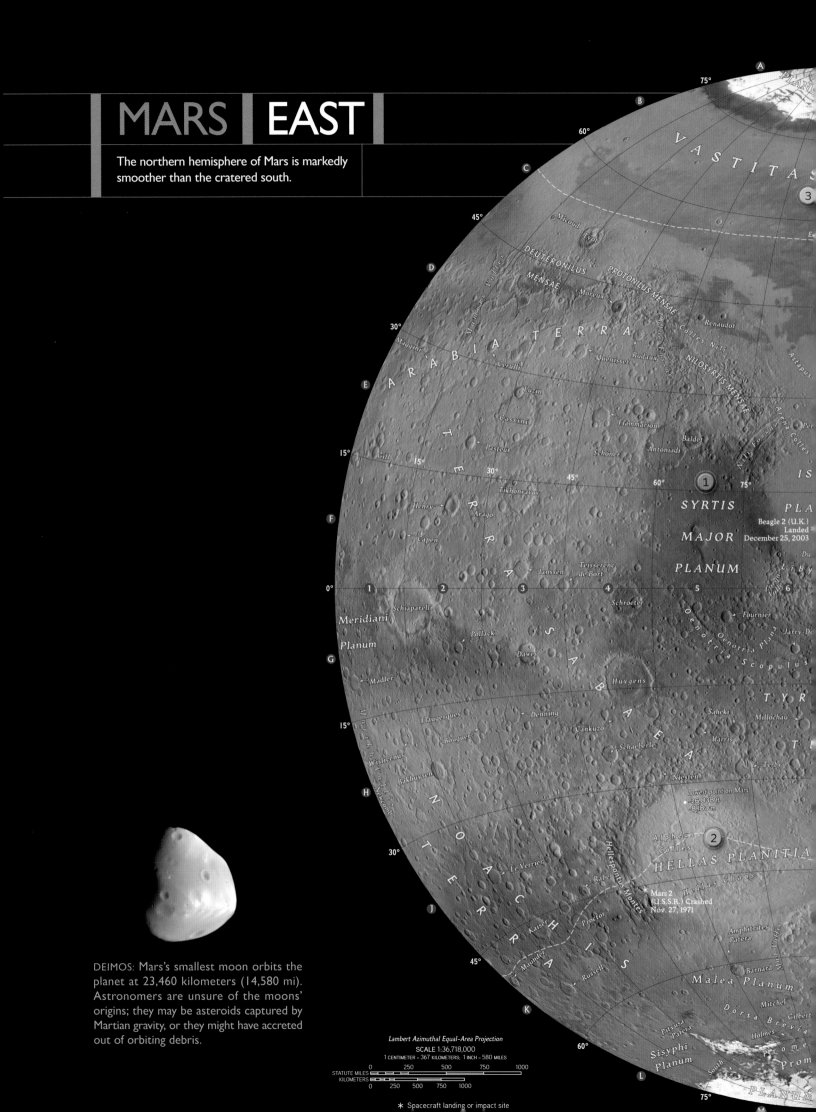

MARS | EAST

The northern hemisphere of Mars is markedly smoother than the cratered south.

A

75°

B

60°

PLANI

VASTITAS

C

3

45°

Miscud

Lyot

DEUTERONILUS

MENSAE

PROTONILUS MENSAE

Colles Nili

Moreux

Renaudot

NILOSYRTIS MENSAE

Astapus

30°

Magging

A R A B I A   T E R R A

Cerulli

Quenisset

Rudaux

Nili Fossae

Arena Colles

Per

Luzin

Flammarion

E

Cassini

Baldet

IS

15°

Gill

15°

Pasteur

Schöner

Antoniadi

60°

1

75°

T

Tikhonravoy

45°

SYRTIS

PLA

Henry

Arago

Beagle 2 (U.K.)
Landed
December 25, 2003

Capen

F

MAJOR

Du

E

Teisserenc
de Bort

Liby

Janssen

PLANUM

Turno Valles

R

0°

1

2

3

4

5

6

Schroeter

Schiaparelli

Fournier

Meridiani

A

Oenotria Plana

Jarry-De

Pollack

Oenotria

Scopulus

Planum

Dawes

G

Huygens

Mädler

B

T Y R

Flaugergues

Denning

A

Saheki

Millochau

T

15°

Bouguer

Cankuzo

Harris

Schaeberle

E

Terby

Wislicenus

A

Niesten

Bakhuysen

Newcomb

Lowest point on Mars
-26,838 ft
-8,180 m

H

Alpheus
Colles

2

N

Le Verrier

HELLAS PLANITIA

O

A

Rabe

Hellas Chaos

30°

Mars 2
(U.S.S.R.) Crashed
Nov. 27, 1971

Hellespontus Montes

C

Kaiser

Proctor

Amphitrites
Patera

H

Russell

J

I

Mätinej

Malea Planum

45°

S

Barnard

Mitchel

Dorsa Brevia

Gilbert

K

Pityusa
Patera

Holmes

Prom

Sisyphi
Planum

South

Prom

60°

PLANUM

L

75°

DEIMOS: Mars's smallest moon orbits the planet at 23,460 kilometers (14,580 mi). Astronomers are unsure of the moons' origins; they may be asteroids captured by Martian gravity, or they might have accreted out of orbiting debris.

Lambert Azimuthal Equal-Area Projection
SCALE 1:36,718,000
1 CENTIMETER = 367 KILOMETERS; 1 INCH = 580 MILES

0   250   500   750   1000
STATUTE MILES
KILOMETERS
0   250   500   750   1000

✳ Spacecraft landing or impact site

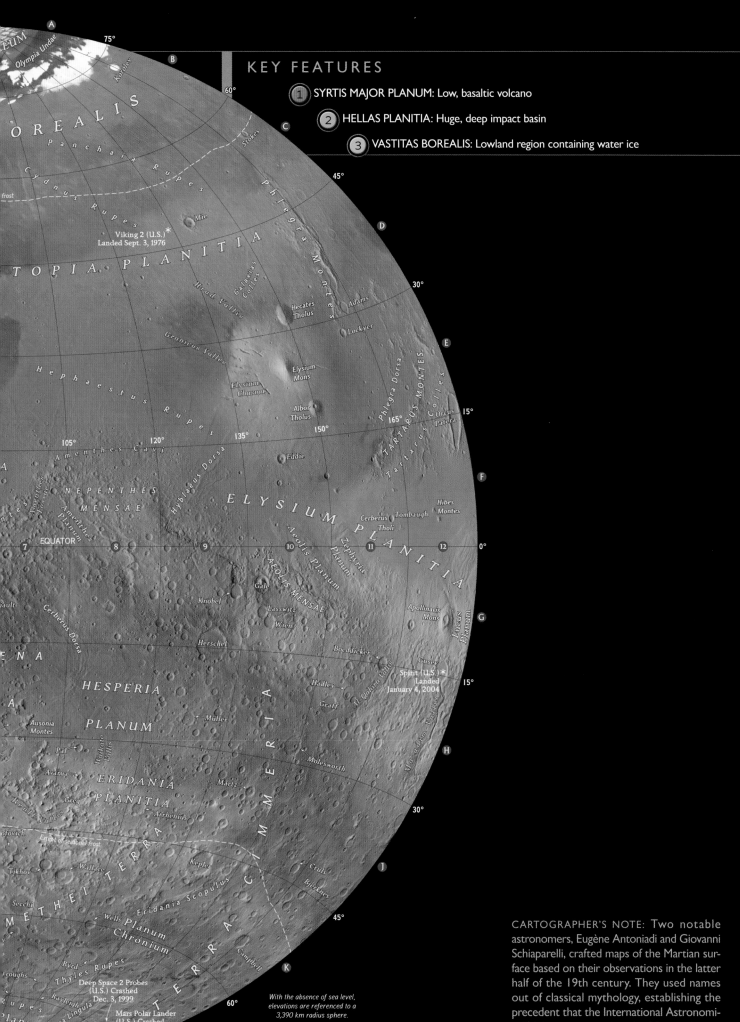

KEY FEATURES

(1) SYRTIS MAJOR PLANUM: Low, basaltic volcano

(2) HELLAS PLANITIA: Huge, deep impact basin

(3) VASTITAS BOREALIS: Lowland region containing water ice

Olympia Undae

BOREALIS

Panchaia Rupes

Cydnus Rupes

frost

Korolev

Stokes

Phlegra Montes

Mie

Viking 2 (U.S.) ✱
Landed Sept. 3, 1976

UTOPIA PLANITIA

Galaxias Colles

Hrad Vallis

Hecates Tholus

Adams

Lockyer

Granicus Valles

Hephaestus Rupes

Elysium Chasma

Elysium Mons

Phlegra Dorsa

TARTARUS MONTES

Tartarus Colles

Orcus Patera

Albor Tholus

Amenthes Cavi

Eddie

NEPENTHES MENSAE

Amenthes Planum

Hyblaeus Dorsa

ELYSIUM PLANITIA

Cerberus Tholi

Tombaugh

Hibes Montes

EQUATOR

Amenthes Fossae

Aeolis Planum

Zephyria Planum

Aeolis Mensae

Gale

Knobel

Lasswitz

Wien

Apollinaris Mons

Lucus Planum

Cerberus Dorsa

Herschel

Boeddicker

Gusev

Spirit (U.S.) ✱
Landed
January 4, 2004

HESPERIA

Hadley

Graff

Ma'adim Vallis

Al-Qahira Vallis

Ausonia Montes

PLANUM

Müller

ERIDANIA PLANITIA

Molesworth

Pal

Avarua

Mendel Tollis

Greg

Martz

Arrhenius

Reull Vallis

Kepler

Cruls

Bjerknes

Tikhov

Wallace

Eridania Scopulus

Secchi

Wells Planum

Chronium

Byrd

Rupes

Thyles Rupes

**Deep Space 2 Probes
(U.S.) Crashed
Dec. 3, 1999**

Rayleigh

Ultima Lingula

**Mars Polar Lander
(U.S.) Crashed
Dec. 3, 1999**

Ultimi Scopuli

AUSTRALE

TERRA CIMMERIA

Campbell

*With the absence of sea level,
elevations are referenced to a
3,390 km radius sphere.*

CARTOGRAPHER'S NOTE: Two notable astronomers, Eugène Antoniadi and Giovanni Schiaparelli, crafted maps of the Martian surface based on their observations in the latter half of the 19th century. They used names out of classical mythology, establishing the precedent that the International Astronomical Union came to adopt for Mars and most of the other bodies in our solar system.

The axis of rotation for Mars is tilted at approximately the same angle as that of Earth, so the red planet has seasons, just as we do. Its "year" is about twice as long as ours, however, so each of its seasons lasts twice as long as those on Earth. During the winter in each hemisphere no sunlight reaches the pole, as on Earth, but on Mars large amounts of carbon dioxide freeze out of the atmosphere to produce a thick layer of frozen carbon dioxide—the stuff we know as dry ice on Earth. This dry ice layer at the poles disappears when the sun returns. Under the dry ice are large, permanent polar caps of water ice. The water ice in Mars's northern ice cap contains a little less than half the ice found in the Greenland ice sheet on Earth.

Mars has a relatively smooth northern hemisphere formed by lava flows and a more complex southern hemisphere displaying old impact craters. Current theories suggest that both hemispheres were covered by oceans early in Mars's history—presumably the water evaporated and, because of the planet's small size, was eventually lost to space. Scientists believe that the most recent ocean was located in the northern hemisphere. Moving southward, we pass through a transitional terrain (one author refers to it as "beachfront property") before we get to the rough, cratered southern hemisphere.

MOUNTAINS AND VALLEYS

Many of the remarkable features on the Martian surface, most visible only in data sent back by spacecraft, are eerily reminiscent of familiar formations on our own planet—only more so. Two deserve special attention. Olympus Mons (Mount Olympus), an extinct volcano, is the largest mountain yet discovered in the solar system. With a height of about 27 kilometers (18 mi), it is over three times as tall as Mount Everest. Valles Marineris (Mariner Valleys) is

a canyon system about 4,000 kilometers (2,600 mi) long and up to 7 kilometers (5 mi) deep. (For reference, the Grand Canyon is about 450 kilometers/300 miles long and up to 2 kilometers/1.3 miles deep.)

THE SEARCH FOR WATER

In 1877 the Italian astronomer Giovanni Schiaparelli produced the first detailed map of the Martian surface. Using a telescope, he saw lines on the surface, which he called *canali* (channels). Unfortunately, this was translated as "canals" in English, which suggested the presence of intelligent life on the planet. Schiaparelli was followed by the American astronomer Percival Lowell, whose book *Mars as the Abode of Life* brought the concept of an inhabited

A cutaway view of Mars. Like the moon, Mars has no tectonic activity. It has a solid core, mostly iron, and a mantle. The crust, with an average thickness of about 50 kilometers (30 mi), is slightly thicker than Earth's.

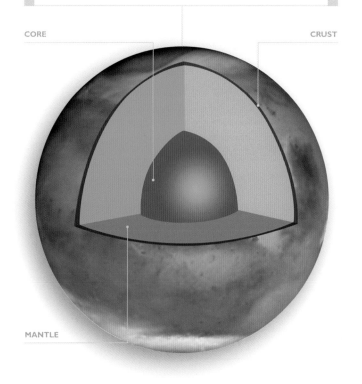

CORE

CRUST

MANTLE

The first intimation that water might once have flowed on the Martian surface was the discovery of gullies like these in the southern highlands, resembling water-formed channels on Earth.

Mars to the attention of the general public. Lowell not only claimed to see canals but also reported on how they filled with water or drained according to the seasons.

The theory promulgated at the time was one of those elegant, beautiful, and wrong ideas that pop up now and again. Mars was the home of a dying civilization, Earth the home of a flourishing civilization, while hot, swampy Venus represented the future. We now know that Lowell's canals were optical illusions, and his conclusions were based on the well-known tendency of humans to see patterns in random assortments of images (think Rorschach test).

The modern exploration of Mars started in 1964, when Mariner 4 flew by the planet, and gained detail in 1971, when Mariner 9 went into orbit. The biggest surprises that these spacecraft produced for scientists were photographs of gullies on Mars, gullies that looked for all the world like ordinary river watersheds on Earth. This was the first genuine intimation we had that there was once liquid water on the Martian surface, an idea that is now well accepted in the scientific community.

The biggest Martian event for the general public, though, was the landing of the Viking 1 and 2 spacecraft in 1976. The resulting photographs from the Martian surface were one of the first big hits on the nascent Internet and were seen in newspapers and magazines around the world.

NASA's Mars Global Surveyor, launched in 1996, was the next important exploration mission. In its ten years of operation while in orbit around Mars, this spacecraft produced a detailed map of the Martian surface. In 1997 the first robotic vehicle—Sojourner—dropped to the Martian ground. This was the first successful landing of a rover on another planet, and it pioneered the technique of surrounding the rover with airbags to cushion the landing, then deflating the airbags to allow the vehicle to move. Originally scheduled for a month of operation, Sojourner sent back data for three, establishing a precedent of longevity for future Mars rover missions.

MODERN EXPLORATION

The 21st century has seen a veritable flotilla of spacecraft, landers, and orbiters launched toward the red planet. In addition to NASA's projects, European Space Agency, Russian, Chinese, and Finnish missions are either in progress or on the drawing boards. The most dramatic missions were those of the Mars exploration rovers Spirit and Opportunity, which landed successfully in 2004. These rovers examined Martian rocks and minerals, establishing in short order that liquid water once existed on the Martian surface.

Expected to operate for only a few months, both rovers functioned for an incredible six *years*—an amazing tribute to the engineers that built them. Scientists link the long lifetime of these vehicles to the fact that storms and dust devils on the Martian surface kept the solar panels free of grit, allowing both rovers to operate at full power. In 2010 Spirit got stuck in deep sand, and, after prolonged attempts to free her, was converted to a stationary observation station. By this time

THE FACE ON MARS

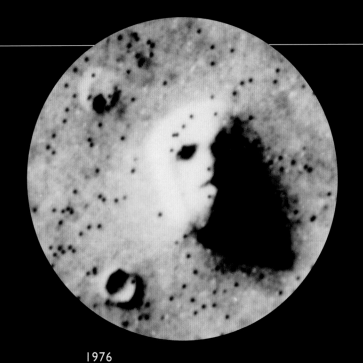

1976

We might as well get down to the only result of the missions to Mars to wind up in tabloids at supermarket checkout counters—the so-called face on Mars. On July 25, 1976, the Viking 1 spacecraft was in orbit around the red planet, photographing possible landing sites for its sister ship, Viking 2. Over a region of the planet's northern hemisphere that marks a kind of coastal zone between the smooth plains of the north and the cratered terrains of the south, a low-resolution photo showed what appeared to be a face staring up at the craft, surrounded by Egyptian-style pyramids. The scientists at control center smiled—clearly it was an optical illusion—but some NASA functionary decided that releasing the photo would be a good way to generate public interest in Martian exploration.

Well, I guess! For years after that decision, the face on Mars became a staple of fringe science. I can recall, for example, a tabloid headline that blared out the news that the face was that of Elvis Presley!

It wasn't until 1998 that the Mars Global Surveyor revisited the site of the face and took photos with much higher resolution (and different lighting). As expected, the face disappeared, a result borne out by several subsequent missions. Though it was fun while it lasted, the face on Mars has joined Lowell's canals in the collection of objects that wishful thinkers

An artist's conception shows what the Mars rover Curiosity will look like on the Martian surface. The most advanced rover ever built by NASA, it was launched in 2011; its main task will be to determine whether Mars is now or ever has been capable of supporting life. It will not look for life directly, but it will search for chemical and mineralogical evidence.

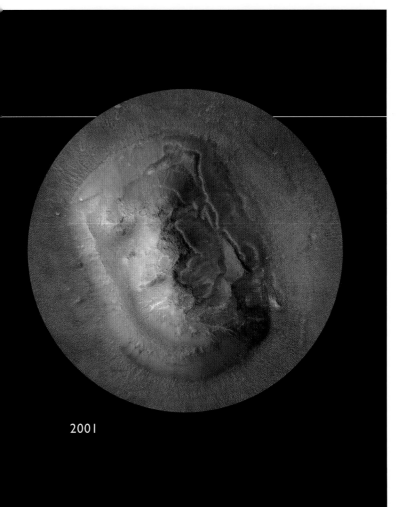

2001

Opportunity had traveled no less than 20 kilometers (14 mi) over the Martian surface.

The next stage of exploration began in 2011, with the launch of the Mars Science Laboratory. A Volkswagen Beetle–size vehicle dubbed Curiosity, this rover will carry scientific instruments from six different countries. An important part of its mission will be to look for organic molecules and to determine whether Mars is now or ever was host to microbial life. It also is expected to provide data that will determine whether Mars is capable of supporting life in the future. This is, of course, a prelude to possible human colonization and the realization of the dreams of all those science-fiction writers.

Dividing the terrestrial and Jovian planets in our solar system is a thin ring of debris known as the asteroid belt. Before we get into a description of the asteroids, we must deal with two popular misconceptions: First, despite the crowded, rock-strewn scenes you may have witnessed in the movies, the asteroid belt is almost completely empty. Spacecraft have flown through the belt without encountering asteroids at all. In fact, the chance of meeting an asteroid during a transit has been estimated to be about one in a billion. Second, the belt is not the remains of an exploded planet—it contains only a tiny fraction of a planet's mass. The "exploded planet" theory was popular in the 19th century, and probably served as the inspiration for Krypton, Superman's fictional home planet.

ASTEROID BELT

[REMNANTS FROM THE SOLAR SYSTEM'S BIRTH]

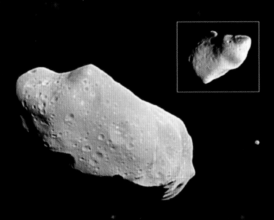

DISCOVERER: GIUSEPPE PIAZZI
DISCOVERY DATE: JANUARY 1, 1801
NAMED FOR: GREEK WORD FOR "STARLIKE"

LOCATION OF MAIN BELT: BETWEEN 2.1 AND 3.3 AU FROM SUN
TOTAL NUMBER: OVER 570,000
DIAMETER OF LARGEST ASTEROIDS:
CERES: 950 KM (640 MI)
VESTA: 580 KM (360 MI)
PALLAS: 540 KM (335 MI)
HYGIEA: 430 KM (270 MI)
NEAR-EARTH ASTEROIDS DISCOVERED TO DATE: 8,484

Artist's conception of Ceres, largest asteroid.
(Inset) Asteroid Ida and its satellite Dactyl

The first asteroid, Ceres, was discovered in 1801. Because the telescopes of that time produced pointlike images of asteroids, the small, dimly reflective bodies resembled stars, hence their name ("asteroid" comes from the Greek for "star-like"). It is no accident that Ceres was the first to be seen: It is the largest body in the asteroid belt at about 950 kilometers (640 mi) across. Ceres is the only object in the belt big enough to be spherical—the mutual gravitational attraction between its constituents pulls it into that shape—and it comprises about a third of the total mass in the asteroid belt. Technically, this makes it a dwarf planet (see page 184). Everything else in the belt is smaller and irregular in shape. It is estimated that there are around 200 asteroids bigger than 100 kilometers (60 mi) across and perhaps more than a million larger than 1 kilometer (0.6 mi).

Not all asteroids are found in the asteroid belt. Some travel in paths that bring them inside the orbits of Mars and Earth—which raises the possibility that they might collide with either planet.

HOW THE BELT TOOK SHAPE

When the solar system was forming, the process of planetesimal accumulation (see pages 45–46) went on in the asteroid belt as it did throughout the inner solar system. Had it not been for the presence of Jupiter, there might well have been a planet where the belt is now. One theory has it that the gravitational pull of the giant planet sped up nearby planetesimals in their orbits, so that they were either knocked out of the asteroid belt by the influence of Jupiter directly or split up when they collided with each other. In either case, it would appear that the influence of Jupiter was able to prevent a planet from forming.

TARGET: EARTH

On a peaceful day 65 million years ago, dinosaurs were going about their daily business when a piece of rock about 12 kilometers (8 mi) across came hurtling out of the sky. Carrying an energy equivalent to many thousands of times the total modern arsenal of nuclear weapons, the asteroid hit near what is now the Yucatán Peninsula in Mexico, blasting out a crater 180 kilometers (112 mi) across and several kilometers deep. The impact and the pulverized rock thrown up by the impact set in motion a train of events that wiped out fully two-thirds of all the species of plants and animals on Earth, including the dinosaurs. Aptly termed a mass extinction event, the extinction of the dinosaurs and all those other species marks just one of many such events in the history of our planet.

More recently, in 1908, a rock some tens of meters across (roughly 50 feet or so) fell out of the sky and exploded in the air over the Tunguska River in Siberia. The impact, estimated to have a thousand times the energy of a World War II–era atomic bomb, leveled trees as far as 70 kilometers (50 mi) from the blast center.

These two events highlight an important truth: Earth is part of the larger solar system, and occasionally an asteroid shows up to remind us of that fact. You can verify this for yourself by watching for shooting stars in the sky, pebble-size rocks burning up in the atmosphere.

In fact, collisions with objects like the one that did in the dinosaurs are expected to occur every 100 million years or so, and smaller collisions are expected to be more frequent—a kilometer-size (1,000 yd) object will hit every 70,000 years, and a rock 140 meters (153 yd) across every 30,000 years. A Tunguska-size event could happen every few centuries. Asteroids in the size range of 5 to 10 meters (15 to 30 ft) in diameter enter Earth's atmosphere about once a year, but typically explode at high altitude and cause little (if any) damage.

In 2005, recognizing the danger of these impacts, Congress directed NASA to catalog 90 percent of all detectable asteroids, comets, and other potentially dangerous objects near Earth by 2020. (Funding shortfalls make it unlikely that this deadline will be met.) This project is important, because we can't assume we'll always be lucky enough to have the serious impacts occur only in sparsely settled areas. As to what we'd do if we found an asteroid out there with our name on it: We're not sure. Our response to a killer asteroid is still largely a matter of speculation, despite what we see in the movies.

Indeed, computer models suggest that much of the material in the original asteroid belt was ejected in the first million years or so of the solar system's history, with a good portion of the rest being expelled during the Late Heavy Bombardment. During that early period, the minerals in asteroids were subject to influences such as heating from collisions, the radioactive decay of nuclei, and, in some cases, the same process of differentiation that produced the structure of Earth (see page 76). Studying these ancient bodies should cast some light on the way the solar system formed.

MISSIONS TO THE ASTEROIDS

Beginning in 1972, a number of spacecraft—the Pioneers, the Voyagers, and Ulysses—passed through the asteroid belt without mishap, but none of them tried to image the asteroids they passed. Since then, we have obtained flyby images from a number of spacecraft bound for other targets—Galileo on its way to Jupiter, for example, and Cassini traveling to Saturn. The probe NEAR (Near Earth Asteroid Rendezvous) went into orbit around the near-Earth asteroid Eros in 2000. In 2010 the Japanese probe Hayabusa returned after a seven-year, five-billion-kilometer (3 billion mi) landing mission to a small asteroid named Itokawa. Although the spacecraft burned up on entry into Earth's atmosphere, the sample return packet, containing small particles of the asteroid, landed safely in Australia. Analysis showed that the asteroid indeed dated to the early days of the solar system.

NASA's DAWN mission, launched in 2007, went into orbit around the massive asteroid Vesta in 2011; in 2015 it will move on to visit Ceres. Its mission will be to study each of these bodies in some detail, with the hope of learning more about the formation of the solar system.

N E P T U N E

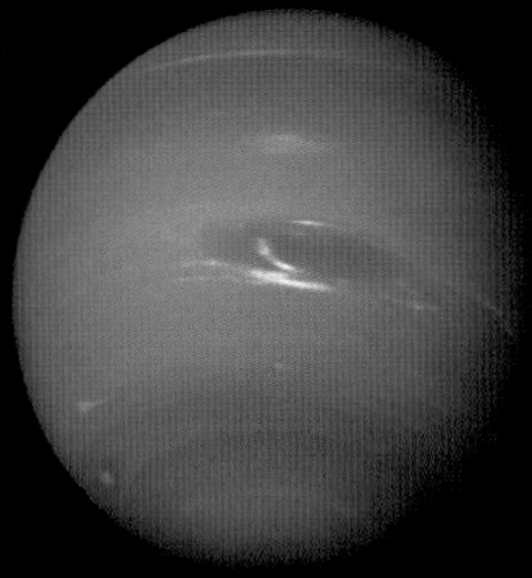

Avg. distance from the sun:	4,495,100,000 km
Perihelion:	4,444,450,000 km
Aphelion:	4,545,670,000 km
Revolution period:	163.84 years
Average orbital speed:	5.4 km/s
Average temperature:	-200°C
Rotation period:	16.1 hours
Equatorial diameter:	49,528 km
Mass (Earth=1):	17.1
Density:	1.64 g/cm^3
Surface gravity (Earth=1):	1.12
Known satellites:	13
Largest satellite:	Triton

Image by: Voyager II

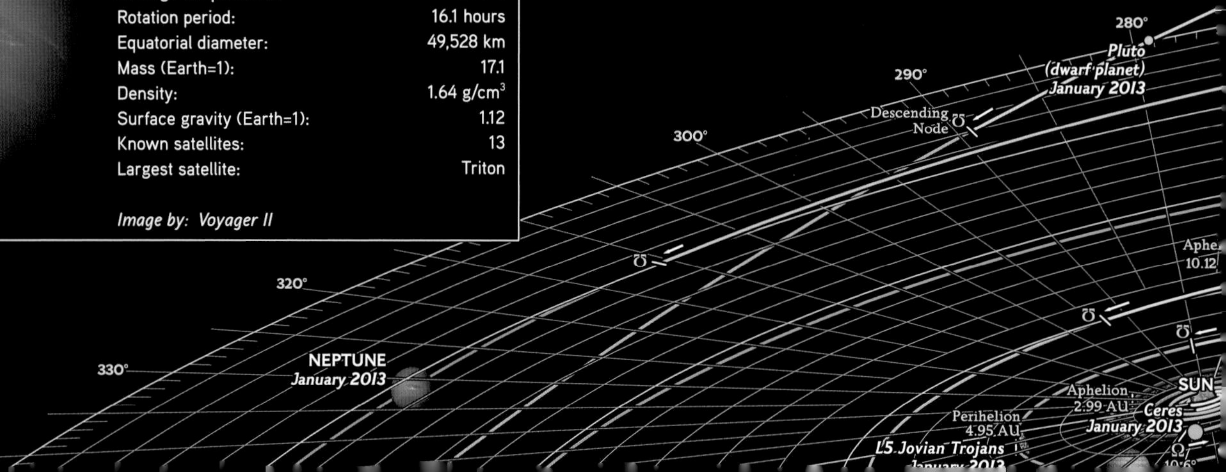

280°

Pluto
(dwarf planet)
January 2013

290°

Descending
Node

300°

Aphe.
10.12

NEPTUNE
January 2013

320°

330°

Aphelion
2.99 AU

SUN

Perihelion
4.95 AU

Ceres
January 2013

340°

L5 Jovian Trojans
January 2013

Among the outer planets, Jupiter and Saturn are well studied. The Galileo probe introduced observers to each of the Jovian moons and revealed the largest planet's character. The Cassini-Huygens ongoing mission continues to relay data on Saturn and its satellites, with special attention given to Titan. The other outer planets, visited only by the Voyager spacecraft, have many questions left unanswered.

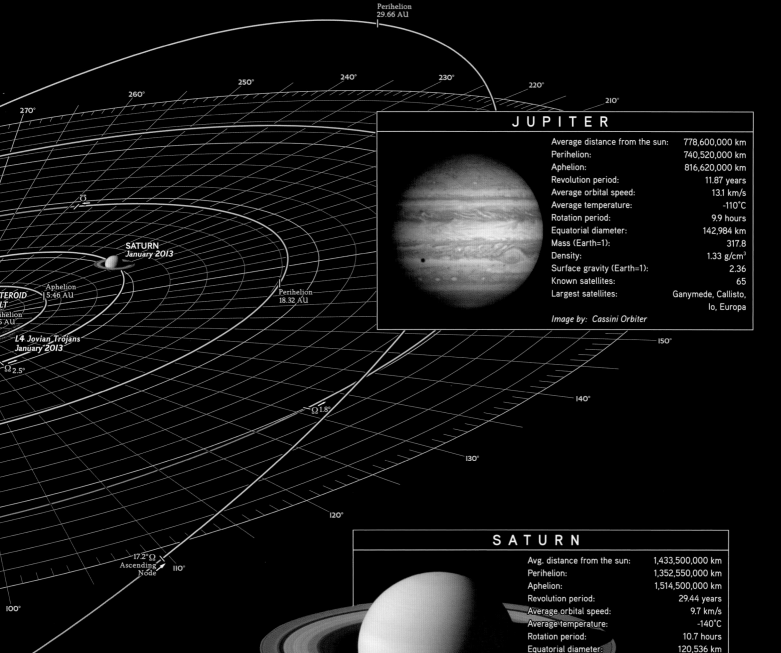

Perihelion
29.66 AU

270°   260°   250°   240°   230°   220°   210°

SATURN
January 2013

Aphelion
5.46 AU

TEROID
LT
phelion
AU

L4 Jovian Trojans
January 2013

Ω 2.5°

Ω 1.8°

Perihelion
18.32 AU

17.2° Ω
Ascending
Node

110°

100°

120°

130°

140°

150°

JUPITER

Average distance from the sun:	778,600,000 km
Perihelion:	740,520,000 km
Aphelion:	816,620,000 km
Revolution period:	11.87 years
Average orbital speed:	13.1 km/s
Average temperature:	-110°C
Rotation period:	9.9 hours
Equatorial diameter:	142,984 km
Mass (Earth=1):	317.8
Density:	1.33 g/cm^3
Surface gravity (Earth=1):	2.36
Known satellites:	65
Largest satellites:	Ganymede, Callisto, Io, Europa

Image by: Cassini Orbiter

SATURN

Avg. distance from the sun:	1,433,500,000 km
Perihelion:	1,352,550,000 km
Aphelion:	1,514,500,000 km
Revolution period:	29.44 years
Average orbital speed:	9.7 km/s
Average temperature:	-140°C
Rotation period:	10.7 hours
Equatorial diameter:	120,536 km

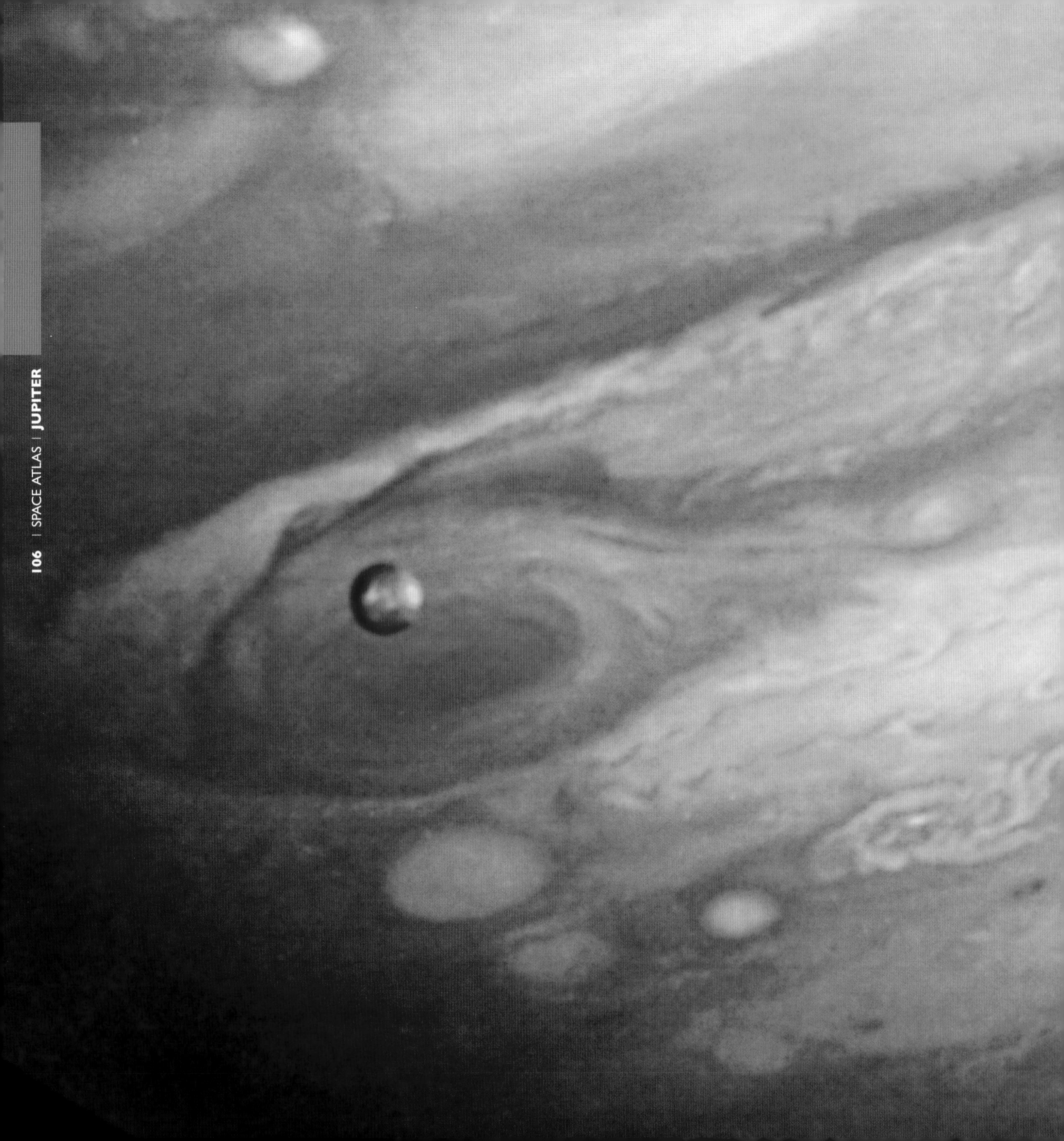

The fifth planet from the sun, Jupiter is also the largest, packing over two and a half times the mass of all the other planets combined. As seen through a telescope, it is a beautiful thing, with alternating stripes of different colors, surrounded by a swarm of small moons. • Jupiter, the first of the gas giants, exhibits some unusual traits that appear in all the outer planets. In the first place, Jupiter doesn't really have a "surface." Dropping down into Jupiter's atmosphere would be like sinking into a milkshake, with the density increasing as the surroundings changed from gas to liquid to slush. We wouldn't encounter anything we would call a solid surface until, perhaps, we got near the very center.

JUPITER

[KING OF THE GAS GIANTS]

DISCOVERER: UNKNOWN
DISCOVERY DATE: PREHISTORIC
NAMED FOR: KING OF THE ROMAN GODS

MASS: 317.82 × EARTH'S
VOLUME: 1,321.34 × EARTH'S
MEAN RADIUS: 69,911 KM (43,441 MI)
EFFECTIVE TEMPERATURE: -148°C (-234°F)
LENGTH OF DAY: 9.92 HOURS
LENGTH OF YEAR: 11.86 EARTH YEARS
NUMBER OF MOONS: 65 (51 NAMED)
PLANETARY RING SYSTEM: YES

The moon Io orbits Jupiter. (Inset) Jupiter, with shadow of Europa

JUPITER

Unlike the inner planets, Jupiter does not have a rocky surface with craters, mountains, or valleys. Its complex, stormy atmosphere changes over time.

Usually some shade of reddish-brown, warmer belts are composed of regions where low-pressure gasses are beginning to sink through the atmosphere. Where belts and zones collide, great areas of turbulence ensue, giving rise to the many storm systems that speed around Jupiter's atmosphere.

Jupiter's magnetosphere—the magnetic field that surrounds the planet—while similar to Earth's, is about 20,000 times more intense. Its pull captures charged particles from solar wind, forming vast and intense bands of radiation. The effect is similar to Earth's Van Allen belts but immensely more powerful. Any unprotected spacecraft would quickly receive a destructive dose.

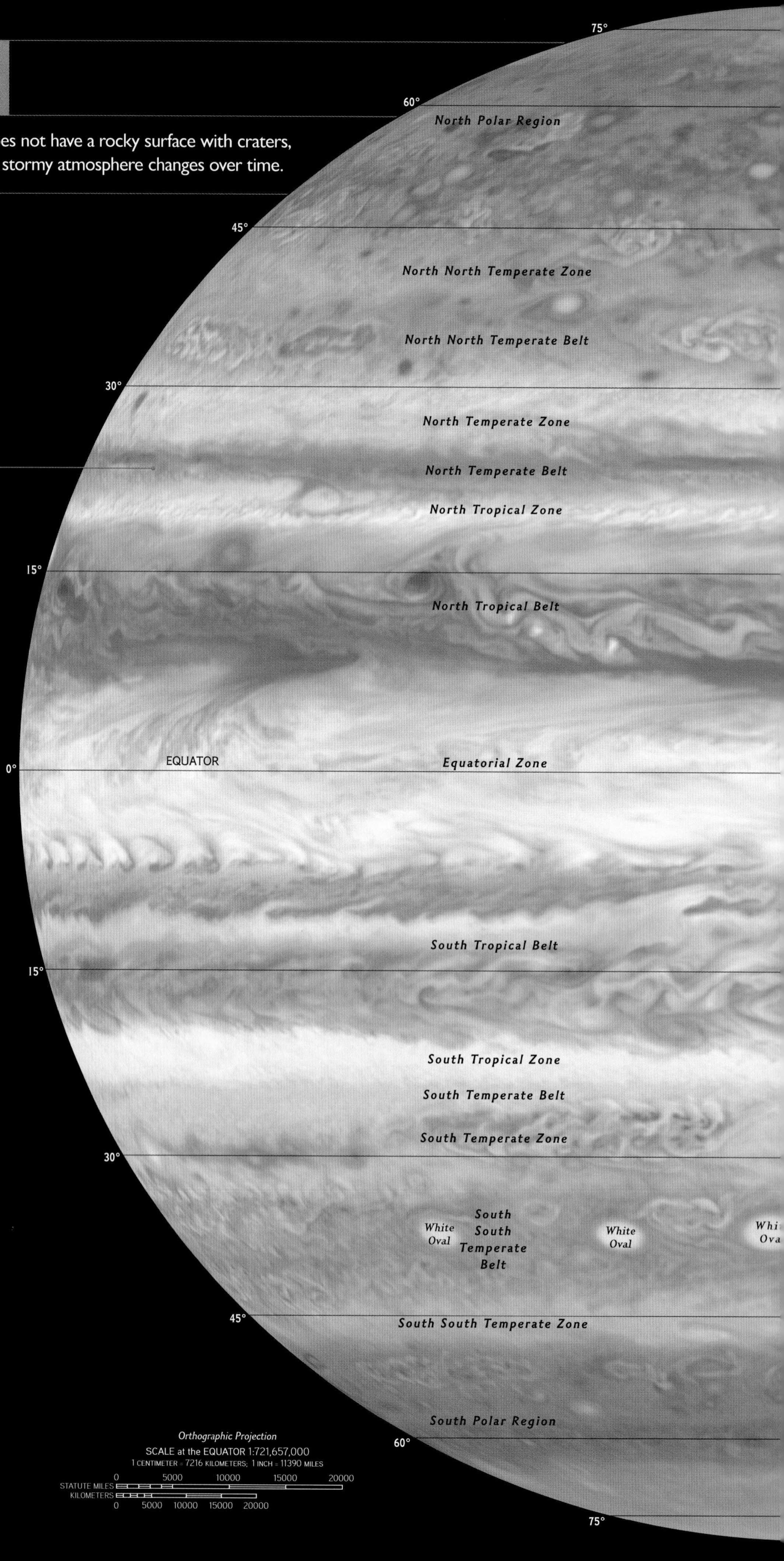

75°

60°
North Polar Region

45°

North North Temperate Zone

North North Temperate Belt

30°

North Temperate Zone

North Temperate Belt

North Tropical Zone

15°

North Tropical Belt

0° EQUATOR   Equatorial Zone

South Tropical Belt

15°

South Tropical Zone

South Temperate Belt

South Temperate Zone

30°

White
Oval

South
South
Temperate
Belt

White
Oval

Whi
Ova

45°
South South Temperate Zone

South Polar Region

60°

75°

Orthographic Projection
SCALE at the EQUATOR 1:721,657,000
1 CENTIMETER - 7216 KILOMETERS; 1 INCH = 11390 MILES

0   5000   10000   15000   20000
STATUTE MILES
KILOMETERS
0   5000   10000 15000 20000

North Polar Region

60°

1   NORTH TROPICAL BELT: The darker belts have warmer, descending air.

2   SOUTH TROPICAL ZONE: The lighter zones have colder, upwelling air.

3   GREAT RED SPOT: A long-lived anticyclone larger than Earth

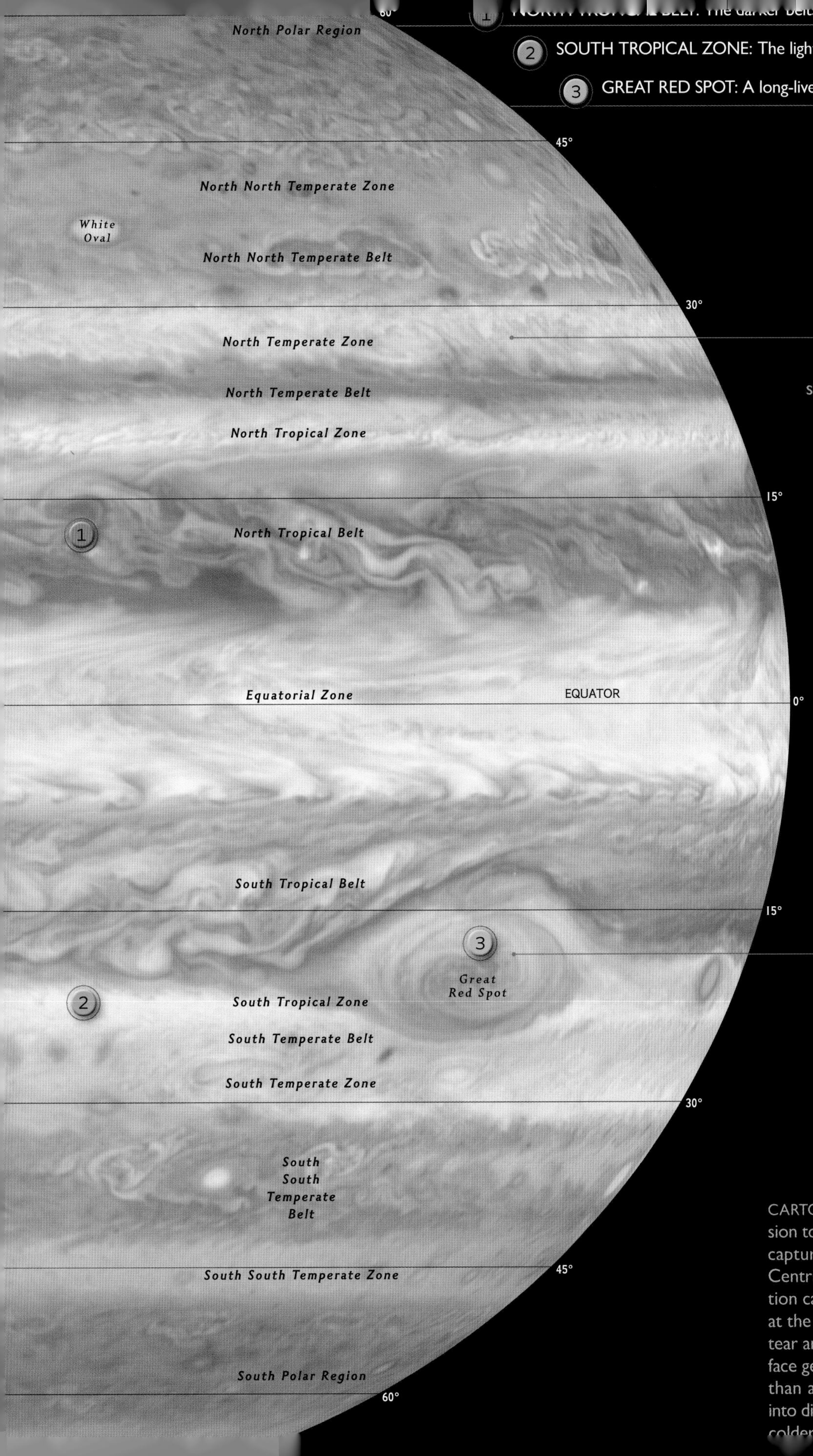

45°

North North Temperate Zone

White Oval

North North Temperate Belt

30°

North Temperate Zone

North Temperate Belt

Light in color and cooler than their surroundings, Jupiter's zones represent high-pressure rising gasses.

North Tropical Zone

15°

1

North Tropical Belt

Large enough to fit three Earths, the Great Red Spot is similar to a hurricane. Whereas storms on our planet are the result of low-pressure rising air, this maelstrom is formed by high-pressure gases spinning downward. Driven by the massive planet's internal heat, and never encountering solid terrain that would destabilize it, the storm has been in existence since telescopic observation of Jupiter began over 400 years ago.

Equatorial Zone   EQUATOR   0°

South Tropical Belt

15°

3

Great Red Spot

2

South Tropical Zone

South Temperate Belt

South Temperate Zone

30°

South South Temperate Belt

South South Temperate Zone   45°

CARTOGRAPHER'S NOTE: En route to its mission to study Saturn, NASA's Cassini spacecraft captured this ultradetailed mosaic of Jupiter. Centrifugal force from the planet's quick rotation causes the gaseous atmosphere to flatten at the poles and expand at the equator. Winds tear around the planet and, unimpeded by surface geography, they reach speeds much higher than anything observed on Earth. Organized into distinct bands—dark warmer belts and light colder zones—where they abut, titanic storms

South Polar Region

60°

The second thing to know about gas giants is that at the enormous pressures that exist inside a planet this size, matter can be forced into some pretty unusual forms, which means that the interior structures of the gas giant planets aren't like anything we've seen so far. Neither are the temperatures: These range from minus 148°C (-234°F) at the top of Jupiter's cloud layer to around 24,000°C (43,000°F)—hotter than the surface of the sun—in the core!

THE METAL OCEAN

Gravitational measurements suggest that Jupiter may have a small rocky core, possibly with as much as 20 to 40 times the mass of Earth. Surrounding this core is a layer of a strange material called metallic hydrogen. Again, we are used to thinking of hydrogen as a gas, or perhaps, at very low temperatures, as an ordinary liquid. Pressures in the interior of Jupiter are so high, however, that the atoms are forced into a state that, while still a liquid, has the properties of a metal. (If you have trouble thinking of a liquid as a metal, picture mercury.) Metallic hydrogen, an exotic rarity on Earth, actually comprises a significant fraction of the mass of Jupiter. The layer

A cutaway view of our current idea of the structure of Jupiter. Jupiter is unusual because its rocky core is surrounded by a layer of metallic hydrogen, a material not found naturally on Earth. It exists on Jupiter because of the extremely high pressures in the interior.

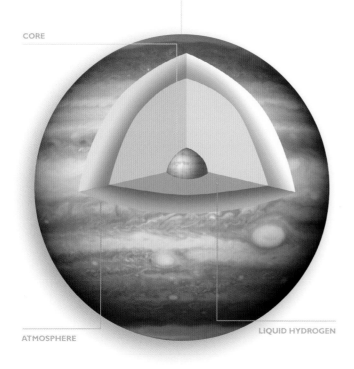

CORE

ATMOSPHERE

LIQUID HYDROGEN

A close-up view of Jupiter's swirling clouds, produced by Voyager 1, highlights the enormous storm known as the Great Red Spot.

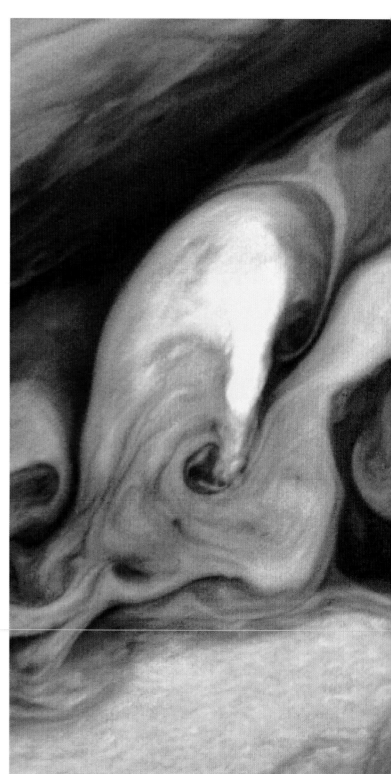

of metallic hydrogen is covered by a layer of ordinary liquid hydrogen: hydrogen atoms forced together by the tremendous pressures in the planetary interior. There probably isn't a sharp transition between the two states of hydrogen—certainly nothing we would call a surface.

BELTS AND ZONES

The upper atmosphere, of course, is what we actually see when we look at Jupiter. This outermost layer of the planet is composed almost entirely of hydrogen and helium (75 percent to 24 percent by weight). The colorful bands result from a rather complex cloud structure in the top 50 kilometers (35 mi). There are two main layers of clouds. Upwelling materials from farther down in Jupiter's atmosphere mix with the lower layer and change color when they encounter the ultraviolet rays of the sun. These are the dark bands, or "belts," that we see. The lighter bands, or "zones," are clouds of crystallized ammonia carried upward, hiding the darker lower layer from view. The deepest layers, seen only occasionally, appear blue.

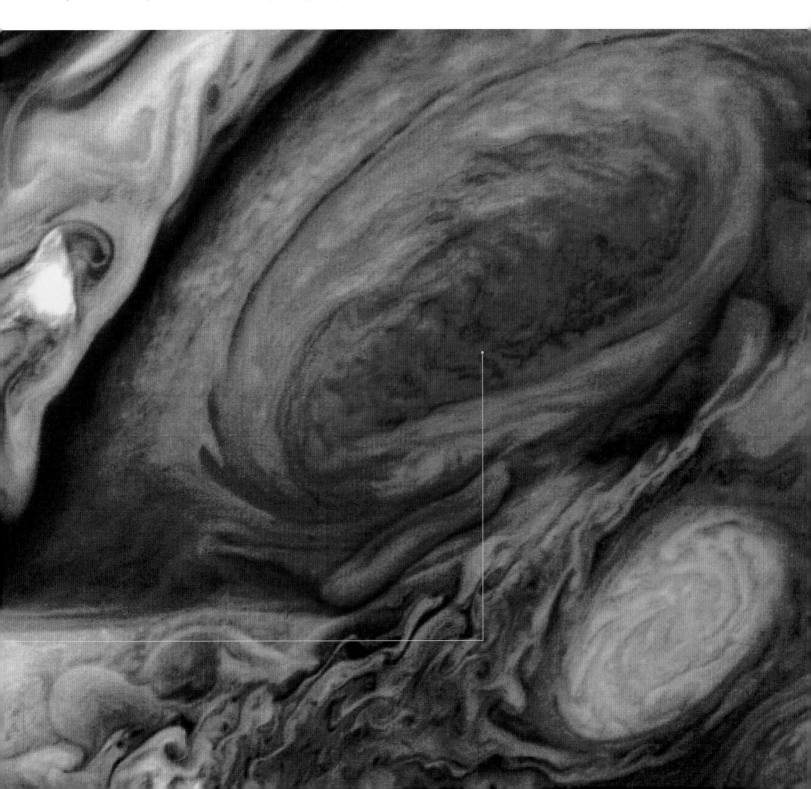

We are used to the fact that on Earth the prevailing winds blow in different directions at different latitudes—the tropical trade winds blowing from east to west, for example, while winds in higher latitudes go from west to east. The same thing happens on Jupiter, but its fast rotation (its "day" is about ten hours long) and greater size produce many more counterrotating bands than on Earth. This together with the complicated cloud dynamics discussed above is what gives rise to the colored stripes we see when we look at the planet.

As befits its role as king of the planets, Jupiter has a strong magnetic field—20,000 times as strong as Earth's—most likely produced by motions in its metallic core. This strong field has the effect of diverting charged particles streaming outward from the sun away from the planet, creating what scientists call a bow shock. The largest four moons of Jupiter have their orbits inside this protected zone.

Perhaps the most striking feature on Jupiter is the Great Red Spot, a storm in the southern hemisphere. This spot may have been seen as early as 1665, and it was sketched by astronomers in 1831. It is so big that the entire Earth could be dropped into it with no trouble. Some theorists think it may actually be a permanent feature of the planet. Interestingly enough, at the end of the 20th century, astronomers observed what may be the start of a similar but smaller storm on Jupiter, which they nicknamed Red Spot Junior.

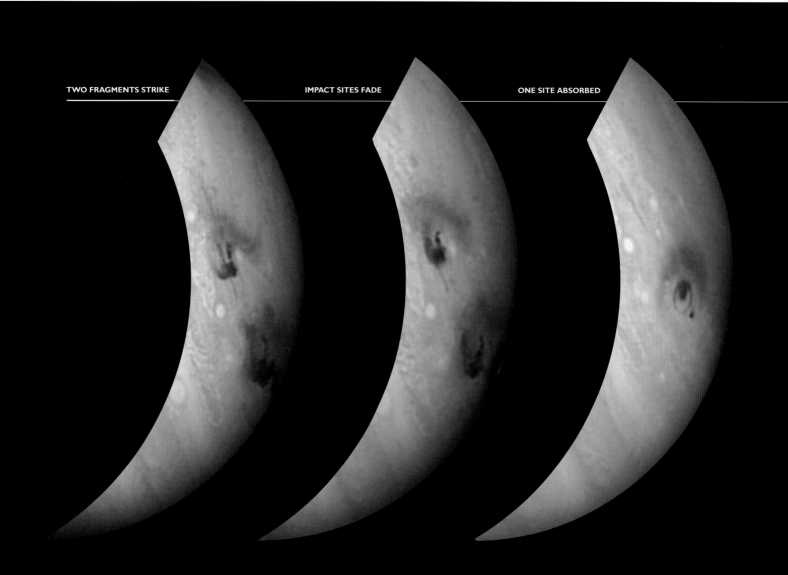

TWO FRAGMENTS STRIKE **IMPACT SITES FADE** **ONE SITE ABSORBED**

VISITING JUPITER

Jupiter is bright in the night sky. Galileo was the first to observe the planet with a telescope in 1610 and the first to document the fact that the planet has moons. Many spacecraft have flown by Jupiter on their way to other destinations, including New Horizons in 2007, while on its way to Pluto. The Voyager spacecraft in 1979 discovered that Jupiter, like all of the Jovian planets, has a system of rings. In Jupiter's case, there are three rings that appear to be made out of dust ejected from nearby moons.

The main space probe associated with Jupiter, however, was the Galileo mission, launched in 1989 and inserted into orbit around the planet in 1995. Later in 1995 it dropped a probe into the Jovian atmosphere.

The probe descended and returned data for almost an hour before it was crushed by the pressure 153 kilometers (95 mi) down. For seven years, the orbiting Galileo spacecraft gathered a trove of data on Jupiter and its moons. In 2003 the spacecraft met its end when it was deliberately steered into the Jovian atmosphere to eliminate any possibility that it would contaminate the moon Europa—a place where, as we shall see, scientists believe that we might find life.

In August 2011 NASA launched its most recent Jovian mission, named Juno. Due to reach Jupiter in 2016, the solar-powered spacecraft will circle the giant planet 33 times in a polar orbit, studying its turbulent atmosphere, its magnetic field, and its powerful gravitational field.

JUPITER TAKES A HIT

One of the most spectacular events in astronomical history took place during July 1994. A comet named Shoemaker-Levy (comets are customarily named after the people who discover them) collided with Jupiter, giving observers on Earth a ringside seat to a cosmic show.

The comet was first seen in orbit around Jupiter in 1993. Calculations indicated it had been captured by the planet's gravitational field and broken into fragments a year earlier. As it became clear that that the fragments were going to collide with Jupiter, all of the astronomical instruments available to astronomers—land-based and in space—were turned toward the giant planet. The hope was that the impacts would churn up the Jovian atmosphere and give scientists a look at what lay between the outer cloud layers.

For six days after July 16, sky-watchers observed no fewer than 21 impacts. The Galileo spacecraft (see above) was in a position to see the actual collisions, which took place on the far side of the planet, and the Hubble Space Telescope (see sidebar page 275) took spectacular pictures. The initial blows produced fireballs and churned up areas of the Jovian atmosphere as wide as the radius of Earth. Elements such as sulfur were seen in the aftermath of the collisions but, against expectations, very little water. Scientists are still poring over the data taken during the collisions, refining their theories about the structure of the largest planet in the solar system.

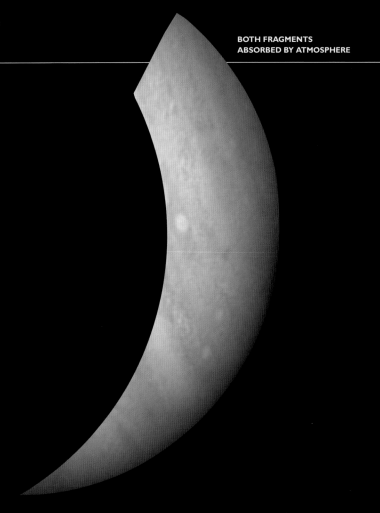

BOTH FRAGMENTS ABSORBED BY ATMOSPHERE

Jupiter has 63 moons and counting. Most of them are rocky, irregular lumps, some only a few miles across. The consensus is that these moons are captured asteroids. In fact, only eight of the Jovian moons fit the standard image of what a moon should be—spherical bodies in equatorial orbits around their planet—and of these, four are small bodies with orbits close to Jupiter. These four inner moons may be the source of the dust that makes up Jupiter's rings. From our point of view, the most important moons of Jupiter are the remaining four—the so-called Galilean moons.

JUPITER'S MOONS

[WATER WORLDS, VOLCANO WORLDS, AND ROCKS]

DISCOVERER: FIRST FOUR DISCOVERED BY GALILEO GALILEI
DISCOVERY DATE: JANUARY 1610
NAMED FOR: JUPITER'S LOVERS AND DESCENDANTS

....

LARGEST MOONS, BY RADIUS:
GANYMEDE: 2,631 KM (1,635 MI)
CALLISTO: 2,410 KM (1,498 MI)
IO: 1,822 KM (1,132 MI)
EUROPA: 1,561 KM (970 MI)
HIMALIA: 85 KM (53 MI)
AMALTHEA: 83 KM (52 MI)
THEBE: 49 KM (30 MI)

The volcanic surface of Jupiter's moon Io. (Inset, left to right)
The moons Io, Europa, Ganymede, and Callisto

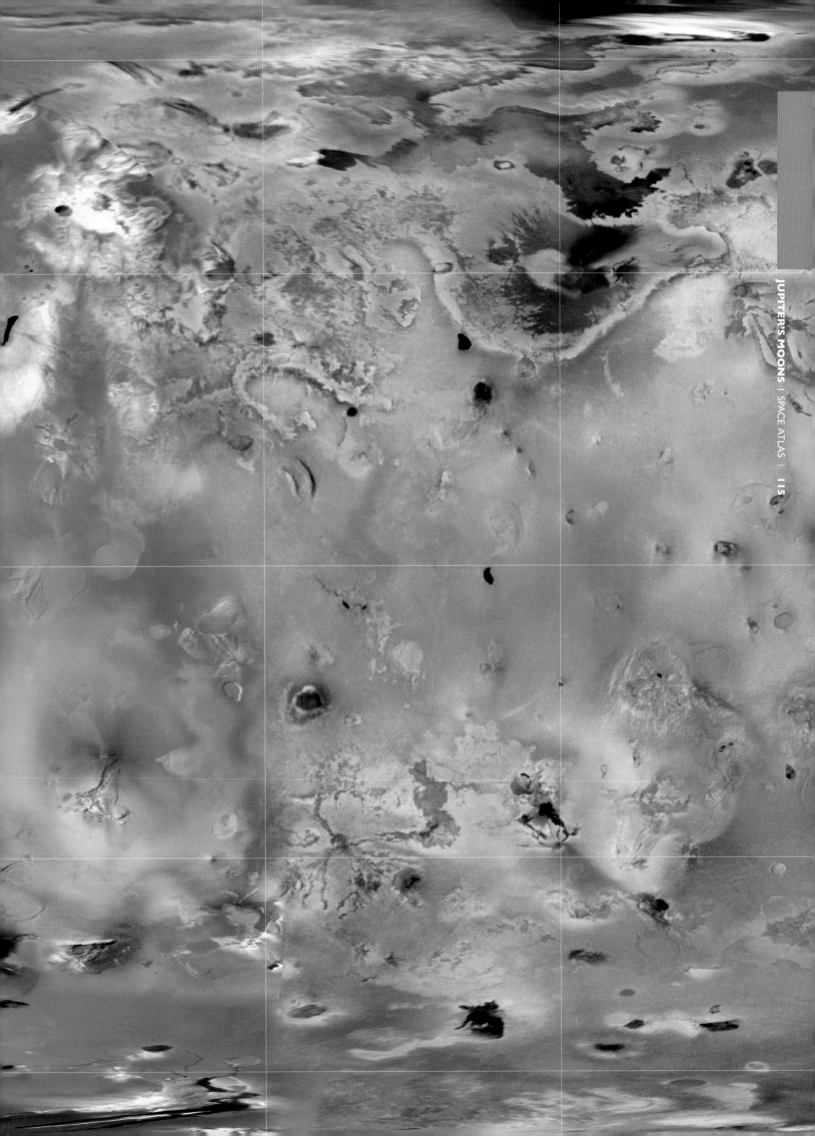

Stretched and pulled between Jupiter's massive gravity and the other Galilean moons, tidal heating makes Io the most volcanic body in the solar system.

WESTERN HEMISPHERE

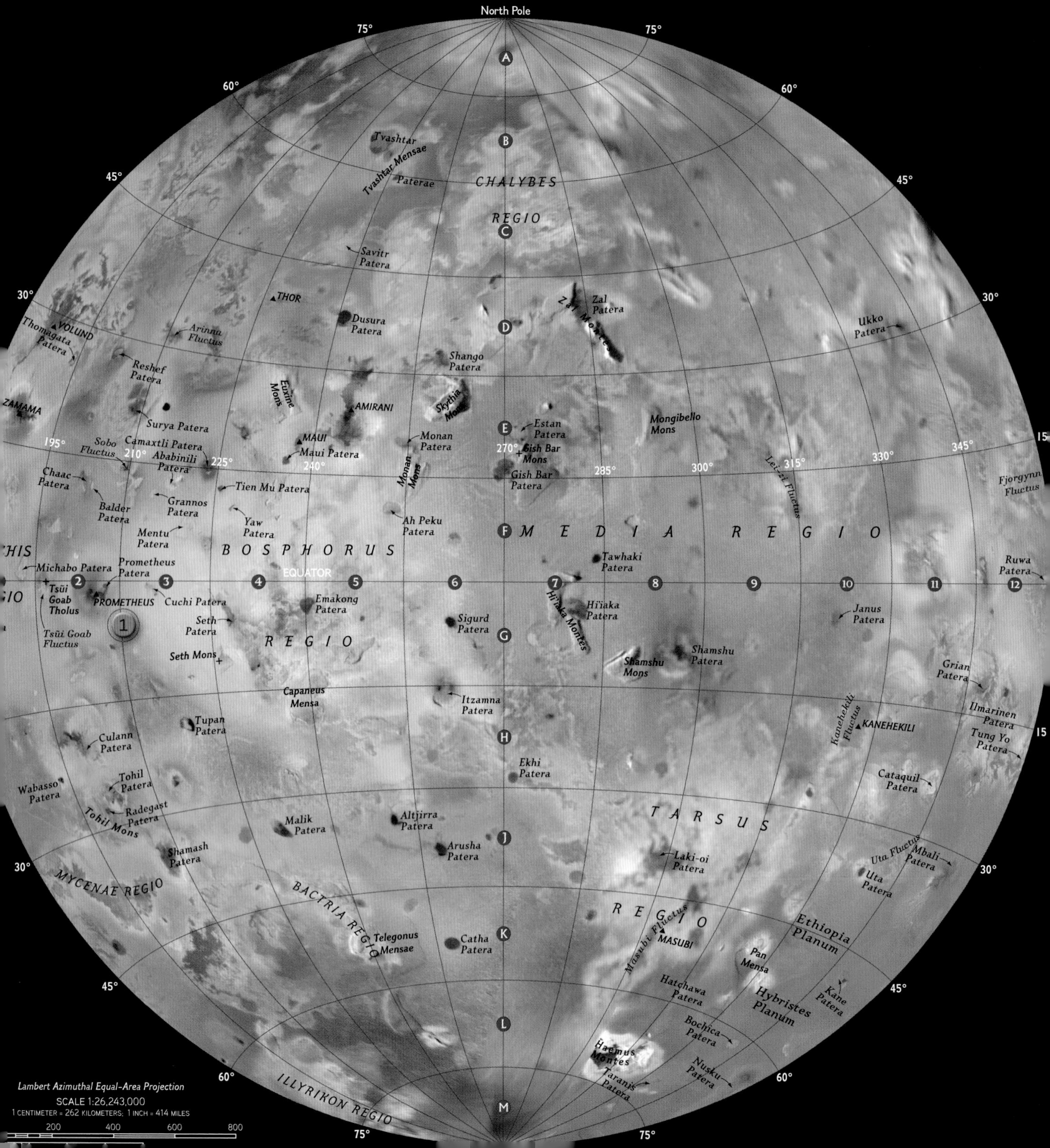

North Pole

75° 75°
60° 60°
45° 45°
30° 30°

Tvashtar

Tvashtar Mensae

Paterae

CHALYBES

REGIO

Savitr Patera

THOR

Dusura Patera

Zal Montes Zal Patera

Ukko Patera

VOLUND
Thomagata Patera

Arinna Fluctus

Shango Patera

ZAMAMA

Reshef Patera

Eubine Mons

AMIRANI

Skythia Mons

Monan Patera

Estan Patera

Mongibello Mons

Surya Patera

MAUI
Maui Patera

Monan Mons

Gish Bar Mons

Gish Bar Patera

Leizi Fluctus

Fjorgynn Fluctus

195° 210° 225° 240° 270° 285° 300° 315° 330° 345° 15°

Sobo Fluctus

Camaxtli Patera

Ababinili Patera

Tien Mu Patera

Ah Peku Patera

MEDIA REGIO

Ruwa Patera

Chaac Patera

Grannos Patera

Yaw Patera

Tawhaki Patera

Balder Patera

BOSPHORUS

Mentu Patera

Prometheus Patera

EQUATOR

Hi'iaka Patera

THIS

Michabo Patera

2 3 4 5 6 7 8 9 10 11 12

Tsüi Goab Tholus

PROMETHEUS

Cuchi Patera

Emakong Patera

Sigurd Patera

Hi'iaka Montes

Janus Patera

REGIO

1

Seth Patera

REGIO

Shamshu Mons

Shamshu Patera

Grian Patera

Tsüi Goab Fluctus

Seth Mons

Capaneus Mensa

Itzamna Patera

Kanehekili Fluctus

KANEHEKILI

Ilmarinen Patera

Tung Yo Patera

Tupan Patera

Cataquil Patera

Culann Patera

Ekhi Patera

Wabasso Patera

Tohil Patera

TARSUS

Uta Fluctus Mbali Patera

Radegast Patera

Malik Patera

Altjirra Patera

Uta Patera

Tohil Mons

Arusha Patera

Laki-oi Patera

Ethiopia Planum

Shamash Patera

REGIO

Hybristes Planum

MYCENAE REGIO

BACTRIA REGIO

Catha Patera

MASUBI

Pan Mensa

Kane Patera

Telegonus Mensae

Masubi Fluctus

Hatchawa Patera

Lambert Azimuthal Equal-Area Projection
SCALE 1:26,243,000
1 CENTIMETER = 262 KILOMETERS; 1 INCH = 414 MILES

200 400 600 800

ILLYRIKON REGIO

Haemus Montes

Taranis Patera

Bochica Patera

Nusku Patera

A B C D E F G H J K L M

KEY FEATURES

1. **PROMETHEUS:** Large, active volcano
2. **LOKI PATERA:** Volcanic depression
3. **MAFUIKE PATERA:** Orange coloration on Io caused by sulfur compounds

EASTERN HEMISPHERE

△ Eruptive center

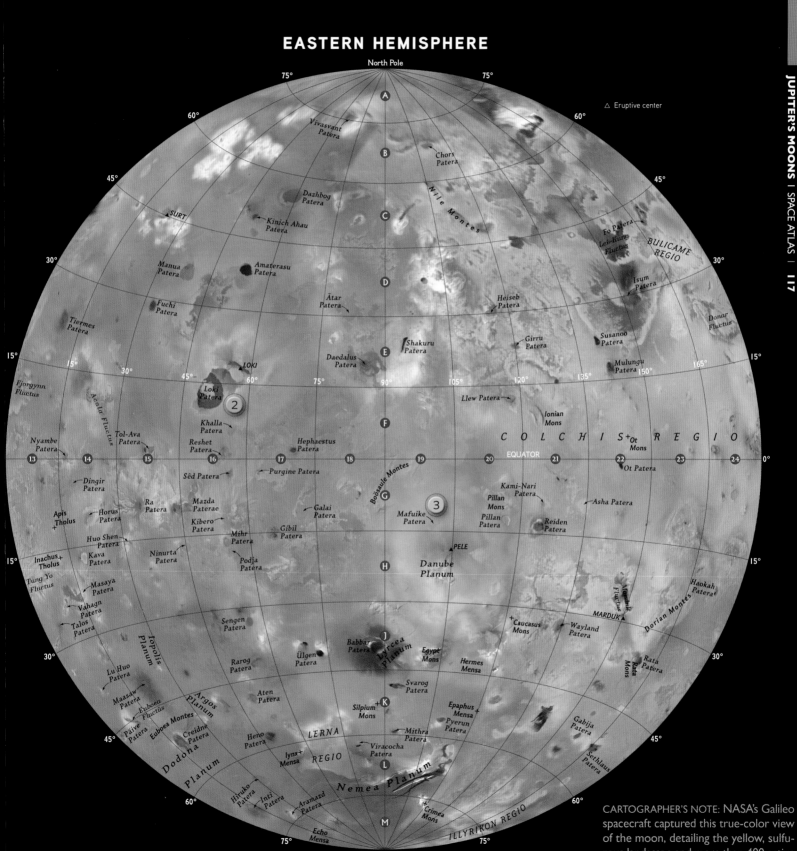

North Pole

A
B
C
D
E
F
G
H
J
K
L
M

South Pole

Vivasvant Patera
Chors Patera
Dazhbog Patera
Kinich Ahau Patera
SURT
Manua Patera
Amaterasu Patera
Ätar Patera
Fuchi Patera
Tiermes Patera
Shakuru Patera
Daedalus Patera
LOKI
Loki Patera
Fjorgynn Fluctus
Acala Fluctus
Khalla Patera
Nyambe Patera
Tol-Ava Patera
Reshet Patera
Hephaestus Patera
Purgine Patera
Dingir Patera
Sêd Patera
Ra Patera
Mazda Paterae
Horus Patera
Apis Tholus
Kibero Patera
Mihr Patera
Galai Patera
Gibil Patera
Huo Shen Patera
Ninurta Patera
Kava Patera
Podja Patera
Inachus Tholus
Tung Yo Fluctus
Masaya Patera
Vahagn Patera
Talos Patera
Topolis Planum
Sengen Patera
Babbar Patera
Ülgen Patera
Lu Huo Patera
Rarog Patera
Maasaw Patera
Euboea Fluctus
Päive Patera
Euboea Montes
Aten Patera
Creidne Patera
Heno Patera
Argos Planum
Dodona Planum
Iynx Mensa
Hiruko Patera
Inti Patera
Aramazd Patera
Echo Mensa
Viracocha Patera
Silpium Mons
Mithra Patera
Svarog Patera
Lerca Planum
Egypt Mons
Hermes Mensa
LERNA REGIO
Nemea Planum
ILLYRIKON REGIO
Crimea Mons

Nile Montes
Fo Patera
Lei-Kung Fluctus
BULICAME REGIO
Isum Patera
Heiseb Patera
Girru Patera
Susanoo Patera
Donar Fluctus
Mulungu Patera
Llew Patera
Ionian Mons
COLCHIS REGIO
Ot Mons
EQUATOR
Ot Patera
Kami-Nari Patera
Pillan Mons
Pillan Patera
Asha Patera
Mafuike Patera
Reiden Patera
PELE
Danube Planum
Magnik Fluctus
MARDUK
Dorian Montes
Haokah Patera
Caucasus Mons
Wayland Patera
Rata Mons
Rata Patera
Epaphus Mensa
Pyerun Patera
Gabija Patera
Sethlaus Patera

13 14 15 16 17 18 19 20 21 22 23 24

CARTOGRAPHER'S NOTE: NASA's Galileo spacecraft captured this true-color view of the moon, detailing the yellow, sulfurous landscape and more than 400 active volcanoes. Dominating the landscape, great eruptive centers are named for ancient gods of fire, thunder, or the sun.

JUPITER'S MOONS | EUROPA

Scientists believe that the gravitational pull of Jupiter and other moons produces enough heat to allow liquid water to exist below Europa's icy surface.

WESTERN HEMISPHERE

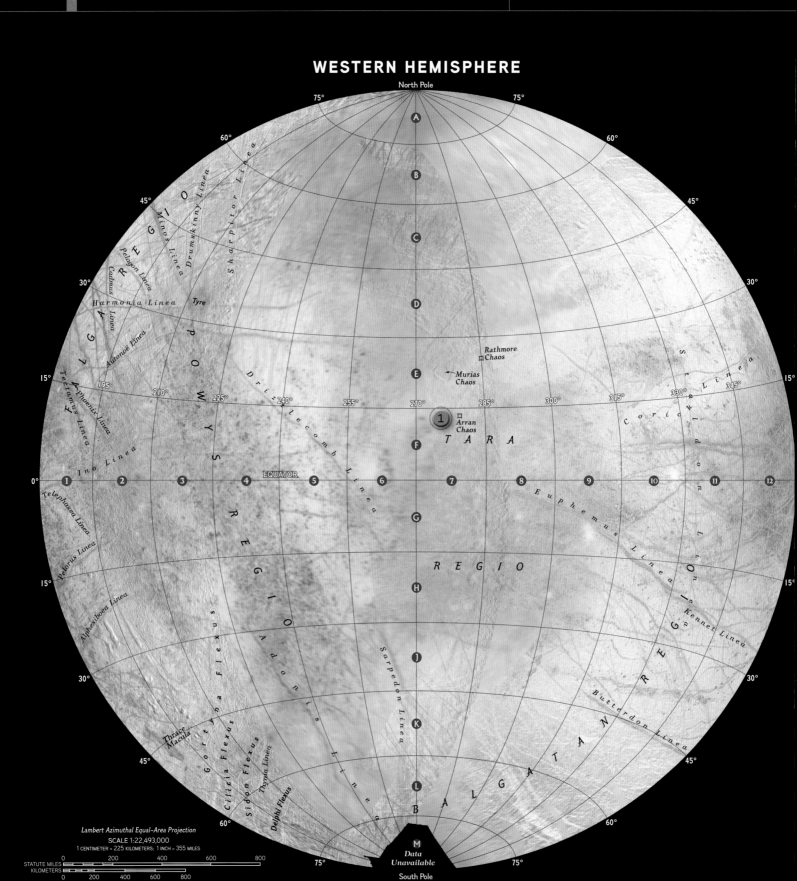

North Pole

Rathmore Chaos

Murias Chaos

Arran Chaos

TARA

EQUATOR

REGIO

Coricka Linea

Euphemus Linea

Kennet Linea

Butterdon Linea

BALGATAN REGIO

Sarpedon Linea

Thynia Linea

Delphi Flexus

Sidon Flexus

Cilicia Flexus

Gortyna Flexus

Thrace Macula

Alphesiboea Linea

Pedarus Linea

Telephassa Linea

Ino Linea

Phoenix Linea

Tectamus Linea

FALGA REGIO

Cadmus Linea

Harmonia Linea

Pelorus Linea

Autonoë Linea

Minos Linea

Drumskinny Linea

Sharptor Linea

Tyre

POWYS REGIO

Dritz Linea

Lecomb Linea

Adonis Linea

Lambert Azimuthal Equal-Area Projection

SCALE 1:22,493,000
1 CENTIMETER = 225 KILOMETERS; 1 INCH = 355 MILES

STATUTE MILES 0 200 400 600 800
KILOMETERS 0 200 400 600 800

Data Unavailable

South Pole

KEY FEATURES

① ARRAN CHAOS: A jumbled region that may sit over a liquid lake

② ANDROGEOS LINEA: One of many cracks in Europa's ice

③ PWYLL: A young impact crater surrounded by white water ice

EASTERN HEMISPHERE

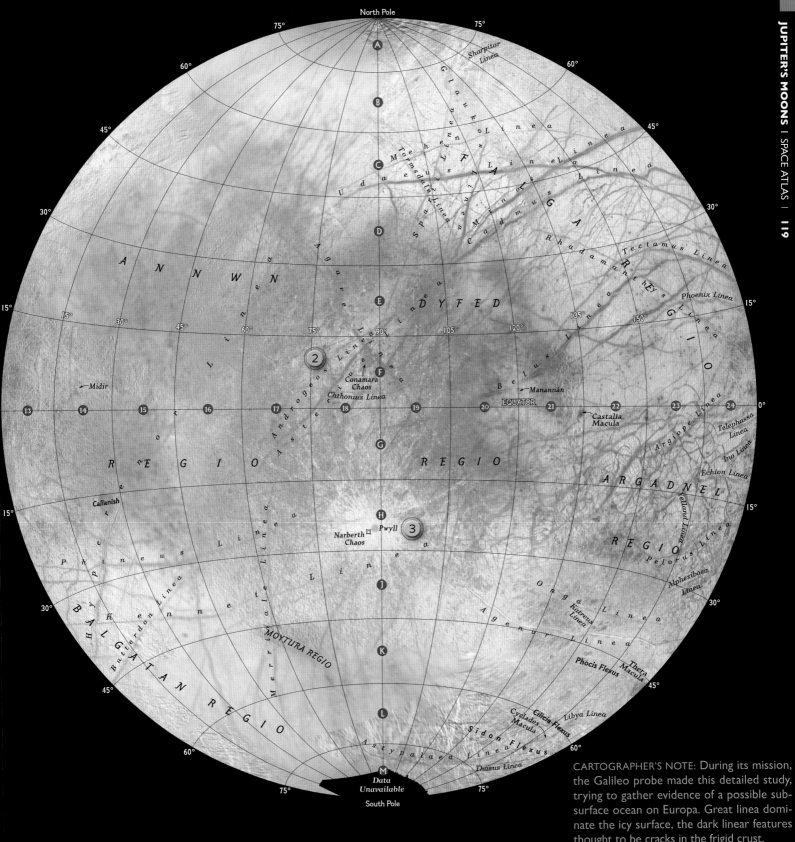

North Pole

75°   75°

60°   60°

45°   45°

30°   30°

15°   15°

15°   15°

30°   30°

45°   45°

60°   60°

75°   75°

South Pole

Sharpitor Linea

Glauko Linea

Tormsdale Linea
Udra Linea
Spaunna Linea
Menena Linea
Harmonia Linea
Cadmus Linea

FALGA REGIO

Rhadamanthys Linea
Belus Linea
Tectamus Linea
Phoenix Linea
RHEY'S GLIO

ANNWN

DYFED

Agave Linea
Asterius Linea
Androgeos Linea
Belus Linea
Argiope Linea
Telephassa Linea
Ino Linea
Echion Linea

Midir

Conamara Chaos
Chthonius Linea
Manannán
Castalia Macula

EQUATOR

13   14   15   16   17   18   19   20   21   22   23   24

REGIO   REGIO

ARGADNEL REGIO

Callanish

Phineus Linea
Belus Linea
Pelorus Linea
Alphesiboea Linea
Yelland Linea

Narberth Chaos   Pwyll

Merriale Linea
Butterdon Linea
MOYTURA REGIO
Agenor Linea
Ongkatreus Linea
Phocis Flexus
Thera Macula

BALGATAN REGIO

Cyclades Macula
Cilicia Flexus
Libya Linea
Sidon Flexus
Astypalaea Linea
Thasus Linea

Data Unavailable

CARTOGRAPHER'S NOTE: During its mission, the Galileo probe made this detailed study, trying to gather evidence of a possible subsurface ocean on Europa. Great linea dominate the icy surface, the dark linear features thought to be cracks in the frigid crust.

JUPITER'S MOONS | GANYMEDE

The third of Jupiter's moons, huge Ganymede is also the largest in the solar system, dwarfing the planet Mercury.

WESTERN HEMISPHERE

North Pole

Data Unavailable

75°   75°
60°   60°
45°   45°
30°   30°
15°   15°

Philae Sulcus
Ur Sulcus
Latpon
Kadi
Halieus
Tettu Facula
Abydos Facula
Ilah
Edfu Facula
Khepri
Siwah Facula
Memphis Facula
Nidaba
Selket
Ninlil
El
Uruk Sulcus
Erech Sulcus
Lagash Sulcus
Ninsum
Gad
Ninki
Mush
Ilus
Irkalla
Menhit
Ashima
Osiris
Thoth
Andjeti
Gilgamesh
Wepwawet
Anubis

Neheh
Adapa
Anzu
Achelous
Ninkasi
Sapas
Aquarius Sulcus
Enki Catena
Sati
Neith
Harakhtes
Kulla
Bigeh Facula
Kibalba Sulcus
Nineveh Sulcus
Bau
Sicyon Sulcus
PERRINE REGIO
PHRYGIA SULCUS
Tros
Ishkur
Danel
Serapis
Mysia Sulci
BARNARD REGIO
Mysia Sulci
Dardanus Sulcus
Cisti
Nah-Hunte
NICHOLSON REGIO
Damkina
Dardanus Sulcus
Nabu

GALILEO REGIO
Lakhmu Fossae
Gir
Gelo

EQUATOR

0° 195° 210° 225° 240° 255° 270° 285° 300° 315° 330° 345° 0°

0 1 2 3 4 5 6 7 8 9 10 11 12

A B C D E F G H J K L M

Data Unavailable

South Pole

Lambert Azimuthal Equal-Area Projection
SCALE 1:37,919,000
1 CENTIMETER = 379 KILOMETERS; 1 INCH = 598 MILES

STATUTE MILES 0 250 500 750 1000
KILOMETERS 0 250 500 750 1000

75°   75°

KEY FEATURES

1. **GALILEO REGIO:** Smooth, icy regions cover 40 percent of Ganymede's surface.

2. **NUN SULCI:** Grooved regions, or sulci, cover 60 percent of Ganymede's surface.

3. **TROS:** A crater; craters on Ganymede are shallow, probably due to the soft, icy surface.

EASTERN HEMISPHERE

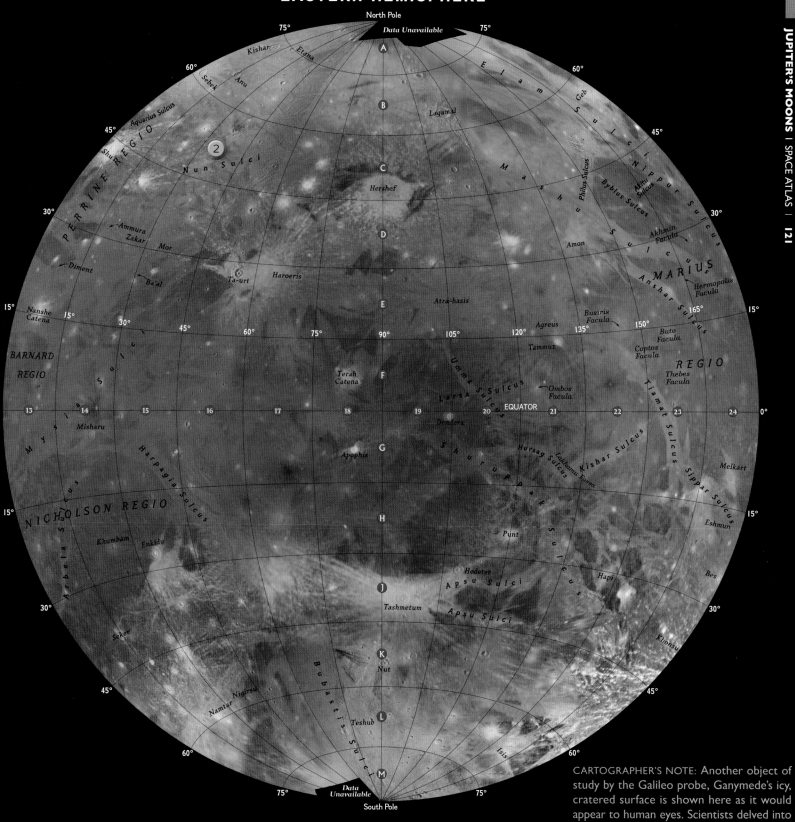

CARTOGRAPHER'S NOTE: Another object of study by the Galileo probe, Ganymede's icy, cratered surface is shown here as it would appear to human eyes. Scientists delved into the mythology of ancient peoples who lived in the Fertile Crescent, from Mesopotamia to the Levant, to name the moon's topography.

JUPITER'S MOONS | CALLISTO

Callisto is the third largest, and most heavily cratered, satellite in the solar system.

WESTERN HEMISPHERE

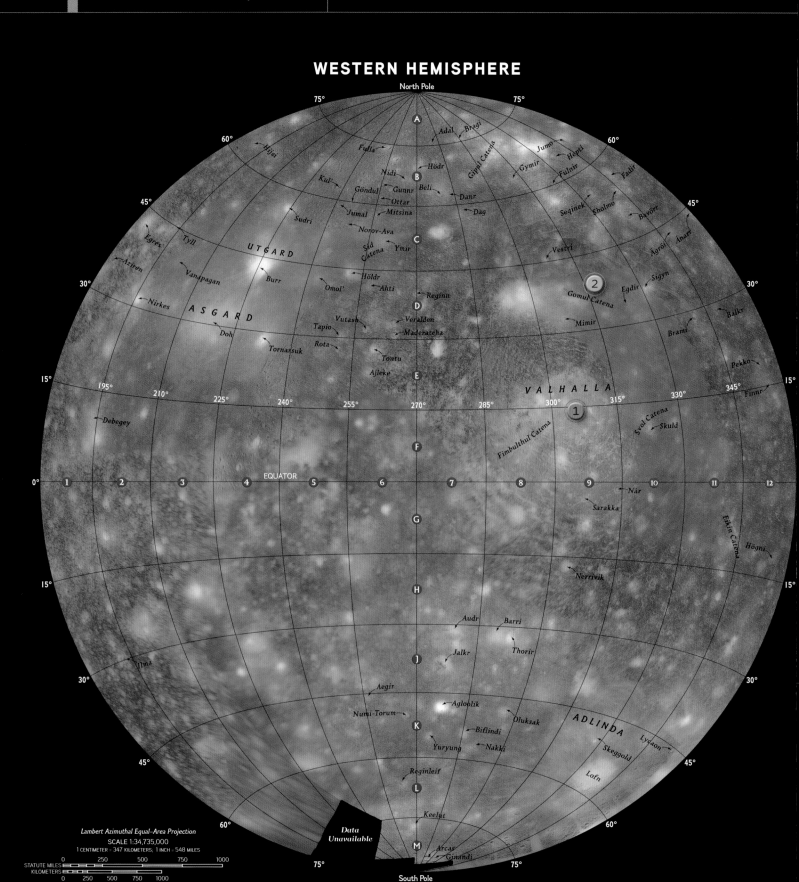

North Pole

South Pole

EQUATOR

UTGARD

ASGARD

VALHALLA

ADLINDA

Lambert Azimuthal Equal-Area Projection
SCALE 1:34,735,000
1 CENTIMETER = 347 KILOMETERS; 1 INCH = 548 MILES

STATUTE MILES
0   250   500   750   1000
KILOMETERS
0   250   500   750   1000

Data Unavailable

Adal Bragi Fulla Hödr Gipul Catena Jumo Hepti
Nidi Beli Gymir Fulnir
Kul' Göndul Gunnr Danr Fadir
Jumal Ottar Mitsina Dag Seqinek Sholmo Bavörr
Sudri Norov-Ava Ägröi Änarr
Egres Tyll Síd Catena Ymir Vestri Sigyn
Aztren Vanapagan Burr Höldr Reginn Egdir Gomul Catena Balkr
Nirkes Omol' Ahti Mimir Brami
Doh Tapio Vutash Veralden Pekko
Tornarsuk Rota Maderakka Finnr
Tontu Svol Catena Skuld
Ajleke
Debegey Fimbulthul Catena
Nár
Sarakka
Eirin Catena
Högni
Nerrivik
Audr Barri
Jalkr Thorir
Ilma Aegir
Agloolik
Numi-Torum Oluksak Lycaon
Biflindi Skeggold
Yuryung Nakki Lofn
Reginleif
Keelut
Arcas Ginandi

Seqinek

KEY FEATURES

① VALHALLA BASIN: Huge, multiringed impact basin

② GOMUL CATENA: Line of craters, possibly created by the same object

③ BRAN: Bright rayed crater

EASTERN HEMISPHERE

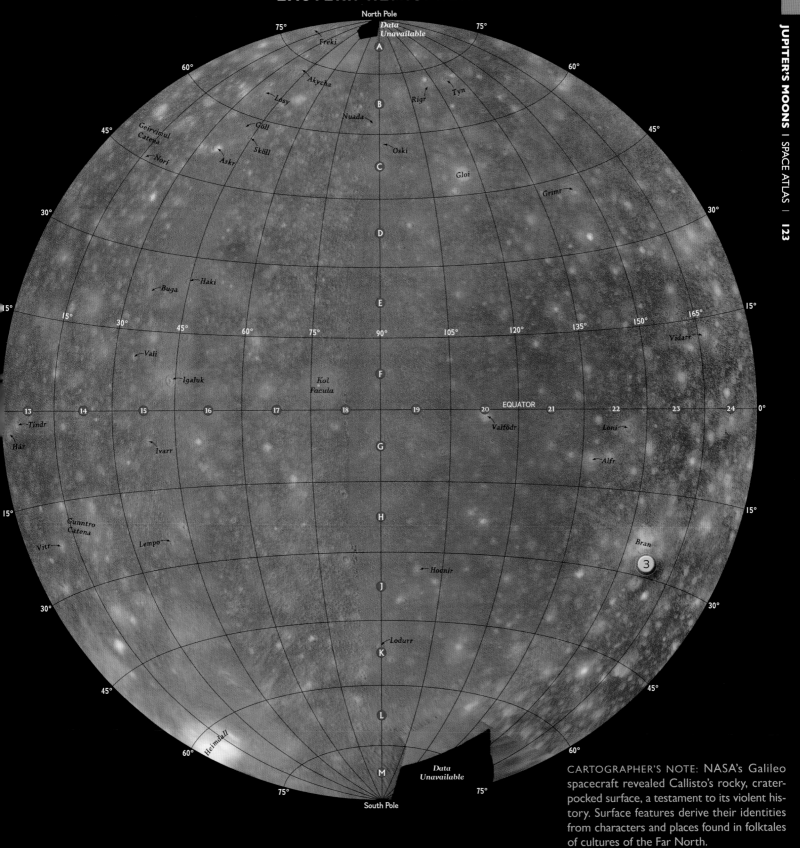

CARTOGRAPHER'S NOTE: NASA's Galileo spacecraft revealed Callisto's rocky, crater-pocked surface, a testament to its violent history. Surface features derive their identities from characters and places found in folktales of cultures of the Far North.

upiter's big moons are called the Galilean moons because the first telescopic sighting of these moons occurred in 1610 when Galileo turned his instrument to the Jovian system. Their discovery played a major role in scientific history: Their existence was a strong argument against the then prevailing Aristotelian view of the universe. Aristotle taught that every material object has an innate property that makes it try to move toward the center of the universe (which, in that system, corresponded to the center of Earth). When a 17th-century philosopher saw an apple fall, then he (and it was always a "he") would ascribe the fall to the apple's innate desire to be at the center of the universe. But here were the Galilean moons, perfectly content to be in orbit around Jupiter, far from that center. Maybe Earth wasn't as important in the grand scheme of things as the ancients had thought!

A cutaway view of Jupiter's moon Io. The moon has an iron-nickel core and a rocky mantle that extends to the surface. Constant flexing in Jupiter's gravitational field makes Io the most volcanically active body in the solar system.

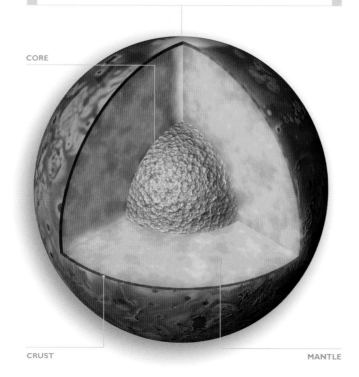

CORE

CRUST

MANTLE

In any case, today we see these moons as examples of the diversity of worlds that exist in our solar system. Each is unique, each has its own story to tell. Indeed, one of them (Ganymede) is bigger than Mercury, and it would be classed as a separate planet if it circled the sun instead of Jupiter.

By convention all of the Jovian moons are named for lovers or children of the god Jupiter. (If you know your mythology, you'll know that Jupiter's activities supply us with many more than 65 possible names in these categories.) For our purposes, though, we'll look at only two of the most interesting moons—Io and Europa.

IO

Io, the innermost of the Galilean moons, is the one that looks like a pizza. It is slightly larger than Earth's moon and the fourth largest moon in the solar system. The mottled appearance of its surface is the result of more than 400 active volcanoes on its surface ejecting various sulfur compounds—compounds that give it its orange and yellow colors. These volcanoes make Io the most geologically active body known.

The existence of volcanoes came as something of a shock to scientists during the flybys of the two Voyager spacecraft in 1979. On Earth volcanoes result from heat rising from deep in the mantle, heat generated in part by radioactive decay. Io is simply too small to generate heat in this way. Scientists quickly realized, however, that there was another way to heat the moon besides radioactivity. Because of the gravitational pull of the other moons, the orbit of Io is not a perfect circle. Instead, the distance of the moon from Jupiter is always changing, so that the

A bluish plume marks a volcanic eruption on the surface of Io. The gas and particles in the plume shoot 100 kilometers (60 mi) above the moon's surface.

IO HAS OVER
400
ACTIVE VOLCANOES

EUROPA HAS AN
OCEAN
UNDER AN ICY SHELL

gravitational force on the moon changes as well. This means that Io is constantly being flexed and distorted, and just as a piece of metal that is bent back and forth will heat up, the moon warms. We see the result of this so-called tidal heating in Io's extensive volcanoes.

EUROPA

At first glance, there doesn't seem to be much about Europa to command our attention. Slightly smaller than Earth's moon, it has a smooth, icy surface crisscrossed by dark streaks.

The most important thing about Europa, the thing that makes it one of the focal points for future space exploration, is the fact that it very likely has an ocean of

Cutaway view of the interior of Europa. The smallest of the four Galilean moons, it has a metallic core and a rocky mantle. These are overlain with water (probably in liquid form) and an outer shell of ice.

METALLIC CORE   ICY CRUST

WATER LAYER   ROCKY INTERIOR

A false-color image of the surface of Jupiter's moon Europa demonstrates that the ice on the moon's surface is contaminated by material (shown in red) brought up from the interior. Icy plains are shown in blue.

liquid water under the thick layer of ice that composes its surface. Like Io, Europa has a slightly eccentric orbit and is affected by tidal heating. Calculations quickly showed that there is enough heat generated by this effect to keep a subsurface layer of water from freezing, despite the fact that the temperature can be as low as minus 220°C (-370°F) at the surface. Because water conducts electricity, the movement of Europa through Jupiter's magnetic field produces changes in that field and in the swarm of particles around the giant planet—effects that were quickly detected by the Galileo orbiter. When you factor in that some large craters on Europa seem to contain smooth areas of recently frozen ice (presumably from upwelling liquid) and include the Galileo mission's measurements of the moon's gravitational field, you have pretty impressive evidence that, like Earth, Europa contains large amounts of liquid water. The current theory is that beneath a layer of ice some tens of kilometers thick is an ocean containing up to twice as much water as is in the oceans of our own planet.

Which brings us, of course, to the question of life. We know that life exists at deep-sea vents on Earth: Presumably there would be similar vents on Europa, which suggests that microbial life might have developed there as well. More interestingly, recent calculations indicate that there might be processes by which the collision of particles with the Europan surface would introduce oxygen into the oceans. This, in turn, raises the possibility of more complex life, perhaps even something like our own fish.

Because of this, there are a number of proposals for sending spacecraft to Europa. Scheduled for launch in 2020, a joint NASA–European Space Agency mission known as Europa Jupiter Systems Mission seems the furthest along. Much further in the future are some blue-sky plans to send a robotic lander to Europa to drill through the overlying ice and explore the ocean itself.

Stay tuned!

The sixth planet from the sun and the farthest visible to the naked eye, Saturn is a gas giant similar in structure to Jupiter. It is believed to have a rocky core, perhaps 10 to 20 times the size of Earth, surrounded by a layer of metallic hydrogen. This, in turn, is enclosed by liquid hydrogen mixed with helium, above which we find the gaseous atmosphere. In appearance, Saturn is a little bland compared to Jupiter, although it has the same (though fainter) banded cloud structure. The clouds appear to consist of water ice, ammonia compounds, and ammonia crystals found in successive tiers tens of kilometers thick. It is these clouds that we see when we look at the planet.

SATURN & ITS MOONS

[GOLDEN GLOBE]

DISCOVERER: UNKNOWN
DISCOVERY DATE: PREHISTORIC
NAMED FOR: ROMAN GOD OF AGRICULTURE

MASS: 95.16 × EARTH'S
VOLUME: 763.59 × EARTH'S
MEAN RADIUS: 58,232 KM (36,184 MI)
EFFECTIVE TEMPERATURE: -178°C (-288°F)
LENGTH OF DAY: 10.66 HOURS
LENGTH OF YEAR: 29.48 EARTH YEARS
NUMBER OF MOONS: 62 (53 NAMED)
PLANETARY RING SYSTEM: YES

Artist's conception of Saturn seen from Mimas.
(Inset) Saturn

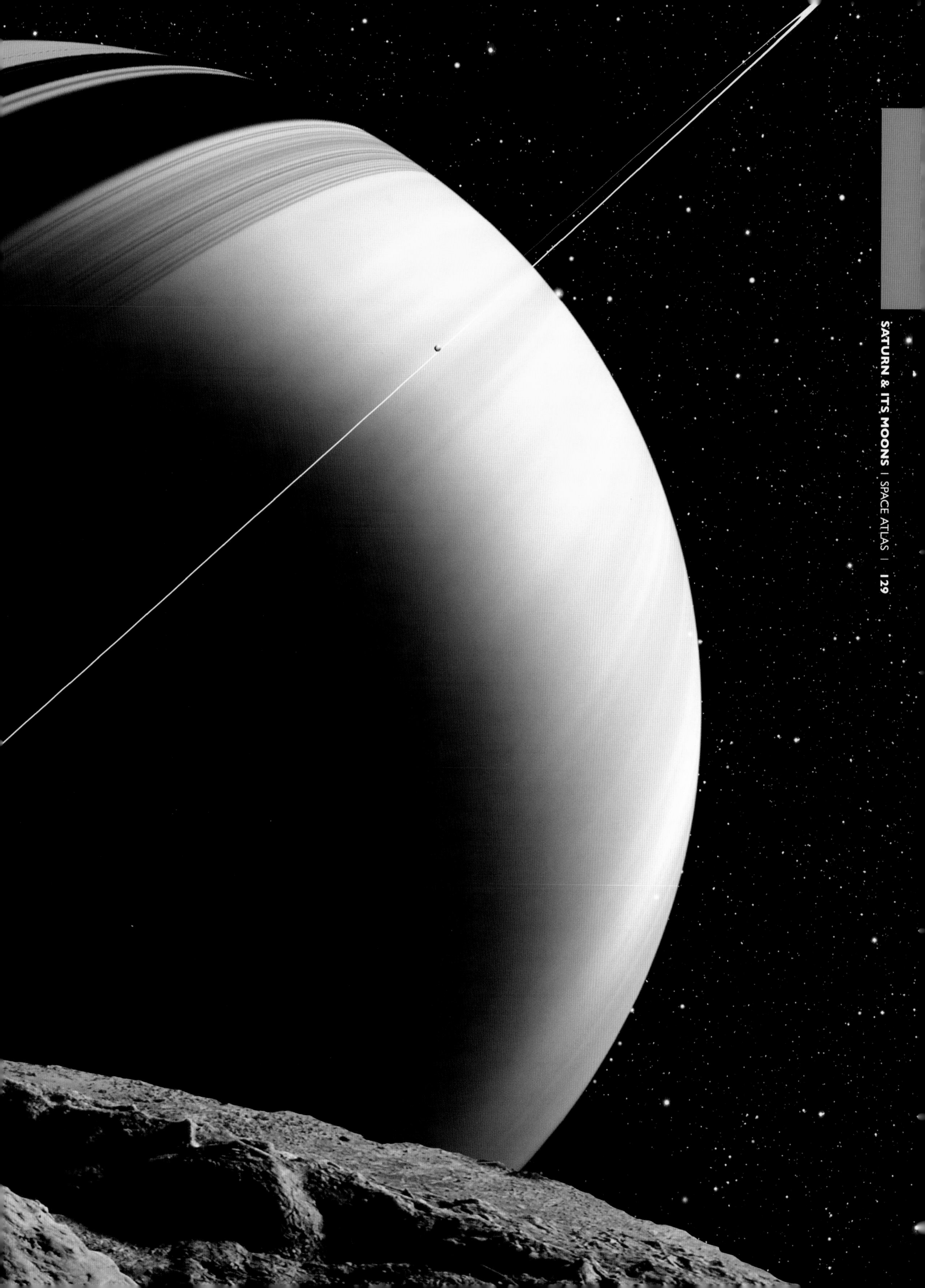

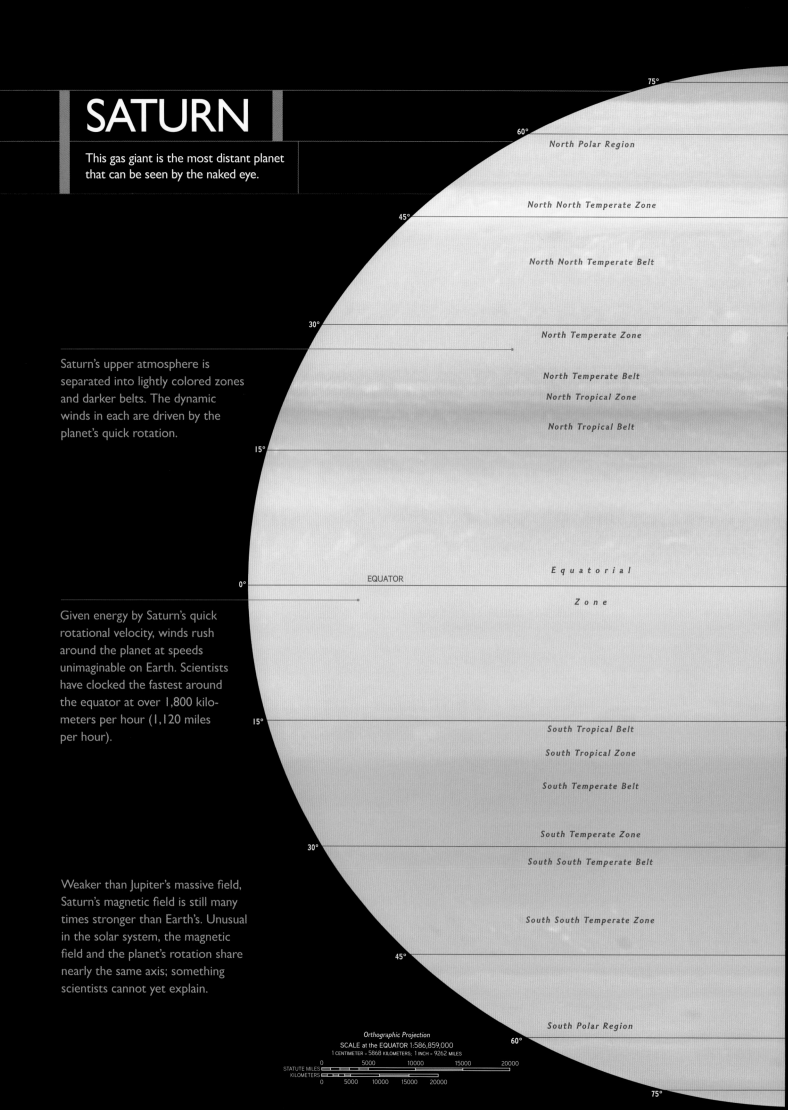

SATURN

This gas giant is the most distant planet that can be seen by the naked eye.

Saturn's upper atmosphere is separated into lightly colored zones and darker belts. The dynamic winds in each are driven by the planet's quick rotation.

Given energy by Saturn's quick rotational velocity, winds rush around the planet at speeds unimaginable on Earth. Scientists have clocked the fastest around the equator at over 1,800 kilometers per hour (1,120 miles per hour).

Weaker than Jupiter's massive field, Saturn's magnetic field is still many times stronger than Earth's. Unusual in the solar system, the magnetic field and the planet's rotation share nearly the same axis; something scientists cannot yet explain.

75°

60°

North Polar Region

North North Temperate Zone

45°

North North Temperate Belt

30°

North Temperate Zone

North Temperate Belt

North Tropical Zone

North Tropical Belt

15°

0° EQUATOR

Equatorial

Zone

15°

South Tropical Belt

South Tropical Zone

South Temperate Belt

South Temperate Zone

30°

South South Temperate Belt

South South Temperate Zone

45°

South Polar Region

60°

75°

Orthographic Projection
SCALE at the EQUATOR 1:586,859,000
1 CENTIMETER = 5868 KILOMETERS; 1 INCH = 9262 MILES

0   5000   10000   15000   20000
STATUTE MILES
KILOMETERS
0   5000   10000   15000   20000

KEY FEATURES

① **NORTH NORTH TEMPERATE ZONE:** Zones represent upwelling air.

② **NORTH NORTH TEMPERATE BELT:** Belts represent descending air.

③ **EQUATOR:** Rapidly rotating Saturn bulges at the equator.

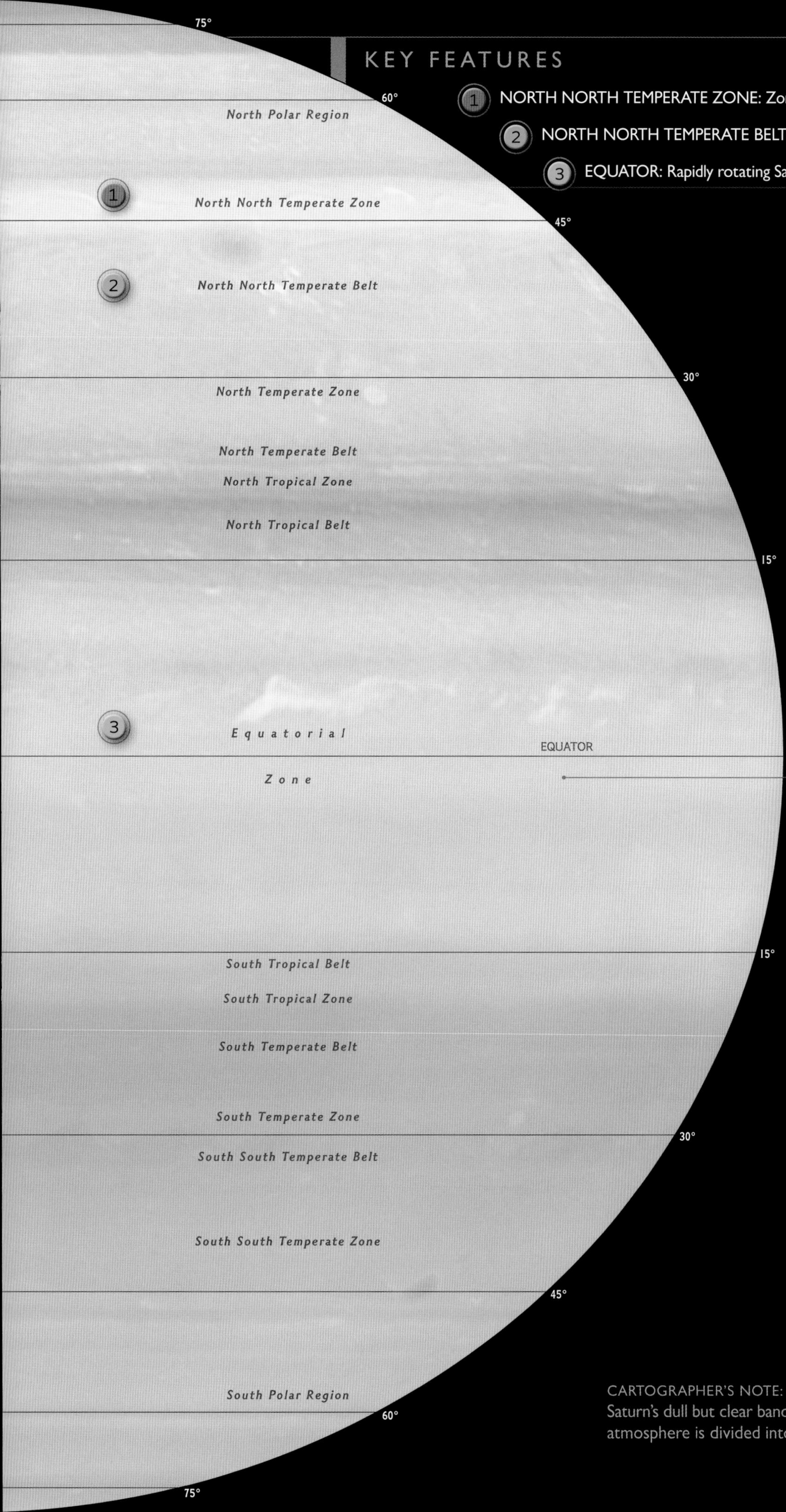

60°

North Polar Region

① *North North Temperate Zone*

45°

② *North North Temperate Belt*

30°

North Temperate Zone

North Temperate Belt

North Tropical Zone

North Tropical Belt

15°

③ *Equatorial*

EQUATOR

Zone

0°

Spinning at over 35,500 kilometers per hour (22,000 miles
per hour) causes the planet to
bulge around its middle. Saturn's
diameter at the equator is nearly
12,000 kilometers greater than
at the poles.

15°

South Tropical Belt

South Tropical Zone

South Temperate Belt

South Temperate Zone

30°

South South Temperate Belt

South South Temperate Zone

45°

South Polar Region

60°

CARTOGRAPHER'S NOTE: Observed by NASA's Cassini mission,
Saturn's dull but clear banding is visible. Like larger Jupiter, Saturn's
atmosphere is divided into darker belts and lighter zones.

75°

SATURN'S MOONS | MIMAS

Heavily cratered, small Mimas is
the innermost of Saturn's moons.

WESTERN HEMISPHERE

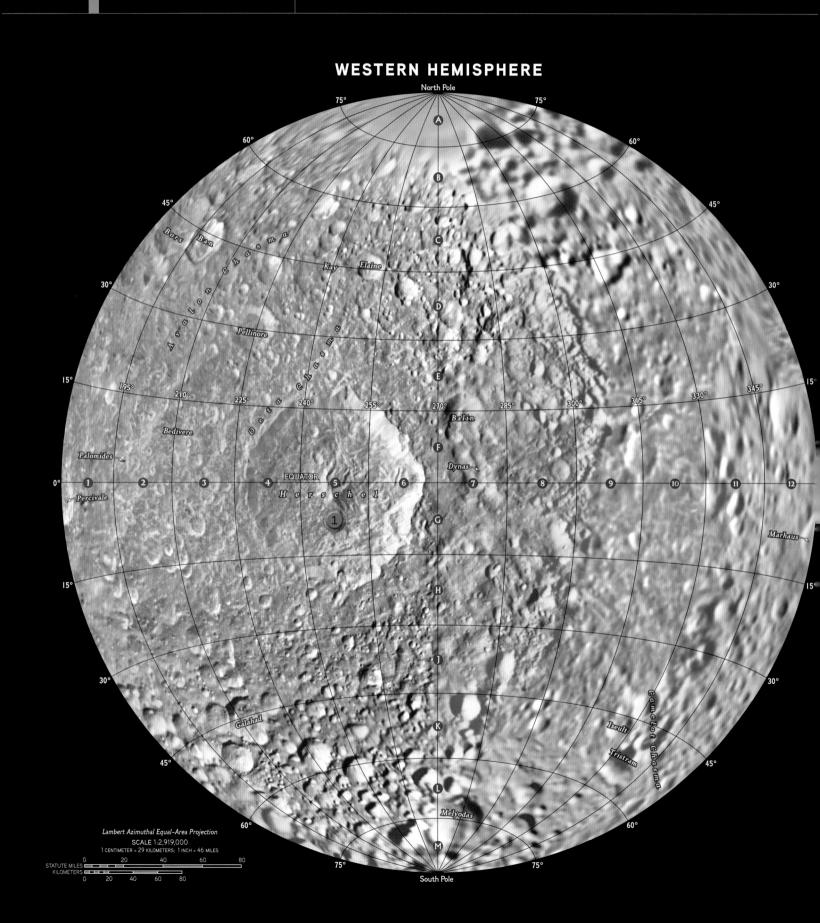

North Pole

75° 75°
60° 60°

A
B

45° 45°

C

Bors Ban

30° 30°

Avalon Chasma

Kay   Elaine

D

Pellinore

45° 45°

E

Oeta Chasma

15° 15°

195° 210° 225° 240° 255° 270° 285° 300° 315° 330° 345°

Balin

Bedivere

F

Palomides

Dynas

EQUATOR

0° 1   2   3   4   5   6   7   8   9   10   11   12 0°

Herschel

Percivale

1

Marhaus

G

15° 15°

H

J

30° 30°

Galahad

K

Iseult

Camelot Chasma

Tristram

45° 45°

L

Meliodas

60° 60°

M

Lambert Azimuthal Equal-Area Projection

SCALE 1:2,919,000
1 CENTIMETER = 29 KILOMETERS; 1 INCH = 46 MILES

STATUTE MILES 0   20   40   60   80
KILOMETERS   0   20   40   60   80

75° 75°

South Pole

EASTERN HEMISPHERE

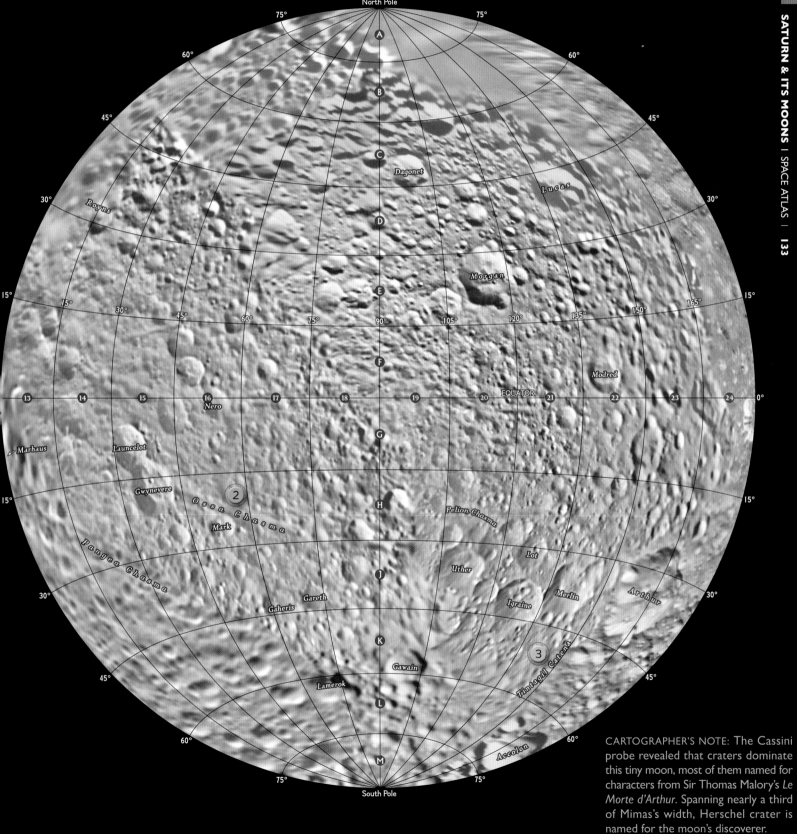

CARTOGRAPHER'S NOTE: The Cassini probe revealed that craters dominate this tiny moon, most of them named for characters from Sir Thomas Malory's *Le Morte d'Arthur*. Spanning nearly a third of Mimas's width, Herschel crater is named for the moon's discoverer.

Like Jupiter's Europa, icy Enceladus may have a body of subsurface liquid water.

WESTERN HEMISPHERE

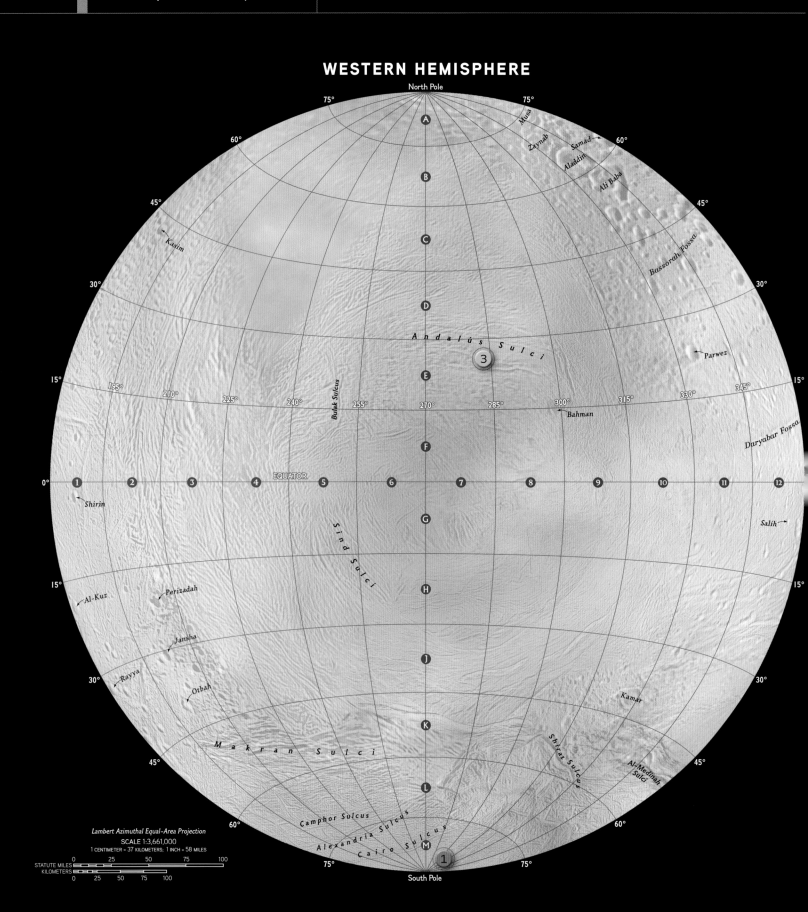

North Pole

75° 75°
Musa
60° Zaynab Samad 60°
Aladdin
Ali Baba

45° 45°

Kasim

Bassorah Fossa

30° 30°

Andalús Sulci

3 Parwez

15° 15°
Bulak Sulcus
195° 210° 225° 240° 255° 270° 285° 300° 315° 330° 345°
Bahman
Daryabar Fossa

EQUATOR
1 2 3 4 5 6 7 8 9 10 11 12
0° 0°
Shirin
Salih
Sind Sulci

15° 15°
Al-Kuz Perizadah
Jansha

30° 30°
Rayya
Otbah Kamar

Makran Sulci Shiraz Sulcus
45° Al-Medinah 45°
Sulci
Camphor Sulcus
60° Cairo Sulcus 60°
Alexandria Sulcus M
75° 1 75°
South Pole

Lambert Azimuthal Equal-Area Projection
SCALE 1:3,661,000
1 CENTIMETER = 37 KILOMETERS; 1 INCH = 58 MILES
STATUTE MILES 0 25 50 75 100
KILOMETERS 0 25 50 75 100

EASTERN HEMISPHERE

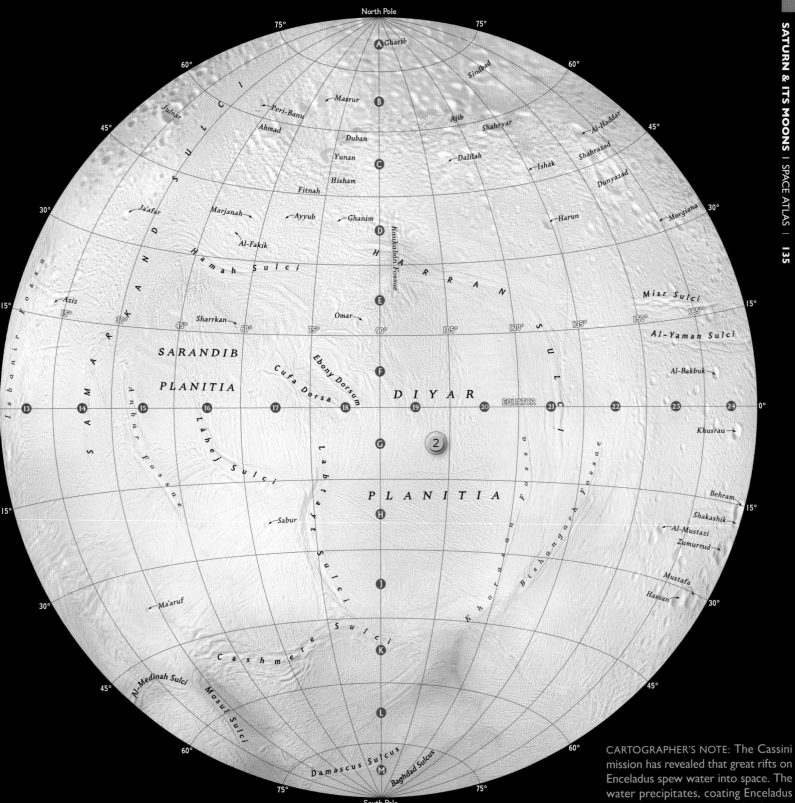

North Pole

Gharib

Sindbad

Masrur

Peri-Banu

Ajib

Shahryar

Al-Haddar

Julnar

Ahmad

Duban

Dalilah

Shahrazad

Ishak

Yunan

Dunyazad

Hisham

Fitnah

Morgiana

Ja'afar

Marjanah

Ayyub

Ghanim

Harun

Al-Fakik

Hamah Sulci

Aziz

Omar

Misr Sulci

Sharrkan

Al-Yaman Sulci

SARANDIB

Al-Bakbuk

PLANITIA

Cufa Dorsa

Ebony Dorsum

DIYAR

EQUATOR

Khusrau

PLANITIA

Sabur

Behram

Shakashik

Al-Mustazi

Zumurrud

Mustafa

Ma'aruf

Hassan

Cashmere Sulci

Al-Medinah Sulci

Mosul Sulci

Damascus Sulcus

Baghdad Sulcus

South Pole

SULCI

SAMARKAND SULCI

Isbanir Fossa

Anbar Fossae

Lahej Sulci

Labtayt Sulci

Kaukaban Fossae

HARRAN SULCI

Khorasan Fossa

Bishangarh Fossae

CARTOGRAPHER'S NOTE: The Cassini mission has revealed that great rifts on Enceladus spew water into space. The water precipitates, coating Enceladus and giving the moon a very bright surface. Geographic names are taken from Sir Richard Burton's *Arabian Nights*.

SATURN'S MOONS | TETHYS

Little Tethys, heavily scarred, is
made primarily of water ice.

WESTERN HEMISPHERE

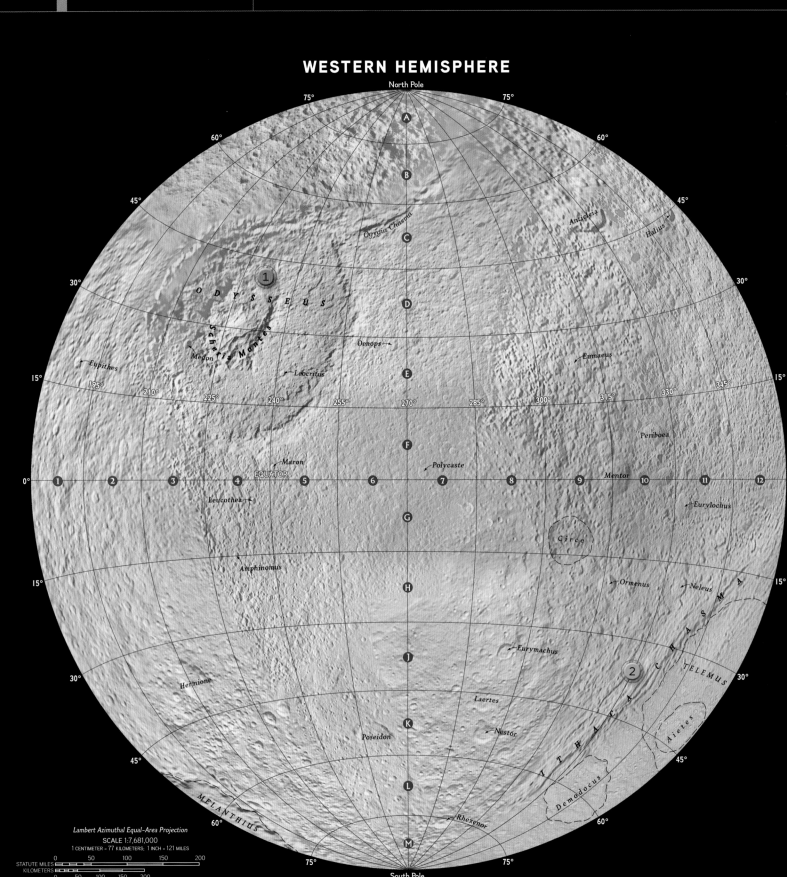

North Pole

75°   75°
60°   60°

Antícleia

Halius

45°   45°

Ogygia Chasma

30°   30°

O D Y S S E U S

Eumaeus

Oenops →

Scheria Montes

Medon

Eupithes

Leocritus

Periboea

195°   210°   225°   240°   255°   270°   285°   300°   315°   330°   345°

15°   15°

Maron

Polycaste

Mentor

0°   ① ② ③ ④ EQUATOR ⑤ ⑥ ⑦ ⑧ ⑨ ⑩ ⑪ ⑫   0°

Leucothea →

Eurylochus

Circe

Amphinomus

Ormenus → ← Neleus

15°   15°

I T H A C A C H A S M A

Eurymachus

TELEMUS

②

30°   30°

Hermione

Laertes

Aietes

Nestor →

Poseidon

Demodocus

45°   45°

MELANTHIUS

Rhexenor

60°   60°

75°   75°

South Pole

Lambert Azimuthal Equal-Area Projection
SCALE 1:7,681,000
1 CENTIMETER = 77 KILOMETERS; 1 INCH = 121 MILES

STATUTE MILES   0   50   100   150   200
KILOMETERS   0   50   100   150   200

KEY FEATURES

① **ODYSSEUS:** Huge impact basin almost two-fifths as large as Tethys

② **ITHACA CHASMA:** Enormous, deep valley, its origins unknown

③ **PENELOPE:** Large impact crater

EASTERN HEMISPHERE

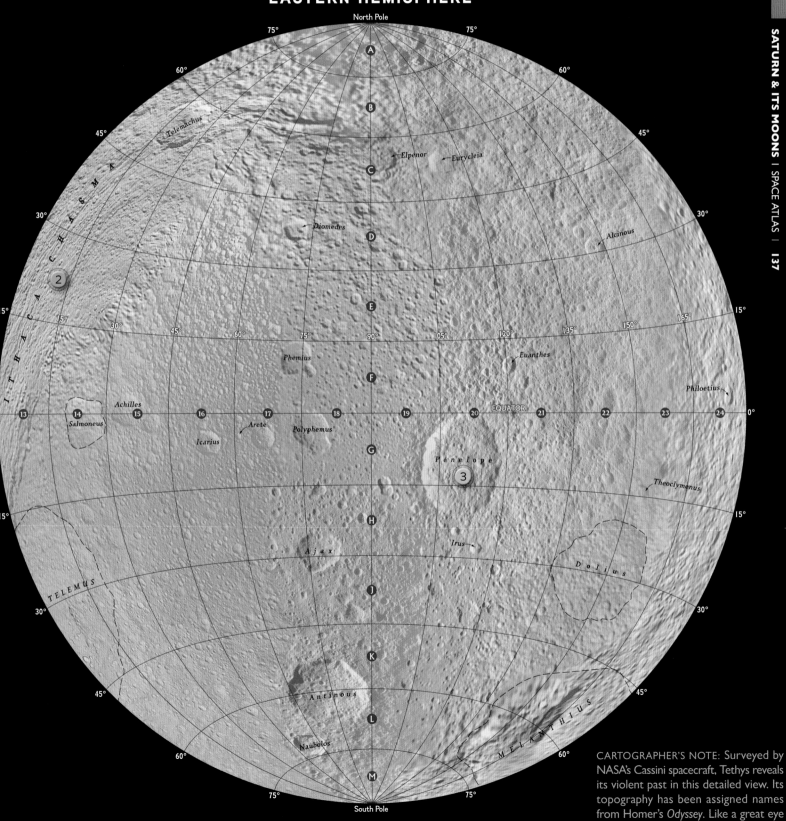

North Pole

75°   75°

A

60°   60°

B

45°   45°

Telemachus

C   Elpenor   Eurycleia

30°   30°

Diomedes   Alcinous

D

ITHACA CHASMA

②

15°   15°

E

15°   30°   45°   60°   75°   90°   105°   120°   135°   150°   165°

Phemius   Euanthes

F

Philoetius

Achilles   EQUATOR

⑬   ⑭   ⑮   ⑯   ⑰   ⑱   ⑲   ⑳   ㉑   ㉒   ㉓   ㉔   0°

Salmoneus   Areté   Polyphemus   Penelope

Icarius   G   ③

Theoclymenus

15°   15°

H

Irus

Ajax   Dolius

TELEMUS   J

30°   30°

K

45°   45°

Antinous   MELANTHIUS

L

60°   60°

Naubolos

75°   M   75°

South Pole

CARTOGRAPHER'S NOTE: Surveyed by NASA's Cassini spacecraft, Tethys reveals its violent past in this detailed view. Its topography has been assigned names from Homer's *Odyssey*. Like a great eye on the surface, the crater Odysseus dominates the northern hemisphere.

SATURN'S MOONS | DIONE

Fractured and cratered, little Dione orbits Saturn every 2.7 Earth days.

WESTERN HEMISPHERE

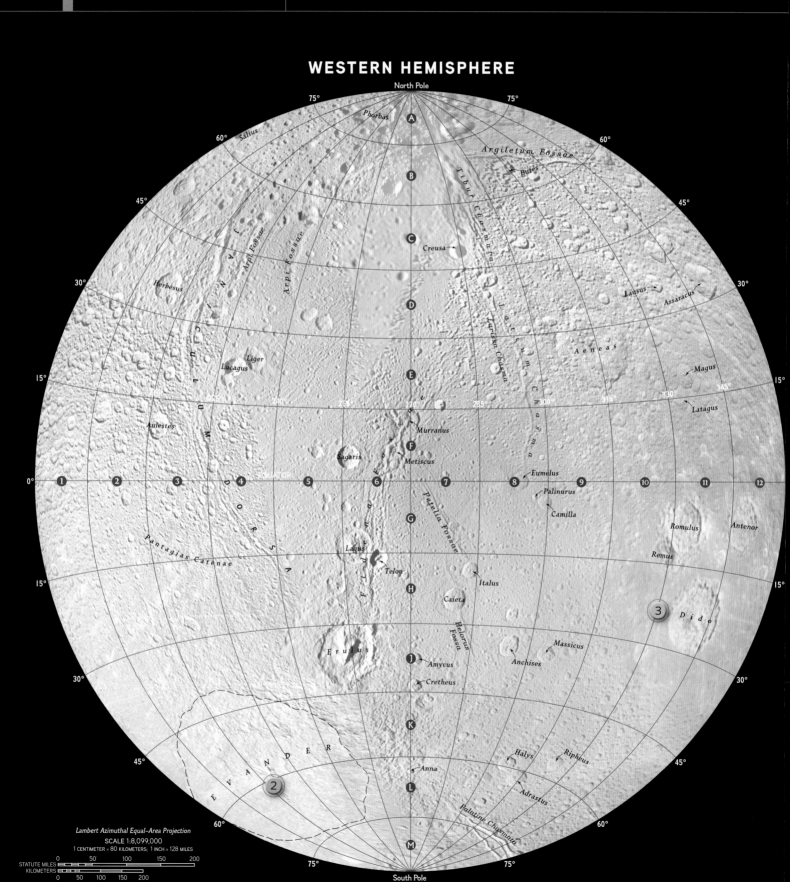

Lambert Azimuthal Equal-Area Projection
SCALE 1:8,099,000
1 CENTIMETER = 80 KILOMETERS; 1 INCH = 128 MILES

STATUTE MILES
0   50   100   150   200

KILOMETERS
0   50   100   150   200

KEY FEATURES

① PALATINE CHASMATA: A system of long, steep-sided depressions

② EVANDER: Largest crater on Dione

③ DIDO: Prominent crater matched by crater Aeneas, above it

EASTERN HEMISPHERE

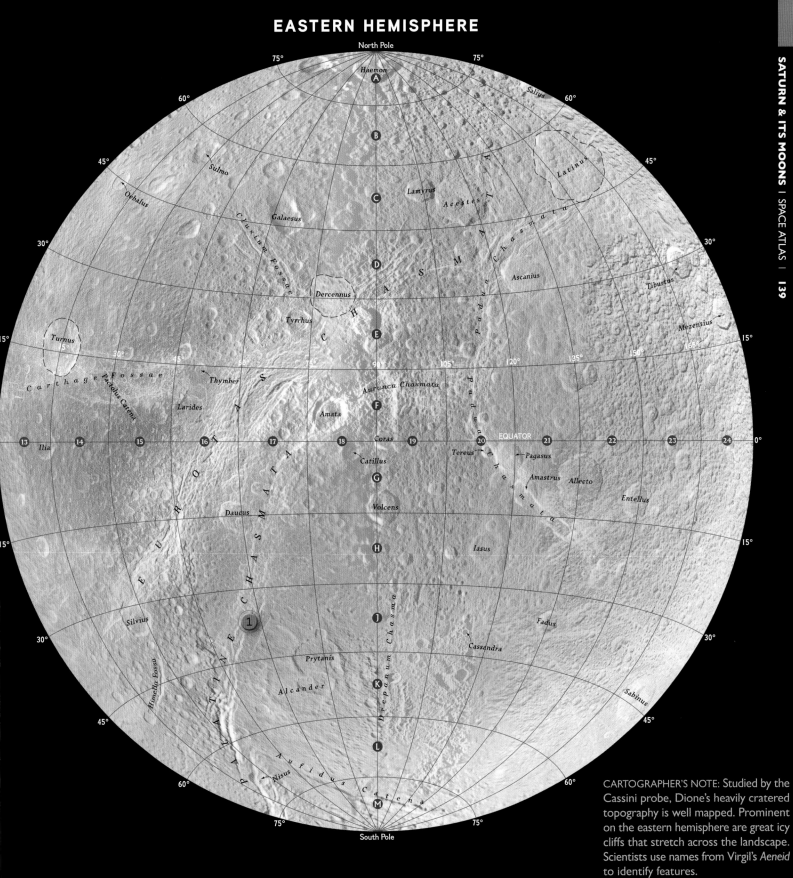

North Pole

Haemon
Saliys
Sulmo
Oebalus
Latinus
Lamyrus
Acestes
Galaesus
Ascanius
Tiburtus
Dercennus
Mezentius
Tyrrhus
PADUA CHASMATA
Turnus
CHASMATA
Carthage Fossae
Pactolus Catena
Thymber
Aurunca Chasmata
Larides
Amata
Coras
EQUATOR
Padua Chasmata
13 Ilia 14 15 16 17 18 19 20 Tereus Pagasus 21 22 23 24
Catillus
Amastrus Allecto
EUROTAS CHASMATA
Daucus
Volcens
Entellus
Iasus
PALATINE CHASMATA
Silvius
Fadus
Drepanum Chasma
Prytanis
Cassandra
Himella Fossa
Alcander
Sabinus
Aufidus Catena
Nisus

South Pole

CARTOGRAPHER'S NOTE: Studied by the Cassini probe, Dione's heavily cratered topography is well mapped. Prominent on the eastern hemisphere are great icy cliffs that stretch across the landscape. Scientists use names from Virgil's *Aeneid* to identify features.

SATURN'S MOONS | RHEA

Saturn's second largest moon, icy
Rhea is scarred by craters and ice cliffs.

WESTERN HEMISPHERE

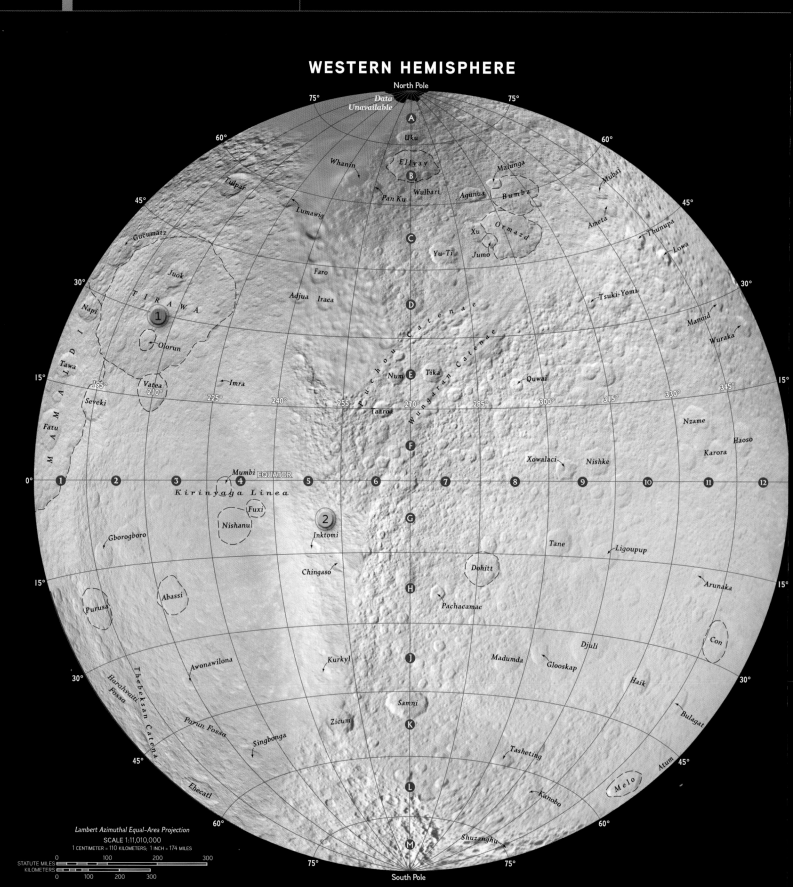

North Pole

Data Unavailable

Uku

Whanin — Ellyay — Malunga — Mubai

Tulpar — Pan Ku — Wulbari — Agunua — Bumba

Lumawig — Ameta — Thunupa

Gucumatz — Xu — Ormazd — Lowa

Juok — Yu-Ti — Jumo

Napi — Faro — Tsuki-Yomi

TIRAWA — Adjua — Iraca — Manoid

① — Wuraka

Olorun — Num — Tika — Quwai

Tawa — Puchou Catenae

Vatea 210° — Imra — Taaroa — Wungaran Catenae — Nzame

Seveki — 195° — Haoso

Fatu — 225° — 240° — 255° — 270° — 285° — 300° — 315° — 330° — 345° — Karora

Xowalaci — Nishke

Mumbi EQUATOR

① — ② — ③ — Kirinyaga Linea — ④ — ⑤ — ⑥ — ⑦ — ⑧ — ⑨ — ⑩ — ⑪ — ⑫

Fuxi — ② — Tane — Ligoupup

Nishanu — Inktomi — Dohitt

Gborogboro

Chingaso — Arunaka

Abassi — Pachacamac

Purusa — Djuli — Con

Awonawilona — Kurkyl — Madumda — Glooskap — Haik

Harahvaiti Fossa — Thebeksan Catena — Bulagat

Parun Fossa — Zicum — Samni — Tasheting

Singbonga — Atum

Ehecatl — Melo

Kanobo

Shuzanghu

South Pole

Lambert Azimuthal Equal-Area Projection
SCALE 1:11,010,000
1 CENTIMETER = 110 KILOMETERS; 1 INCH = 174 MILES

STATUTE MILES 0 100 200 300
KILOMETERS 0 100 200 300

KEY FEATURES

① **TIRAWA:** Large, well-defined impact basin

② **INKTOMI:** Young impact basin with well-defined rays

③ **YAMSI CHASMATA:** Long ice depressions

EASTERN HEMISPHERE

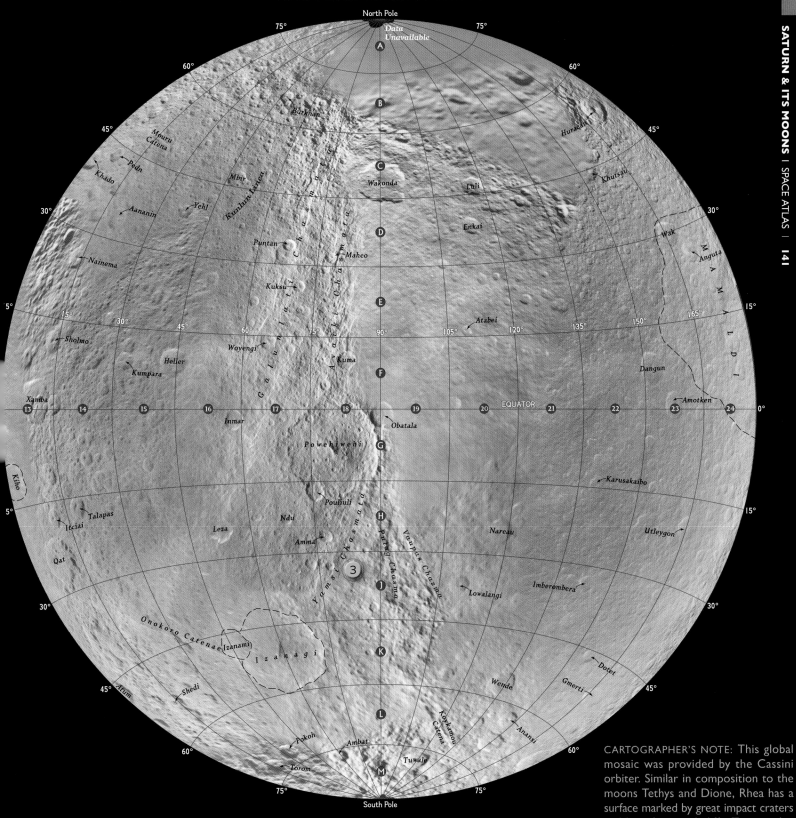

North Pole

75°   75°

A   *Data Unavailable*

60°   60°

B

Burkhan   *Huracan*

45°   45°

Mouru Catene   *Khutsau*

Pedn   C

Khado   *Mbir*   *Wakonda*   *Luli*

30°   *Yehl*   *Enkai*   *Wak*   30°

Aananin   *Puntan*   D   M

Nainema   *Maheo*   *Anguta*   A

Kuksu   E   *Atabei*   M

15°   *Sholmo*   *Woyengi*   *Kuma*   105° 120° 135° 150° 165°   A   15°

Heller   75°   90°   *Dangun*   L

Kumpara   F   D

Xamba   *Inmar*   *Obatala*   *Amotken*   I

13   14   15   16   17   18   19   EQUATOR 20   21   22   23   24   0°

Powehiwehi   G

Kiho   *Karusakaibo*

5°   *Talapas*   *Pouliuli*   H   5°

Itciai   *Ndu*   *Nareau*   *Utleygon*

Leza   *Amma*   *Imberombera*

Qat   ③   *Lowalangi*

Onokoro Catenae (*Izanami*)   J

30°   *Izanagi*   K   *Dotet*   30°

Wende   *Gmerti*

45°   *Atum*   *Shedi*   L   45°

Pokoh   *Koykamou Catena*   *Anansi*

Ambat   *Tuwale*

60°   M   60°

75°   75°

South Pole

CARTOGRAPHER'S NOTE: This global mosaic was provided by the Cassini orbiter. Similar in composition to the moons Tethys and Dione, Rhea has a surface marked by great impact craters and long fractured cliffs. Topographic names originate in the creation legends held sacred by cultures all over the world

SATURN'S MOONS | TITAN

Largest of Saturn's satellites, and the only moon in the solar system with its own atmosphere

WESTERN HEMISPHERE

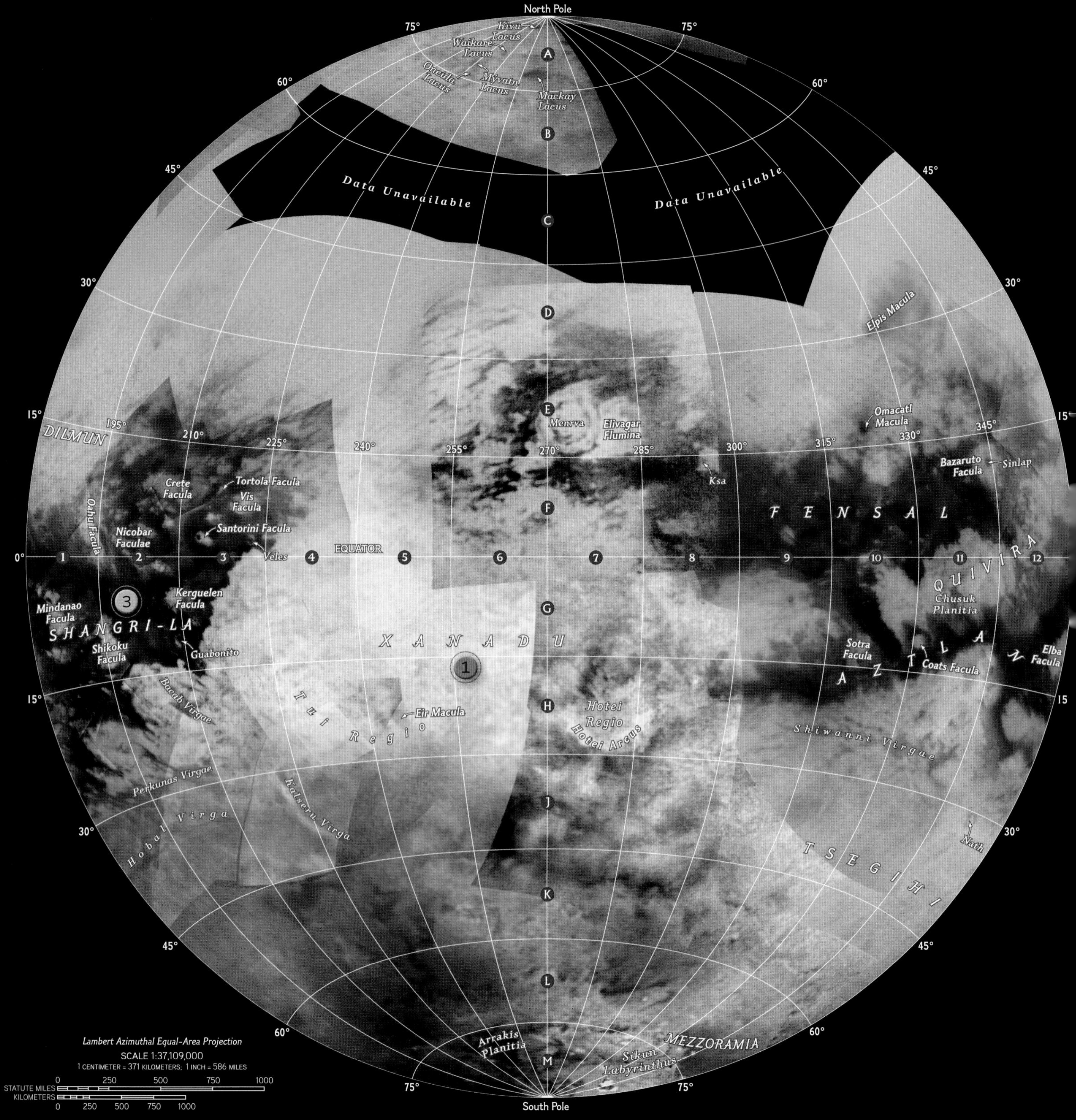

North Pole

75° · 75° · 60° · 60° · 45° · 45° · 30° · 30° · 15° · 15° · 0° · 15° · 30° · 45° · 60° · 75°

Kivu Lacus
Waikare Lacus
Oneida Lacus
Mývatn Lacus
Mackay Lacus

Data Unavailable · Data Unavailable

Elpis Macula

DILMUN

195° · 210° · 225° · 240° · 255° · 270° · 285° · 300° · 315° · 330° · 345°

Menrva · Elivagar Flumina

Omacatl Macula

Bazaruto Facula · Sinlap

Ksa

Crete Facula
Tortola Facula
Vis Facula
Oahu Facula
Nicobar Faculae
Santorini Facula
Veles

FENSAL

QUIVIRA

Kerguelen Facula
Mindanao Facula
SHANGRI-LA
Shikoku Facula
Guabonito

EQUATOR

Chusuk Planitia

Sotra Facula
Coats Facula
AZTLAN
Elba Facula

Barab Virgae

XANADU

Tui Regio

Eir Macula

Hotei Regio
Hotei Arcus

Shiwanni Virgae

Perkunas Virgae

Hobal Virga

Halseru Virga

Nath

TSEGIHI

Arrakis Planitia

MEZZORAMIA

Sikun Labyrinthus

South Pole

A · B · C · D · E · F · G · H · J · K · L · M

1 · 2 · 3 · 4 · 5 · 6 · 7 · 8 · 9 · 10 · 11 · 12

Lambert Azimuthal Equal-Area Projection
SCALE 1:37,109,000
1 CENTIMETER = 371 KILOMETERS; 1 INCH = 586 MILES

STATUTE MILES
KILOMETERS
0 · 250 · 500 · 750 · 1000

KEY FEATURES

1. **XANADU:** Area of hills and valleys
2. **ONTARIO LACUS:** Shallow lake of liquid hydrocarbons
3. **SHANGRI-LA:** Landing place of Huygens probe

EASTERN HEMISPHERE

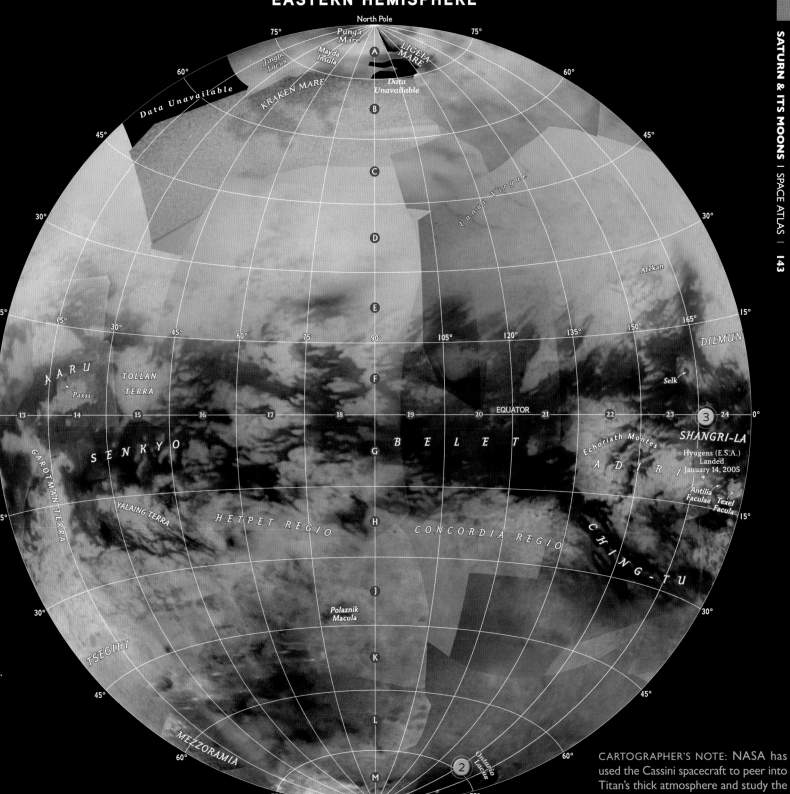

North Pole

75° 75°
Punga Mare
LIGEIA MARE
60° 60°
Jingpo Lacus
Mayda Insula
Data Unavailable
Data Unavailable
KRAKEN MARE
45° 45°
A
B
Uanui Virgae
30° 30°
C
Afekan
D
5° 15°
15° 30° 45° 60° 75° 90° 105° 120° 135° 150° 165° DILMUN
E
AARU Selk
TOLLAN TERRA
Paxsi
F
13 14 15 16 17 18 19 20 EQUATOR 21 22 23 3 24 0°
SHANGRI-LA
SENKYO B E L E T
Echoriath Montes
Hyugens (E.S.A.)
Landed
G A D I R January 14, 2005
GAROTMAN TERRA
YALAING TERRA
15° Antilia Faculae Texel Facula 15°
HETPET REGIO CONCORDIA REGIO
H
5° CHING-TU
Polaznik Macula
J
TSEGIHI
K
30° 30°
L
45° 45°
MEZZORAMIA
60° 60°
M Ontario Lacus
2
75° 75°
South Pole

CARTOGRAPHER'S NOTE: NASA has used the Cassini spacecraft to peer into Titan's thick atmosphere and study the landscape of this unique moon. Surface feature names come from a variety of sources, but the largest—the terrae—are named for sacred, enchanted, or wondrous lands from mythology.

SATURN'S MOONS | IAPETUS

Iapetus's dramatically two-toned surface may result from dark
materials left behind by evaporation on its warmer, leading hemisphere.

WESTERN HEMISPHERE

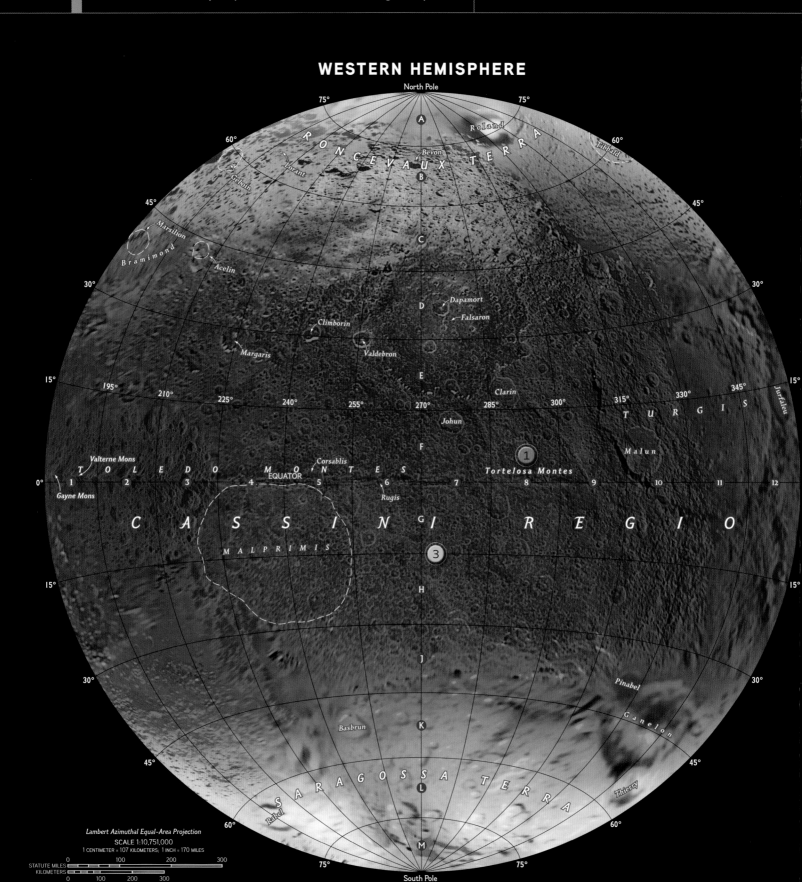

North Pole

RONCEVAUX TERRA

Roland
Tibbald
Bevon
Lorant
Gehoin
Marsilion
Bramimond
Acelin
Dapamort
Climborin
Falsaron
Margaris
Valdebron
Clarin
TURGIS
Jurfaleu
Johun
Malun
Valterne Mons
Corsablis
Tortelosa Montes
TOLEDO MONTES
EQUATOR
Gayne Mons
Rugis
CASSINI REGIO
MALPRIMIS
Pinabel
Ganelon
Basbrun
SARAGOSSA TERRA
Thierry
Rebel

Lambert Azimuthal Equal-Area Projection
SCALE 1:10,751,000
1 CENTIMETER = 107 KILOMETERS; 1 INCH = 170 MILES

STATUTE MILES
KILOMETERS

South Pole

KEY FEATURES

1. **TORTELOSA MONTES:** 10-kilometer-high (6 mi) mountains

2. **ENGELIER:** Impact basin 500 kilometers (300 mi) wide

3. **CASSINI REGIO:** Darker region

EASTERN HEMISPHERE

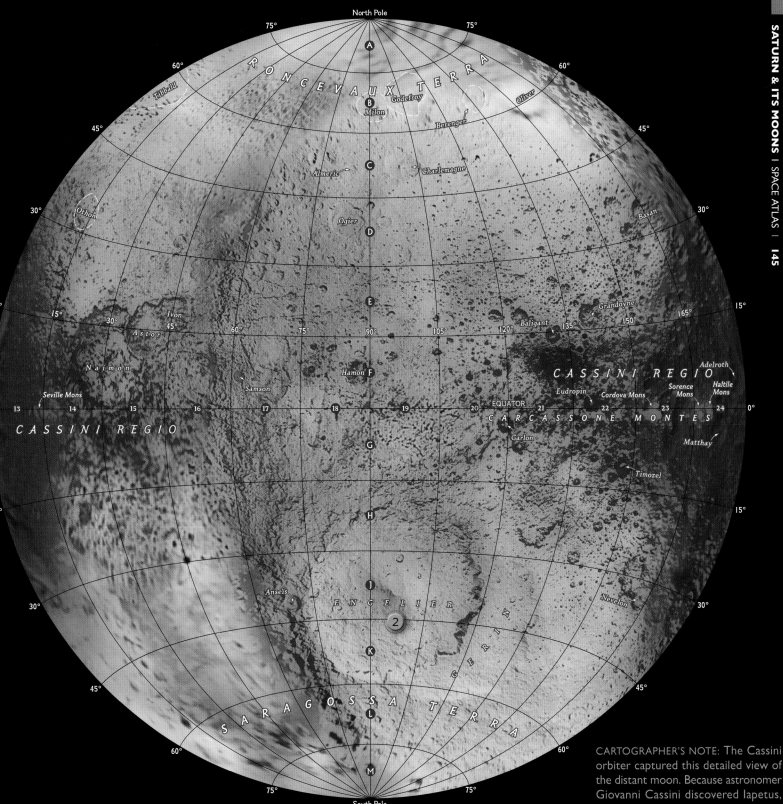

CARTOGRAPHER'S NOTE: The Cassini orbiter captured this detailed view of the distant moon. Because astronomer Giovanni Cassini discovered Iapetus, the dark region bears his name. Other planetary topography is named for characters and places in Dorothy Sayers's translation of *Chanson de Roland*.

Saturn's winds are a striking feature of its atmosphere, registering at up to 1,800 kilometers an hour (1,120 mph)—among the fastest winds anywhere in the solar system. The planet also sports a short-lived storm structure known as the Great White Spot. This structure shows up every Saturn year (about 30 Earth years) at about the time of the summer solstice in Saturn's northern hemisphere. Similar in appearance to Jupiter's Great Red Spot, the Great White Spot was first observed in 1876 and has been seen intermittently since then.

Like Jupiter, Saturn has a large and powerful magnetic field—and like Earth, it has shimmering auroras at both poles, caused by magnetic particles spiraling along the magnetic field lines.

MISSIONS TO SATURN

In addition to observing Saturn by telescope, scientists have sent a number of spacecraft to the Saturnian system. Between 1979 and 1982, Pioneer and Voyager spacecraft flew by the planet on their way out of the solar system, discovering new moons and new aspects of the rings in the process. Then, in 1997, the Cassini spacecraft was launched, reaching and entering its orbit around Saturn in 2004. Most of the detailed information we have about the planet, its rings, and its moons comes from this mission, which is still in orbit and returning data as you read this.

When scientists look at Saturn, however, they seldom concentrate on the planet itself. Instead, they focus on its spectacular ring system (which we will discuss in the next section) and its moons. Like Jupiter, Saturn has a large array of satellites—62 discovered as of this writing, of which 53 have been officially named. As was the case with Jupiter, many of these moons are small—some less than a mile across—and only 13 are more than 50

kilometers (30 mi) in diameter; of these large moons, the two that have attracted the most scientific attention are Titan and Enceladus.

TITAN

Titan was discovered in 1655 by the Dutch astronomer Christiaan Huygens (1629–1695). It is larger than Mercury and the only moon in the solar system to have a significant atmosphere. It is also the one world in the system where we can, perhaps, get a glimpse of the kind of chemistry that led to life on Earth billions of years ago—which is why it has become a center of scientific exploration.

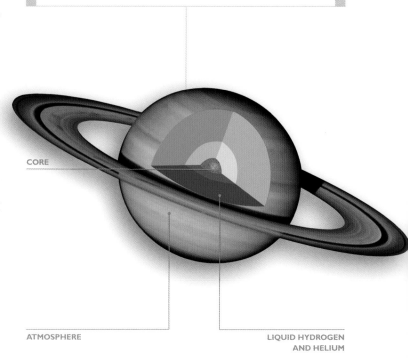

A cutaway view of our best current idea about the structure of Saturn. Like Jupiter, it probably has a rocky core surrounded by a layer of metallic hydrogen, which in turn is surrounded by layers of liquid hydrogen and helium.

CORE

ATMOSPHERE

LIQUID HYDROGEN AND HELIUM

Dwarfed by the bulk of its parent planet, little Mimas (at bottom of image) circles in orbit against the bluish mass of Saturn's northern hemisphere. The dark bands in the picture are shadows cast on Saturn's surface by its rings.

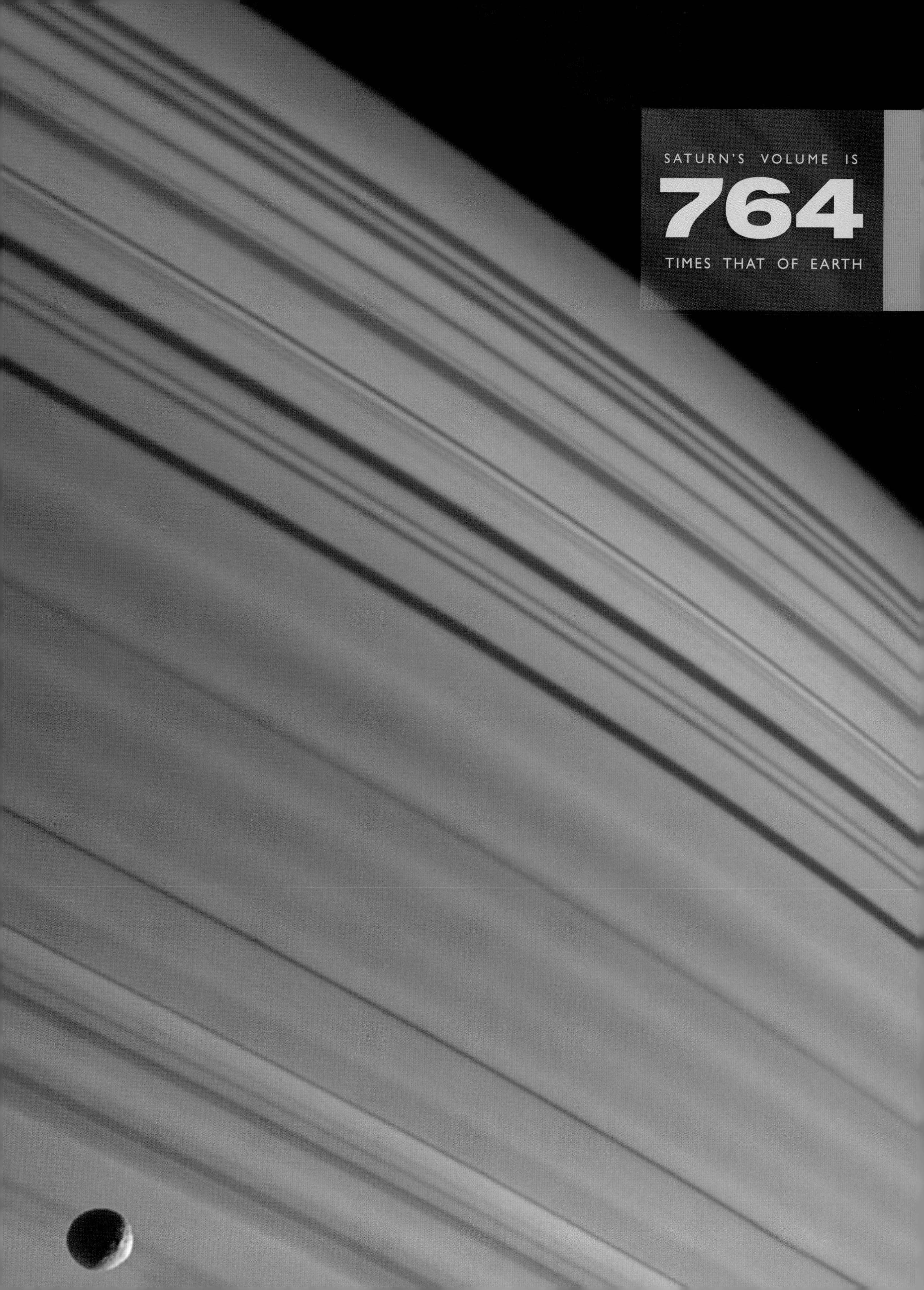

SATURN'S VOLUME IS

764

TIMES THAT OF EARTH

Titan is thought to have a rocky core covered by thick layers of ice, possibly with a water-ammonia ocean beneath the icy surface. Its atmosphere is mostly nitrogen and supports several cloud layers that make it impossible to see the surface from outside. Like Venus, Titan waited for visiting spacecraft before it revealed its surface.

A few months after its arrival at Saturn in July 2004, Cassini dropped a probe, appropriately named Huygens, into Titan's atmosphere. (Eventually, Cassini made more than 60 close approaches to the moon.) Descending by parachute, Huygens sent back the first images of the surface. To the surprise of the observers, the surface of Titan turned out to look almost exactly like the familiar surface of Earth! As one researcher put it, "What was alien about Titan was its eerie familiarity."

In fact, Titan's equatorial region is covered by long sand dunes as big as those in the Sahara, interspersed with rocky hills, while near the poles are large liquid lakes. One is slightly larger than Lake Ontario, after which it is named.

You can understand the similarity of Titan to Earth by noting two things: First, Titan is really cold, with surface temperatures hovering at about minus 180°C (-320°F). Second, at these temperatures familiar materials take on unexpected forms. Water ice is as hard as granite, for example, and methane, which we know as natural gas on our own balmy planet, is a liquid. In essence, what we see on Titan are familiar processes involving unfamiliar materials.

For example, Titan's high clouds, made of hydrocarbon molecules like methane and ethane (methane's chemical cousin), interact with ultraviolet light from the sun to produce a haze that wouldn't be out of place in Los Angeles on a bad day. The hydrocarbons rain down onto the surface, producing, among other things, that "sand" in the equatorial dunes. (One researcher compared the material to a heap of coffee grounds.)

THE FLOATING WORLD

The planets of our solar system exhibit a wide range of densities, from Earth (the densest) to Saturn (the least dense). The density of any object measures how much "stuff" is packed into it, or, technically, its mass divided by its volume. Water's density, 1 gram per cubic centimeter, is used as a standard. Iron, for example, has a density around 7.9, while Earth has a density of about 5.5. Interestingly, ice has a density of about 0.92—less than that of water. This explains why ice stays at the surface of frozen lakes and ponds—any material with a density less than 1 will float in water.

Finding the density of a planet requires that we measure both its mass and its volume. The latter is easy—if we know how far away the planet is and how big its disk appears to be, we can find its radius. From there, simple geometry gets us the volume. Finding the mass is a little more difficult, but if the planet has moons we can deduce the planetary mass by observing the orbits of the moons.

When we apply these methods to Saturn, we find that it is a planet with a volume 764 times bigger and a mass 95 times larger than Earth. This translates into a density of about 0.69—less than the density of water and less, even, than the density of the sun. If you could find an ocean big enough, Saturn would actually float!

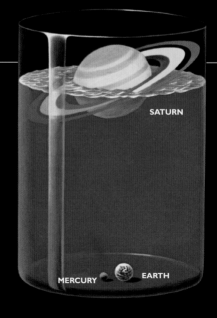

SATURN COULD FLOAT IN WATER.

The lakes are liquid methane, which condenses and falls out of Titan's sky as rain to produce other familiar landscape features just as water does on Earth.

Because hydrocarbons are organic molecules, close relatives of the kinds of molecules we believe produced life on Earth, scientific attention has focused on Titan. We shall see on page 181 that the creation of organic molecules from nonliving material is the first step in the development of living things. The hope, of course, is that by studying the way this process is proceeding on Titan right now we can learn something about how it happened on Earth four billion years ago.

ENCELADUS

The sixth largest moon of Saturn, Enceladus long was known to have a surface made of water ice. In 2005,

An artist's conception depicts the insertion of the spacecraft Cassini into orbit around Saturn. One of the largest and most complex spacecraft ever built, Cassini was launched in 1997 and reached Saturn in 2004. In December 2004 it released the probe Huygens into the atmosphere of Saturn's moon Titan. The product of a collaboration between 16 European countries and NASA, Cassini still orbits Saturn and continues to send back data.

however, Cassini recorded a geyser spewing near the moon's south pole: a geyser that contained liquid water. Like Jupiter's moon Io, Enceladus appears to be heated by the flexing it undergoes in Saturn's gravitational field. Further study suggested that there may be an ocean of liquid water under the moon's icy surface. Although there is still scientific controversy over the existence of the subsurface ocean, if the water exists, Enceladus may join Europa as a possible abode of life in our solar system.

The rings of Saturn are the most spectacular—and least substantial—objects in the solar system. Made up primarily of chunks of water ice in orbit around the planet, they create a stunning display of reflected sunlight visible from Earth through even small telescopes. • Like many objects in the solar system, Saturn's rings were first seen by Galileo when he turned his telescope toward the skies in 1610. Because his instrument was crude by modern standards, he saw the rings as two dots on either side of the main planet and mistook them for moons—at one point he described Saturn as having "ears." It wasn't until 1655 that the Dutch astronomer Christiaan Huygens, using an improved telescope, identified the ears as rings circling the planet.

SATURN'S RINGS

[HOOPS OF GLITTERING ICE]

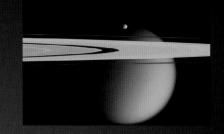

DISCOVERER: GALILEO GALILEI

DISCOVERY DATE: JULY 1610

NAMED FOR: RINGS, GAPS, AND DIVISIONS ARE NAMED ALPHABETICALLY IN ORDER OF DISCOVERY AND FOR SCIENTISTS

LARGEST RINGS AND DISTANCE FROM PLANET'S CENTER:

D: 67,000–74,490 KM (41,630–46,290 MI)

C: 74,490–91,980 KM (46,290–57,150 MI)

B: 91,980–117,500 KM (57,150–73,010 MI)

A: 122,050–136,770 KM (75,840–84,980 MI)

G: 166,000–174,000 KM (103,150–108,120 MI)

E: 180,000–480,000 KM (111,850–300,000 MI)

False-color image of Saturn's rings. (Inset) Saturn's A and F rings stretch in front of Titan and tiny Epimetheus.

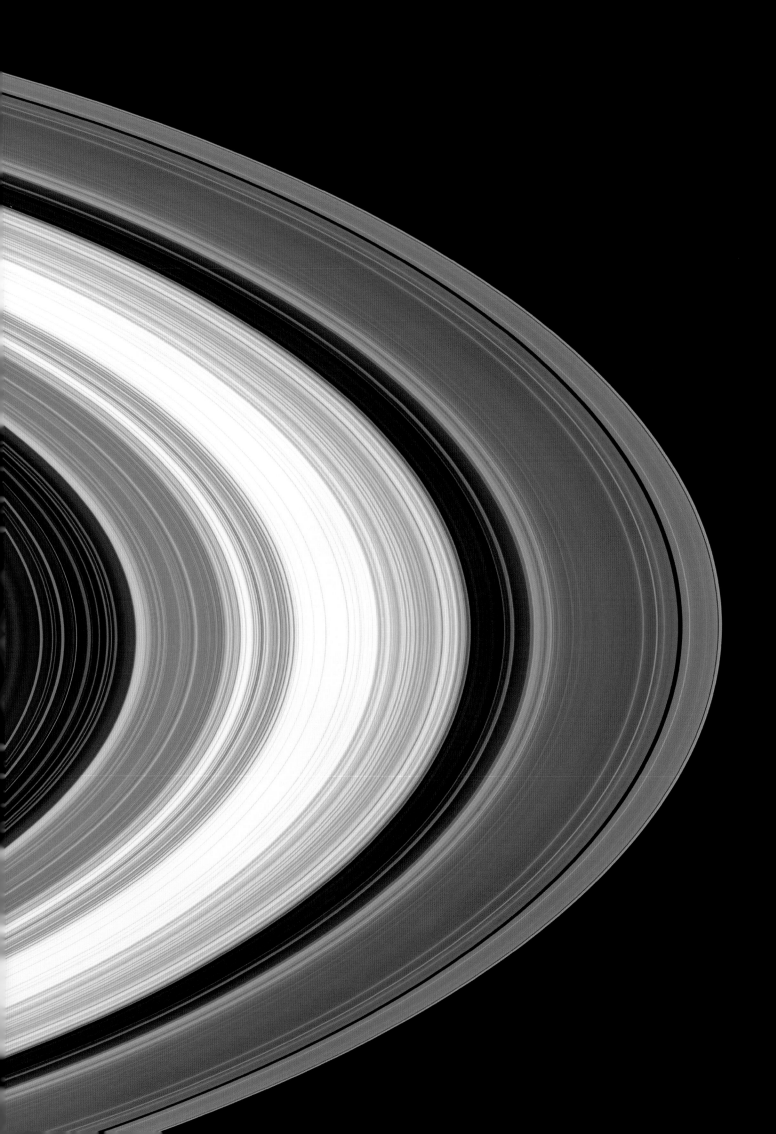

SATURN'S RINGS

Ranging from tiny ringlets to the big A, B, and C rings,
Saturn's rings are broken up by multiple gaps, large and small.

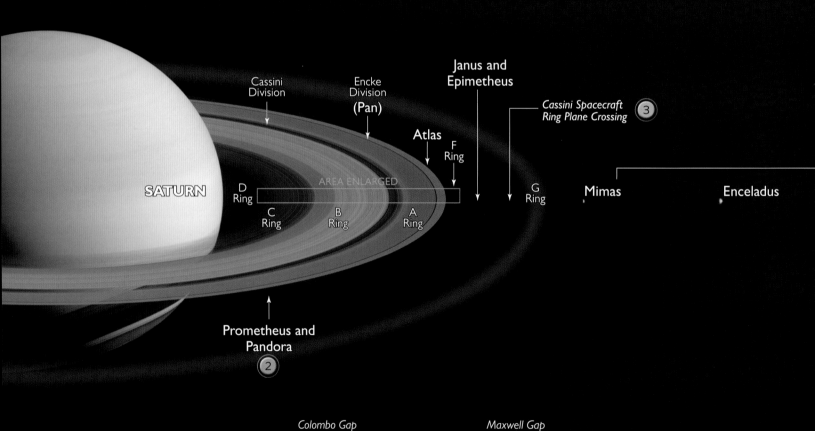

Cassini
Division

Encke
Division
(Pan)

Janus and
Epimetheus

Cassini Spacecraft
Ring Plane Crossing ③

Atlas

F
Ring

SATURN

D
Ring

AREA ENLARGED

G
Ring

Mimas

Enceladus

C
Ring

B
Ring

A
Ring

Prometheus and
Pandora
②

Colombo Gap

Maxwell Gap

D
Ring

C
Ring

74,500 km
(46,300 mi)

92,000 km
(57,200 mi)

KEY FEATURES

(1) CASSINI DIVISION: Largest gap in the ring system

(2) PROMETHEUS AND PANDORA: Tiny "shepherd" moons

(3) CASSINI RING PLANE CROSSING: Gap through which the Cassini orbiter flies

Titan
Hyperion →
Iapetus
Phoebe

E Ring

To Titan

Tethys

Dione

Rhea

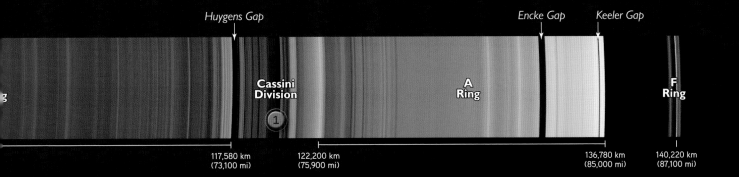

Huygens Gap

Encke Gap   Keeler Gap

Cassini
Division

A
Ring

F
Ring

(1)

117,580 km
(73,100 mi)

122,200 km
(75,900 mi)

136,780 km
(85,000 mi)

140,220 km
(87,100 mi)

Although today we know that all of the giant planets have ring systems, the rings of Saturn remain the largest and most complete set in the solar system. Almost from their discovery, they have stimulated the human imagination. For example, a book published in 1837 by English cleric Thomas Dick, with the grandiloquent title of *Celestial Scenery, or the Wonders of the Planetary System Displayed, Illustrating the Perfections of Deity and a Plurality of Worlds,* estimates the number of human beings living on the rings of Saturn at 8,141,963,826,080!

Unfortunately for the good reverend, in the 18th and 19th centuries scientific calculations established clearly that the rings of Saturn could not be solid, like a cosmic hula hoop, or liquid. In either case, the forces acting in the ring would make it unstable. Early on, then, scientists knew that the rings had to be collections of orbiting particles. In fact, we now know that the constituents of the rings are mainly pieces of water ice, ranging in size from bits a fraction of a centimeter wide to boulders several meters across. The most astonishing thing is that in spite of their optical prominence, the rings are actually quite thin. Estimates put the average thickness of the rings at about 10 meters (30 ft)—an ordinary two-story building would barely stretch from top to bottom.

RINGS, RINGLETS, AND GAPS

At the same time that scientists were gaining a theoretical understanding of the basic dynamics of the rings, astronomers were using telescopes to discover something about their structure. In 1675 the Italian astronomer Giovanni Cassini (1625–1712) found that the rings were not a single unbroken hoop, but had dark divisions between different sections. The largest of these gaps is now called the Cassini division. In fact, as seen from Earth there are two main dark gaps in the

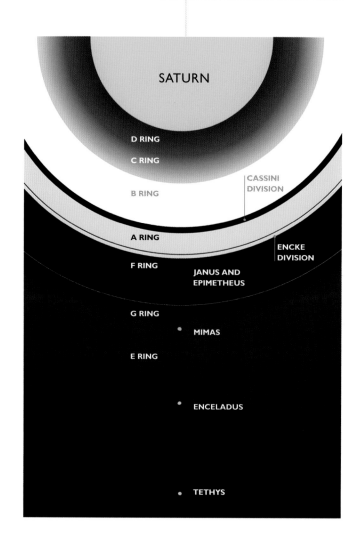

Saturn's rings were named alphabetically in the order of discovery. Gaps and divisions between the rings are named for astronomers; these gaps seem to be maintained by the influence of Saturn's many moons.

SATURN

D RING

C RING

CASSINI DIVISION

B RING

A RING

ENCKE DIVISION

F RING

JANUS AND EPIMETHEUS

G RING

MIMAS

E RING

ENCELADUS

TETHYS

rings, dividing the rings into three parts. These parts were named, with an unusual lack of imagination, the A, B, and C rings. The naming of the many rings discovered since then takes the alphabet up to G, with some also named for the moons that most influence them. Although there are faint rings of dust millions of kilometers from Saturn, the primary rings extend from about 74,000 kilometers (46,000 mi) from the center of the planet out to about 137,000 kilometers (85,000 mi). (For reference, our own moon is 375,000 kilometers/250,000 miles from Earth.)

An artist's conception portrays the icy rubble that makes up one of Saturn's rings. The particles in the rings are actually quite small, ranging from pebble to boulder size. Saturn's moons play a shepherding role, since their gravity keeps these particles in tight orbits.

With the flyby of the Voyager spacecraft in 1980 and, of course, the arrival of the Cassini spacecraft in 2004, our knowledge of the rings has increased considerably. We now understand that the rings are actually complex structures, with thousands of thin gaps separating ringlets. The gaps aren't really empty—they just have less material in them—and seem to be maintained by a variety of gravitational interactions with Saturn's moons. In some cases, close-in moons simply clear out a path, while in other cases the process is more complex, but in the end it is the combination of the moons and their gravity that maintains the structure of the rings.

The one possible exception to this rule is the appearance of phenomena known as spokes, seen intermittently by spacecraft. These are lines running across the rings, bright or dark depending on whether light is being reflected from the spokes or transmitted through them. They are thought to be made of microscopic dust particles, perhaps created in storms on Saturn, that are suspended above the rings by static electricity.

A MOON'S REMAINS?

Finally, we can turn to the question of how such a ring system could form. The most popular theory is that the particles are debris remaining from an ancient moon. The amount of material is about the same as that found in many of Saturn's surviving moons. Different versions of the theory have the moon breaking up because it got too close to Saturn or because of a collision. If the collision theory is correct, then it is likely that the rings of Saturn are one more legacy of the Late Heavy Bombardment (see page 47), that great reshuffling of the solar system that took place four billion years ago.

The last two planets in the solar system are the least known and the least explored. They probably formed closer to the sun than their present orbits, and, although they are similar to each other, they differ significantly from the gas giants Jupiter and Saturn. The atmospheres of Uranus and Neptune contain a lot more of the substances astronomers call "ices"—dense mixtures of water, ammonia, and methane. Thus, they are often referred to as ice giants. • The ice giants are intermediate in size between the terrestrial planets and the gas giants. Neptune, for example, has 17 times the mass of Earth, but only one-nineteenth the mass of Jupiter.

URANUS & NEPTUNE

[ICE GIANTS]

URANUS DISCOVERER: WILLIAM HERSCHEL
URANUS DISCOVERY DATE: MARCH 13, 1781
URANUS NAMED FOR: GREEK GOD OF THE SKY
URANUS MASS: 14.54 × EARTH'S
URANUS VOLUME: 63.09 × EARTH'S

NEPTUNE DISCOVERERS: URBAIN LEVERRIER, JOHN COUCH ADAMS, AND JOHANN GALLE
NEPTUNE DISCOVERY DATE: SEPTEMBER 23, 1846
NEPTUNE NAMED FOR: ROMAN GOD OF THE SEA
NEPTUNE MASS: 17.15 × EARTH'S
NEPTUNE VOLUME: 57.72 × EARTH'S
URANUS AND NEPTUNE MOONS: 27 AND 13
PLANETARY RING SYSTEM: YES FOR BOTH PLANETS

Art of Neptune with rings and moon Triton. (Inset) Uranus

URANUS

Distant and still somewhat mysterious, Uranus has not been visited by a spacecraft since 1986.

With its axis of rotation nearly parallel to the plane of its orbit, Uranus can be thought of as "rolling" through its year. As it revolves, one pole is always lit by the sun, while the other is in darkness. Taking nearly 84 Earth years to complete an orbit, each pole receives about 40 years of day followed by 40 of night.

Methane gas in the atmosphere absorbs the red spectrum of visible light. Only blue-green light is reflected, giving Uranus its characteristic color.

Unlike Earth which has a magnetic field rotated 11 degrees from its geographic poles, Uranus's is shifted 59 degrees. Another baffling feature is that the axis of the magnetic poles does not pass through the center of the planet.

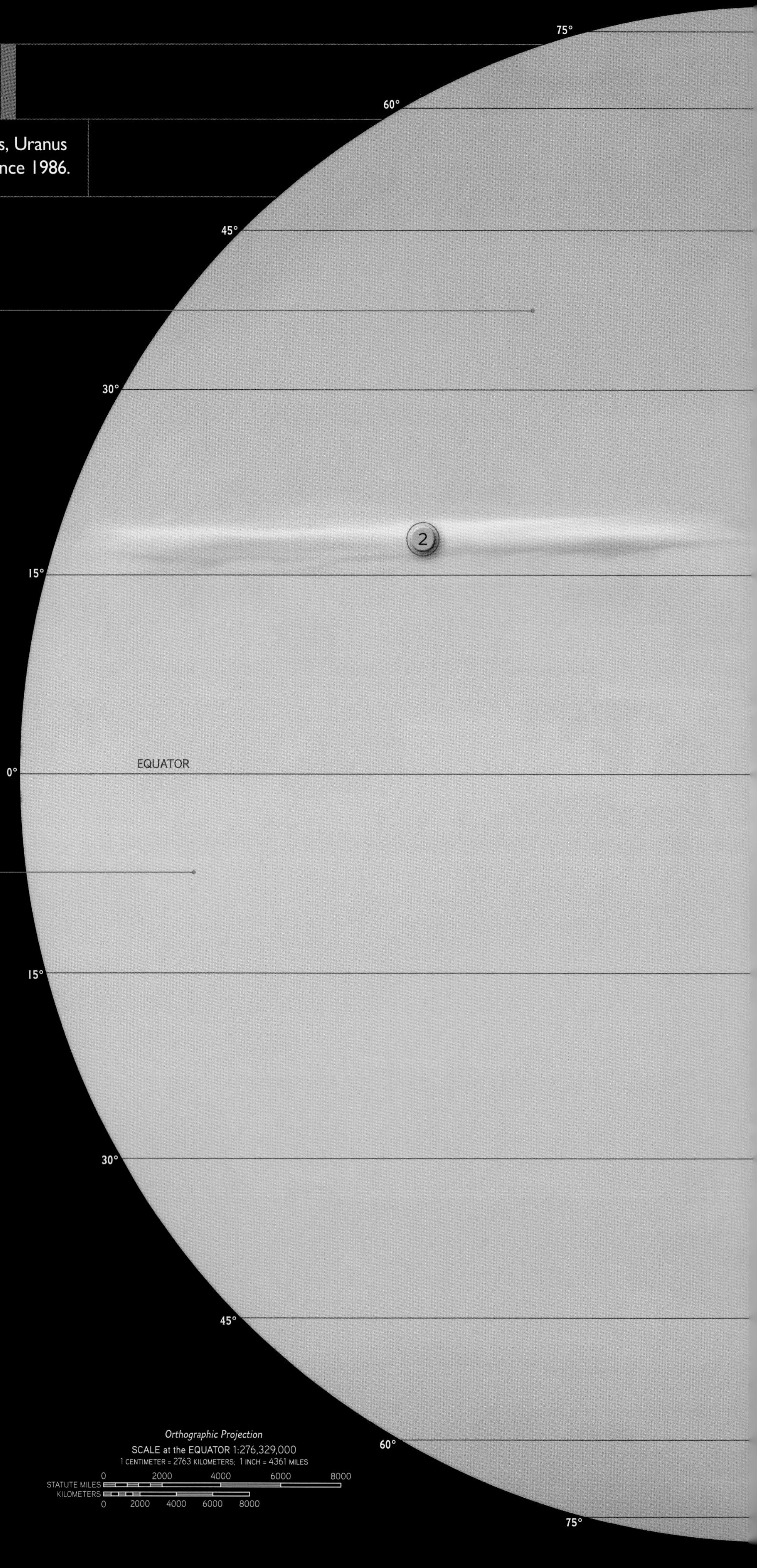

75°
60°
45°
30°
15°

0°   EQUATOR

15°
30°
45°
60°
75°

②

Orthographic Projection
SCALE at the EQUATOR 1:276,329,000
1 CENTIMETER = 2763 KILOMETERS; 1 INCH = 4361 MILES

STATUTE MILES
0   2000   4000   6000   8000

KILOMETERS
0   2000   4000   6000   8000

75°

60°

45°

30°

15°

EQUATOR 0°

15°

30°

45°

60°

75°

KEY FEATURES

1 NORTH POLE: This pole will face the sun directly in 2028.

2 CLOUDS: More clouds are seen as the planet moves into northern summer.

3 ATMOSPHERIC COLOR: Methane absorbs red light, leaving the planet blue-green.

3

Obscured by high-level clouds, Uranus's atmosphere is composed of belts and zones like Jupiter and Saturn. Scientists analyze imagery taken in wavelengths other than visible light to study Uranus's atmosphere.

CARTOGRAPHER'S NOTE: Unique in the solar system, Uranus's axis is tilted nearly 98 degrees, perhaps the result of an immense impact with an Earth-size object in the distant past. The blue-green hue of this Voyager 2–based global mosaic is caused by the presence of methane in the atmosphere.

URANUS'S MOONS

Little Miranda has one of the most shattered surfaces in the solar system; Ariel may have had recent geological activity.

MIRANDA–SOUTHERN HEMISPHERE

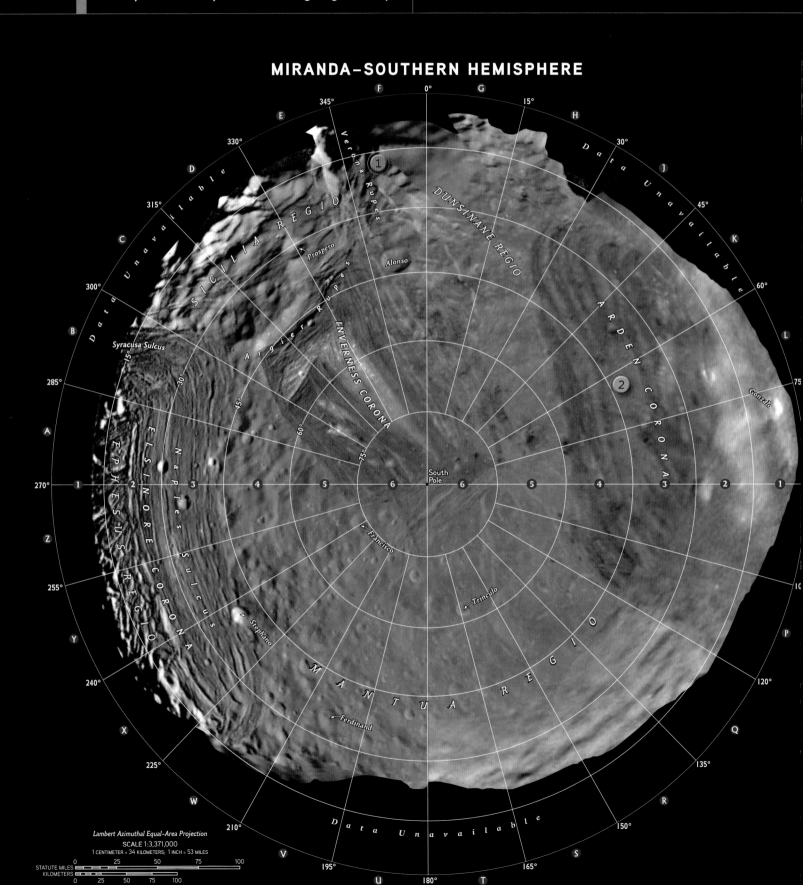

Lambert Azimuthal Equal-Area Projection

SCALE 1:3,371,000

1 CENTIMETER = 34 KILOMETERS; 1 INCH = 53 MILES

STATUTE MILES 0   25   50   75   100

KILOMETERS 0   25   50   75   100

KEY FEATURES

1. **VERONA RUPES:** One of many sharp cliffs and canyons
2. **ARDEN CORONA:** Grooved landscape, possibly over warmer ice
3. **KACHINA CHASMATA:** Canyons possibly formed by faulting

ARIEL–SOUTHERN HEMISPHERE

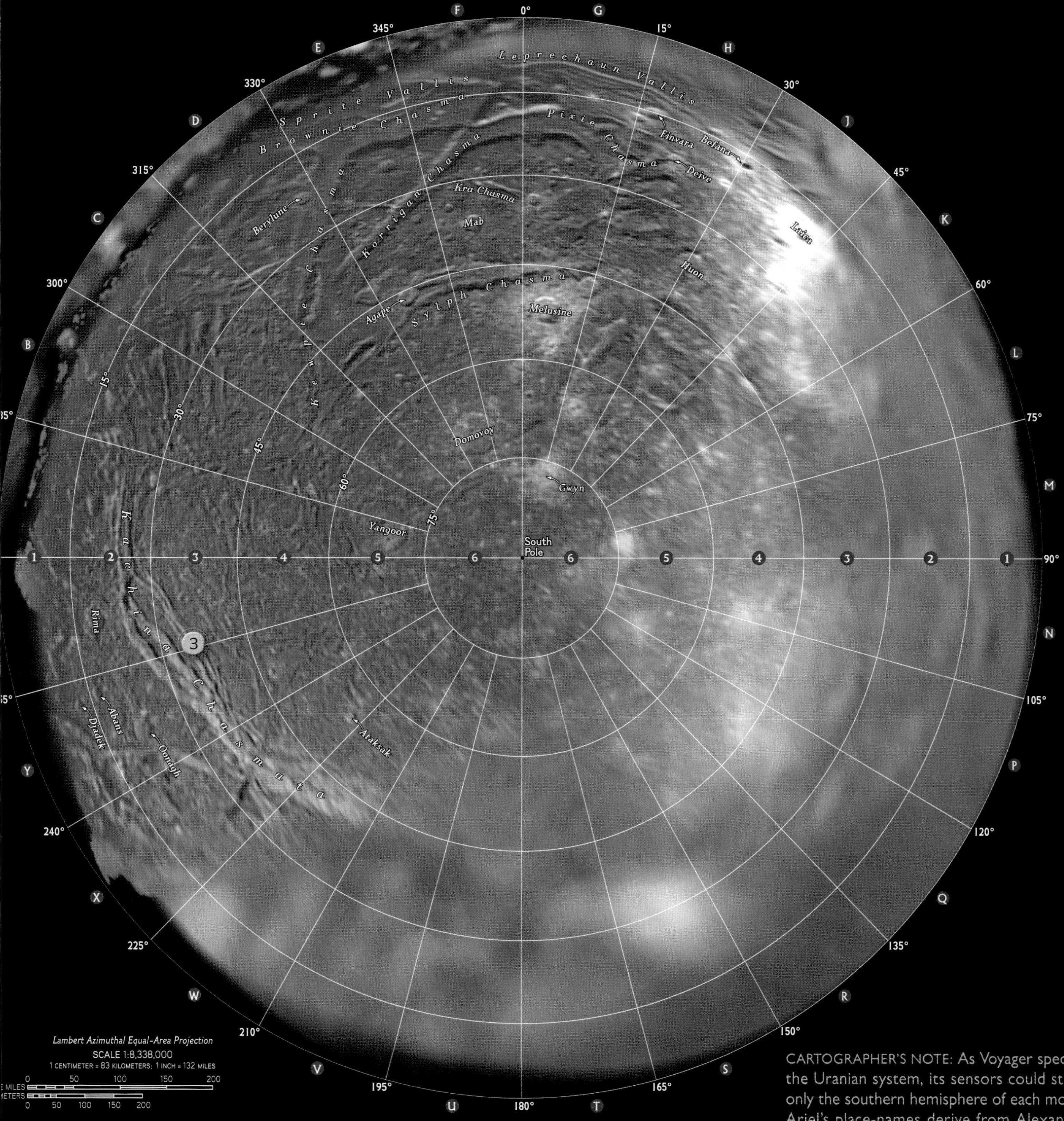

F 0° G
345° 15°
E H
330° 30°
D J
315° 45°
C K
300° 60°
B L
75°
90°
105°
M
N
P
Q
R
S
T U
165° 180°
195° 210°
225°
240°
135°
150°
120°

Leprechaun Vallis
Sprite Vallis
Brownie Chasma
Pixie Chasma
Finvara
Befana
Deive
Berylune
Korrigan Chasma
Kra Chasma
Mab
Laica
Mephistopheles Chasma
Agape
Sylph Chasma
Melusine
Huon
Domovoy
Gwyn
Yangoor
South Pole
Rima
Kachinda Chasmata
Djadek
Albans
Oonagh
Ataksak

Lambert Azimuthal Equal–Area Projection
SCALE 1:8,338,000
1 CENTIMETER = 83 KILOMETERS; 1 INCH = 132 MILES

0 50 100 150 200
MILES
METERS
0 50 100 150 200

CARTOGRAPHER'S NOTE: As Voyager sped by the Uranian system, its sensors could study only the southern hemisphere of each moon. Ariel's place-names derive from Alexander Pope's *Rape of the Lock*, while Miranda's are taken from Shakespeare's *Tempest*.

URANUS'S MOONS | UMBRIEL & TITANIA

Mysteriously dark, Umbriel has an ancient surface. Titania, largest of the planet's moons, may be geologically active.

UMBRIEL–SOUTHERN HEMISPHERE

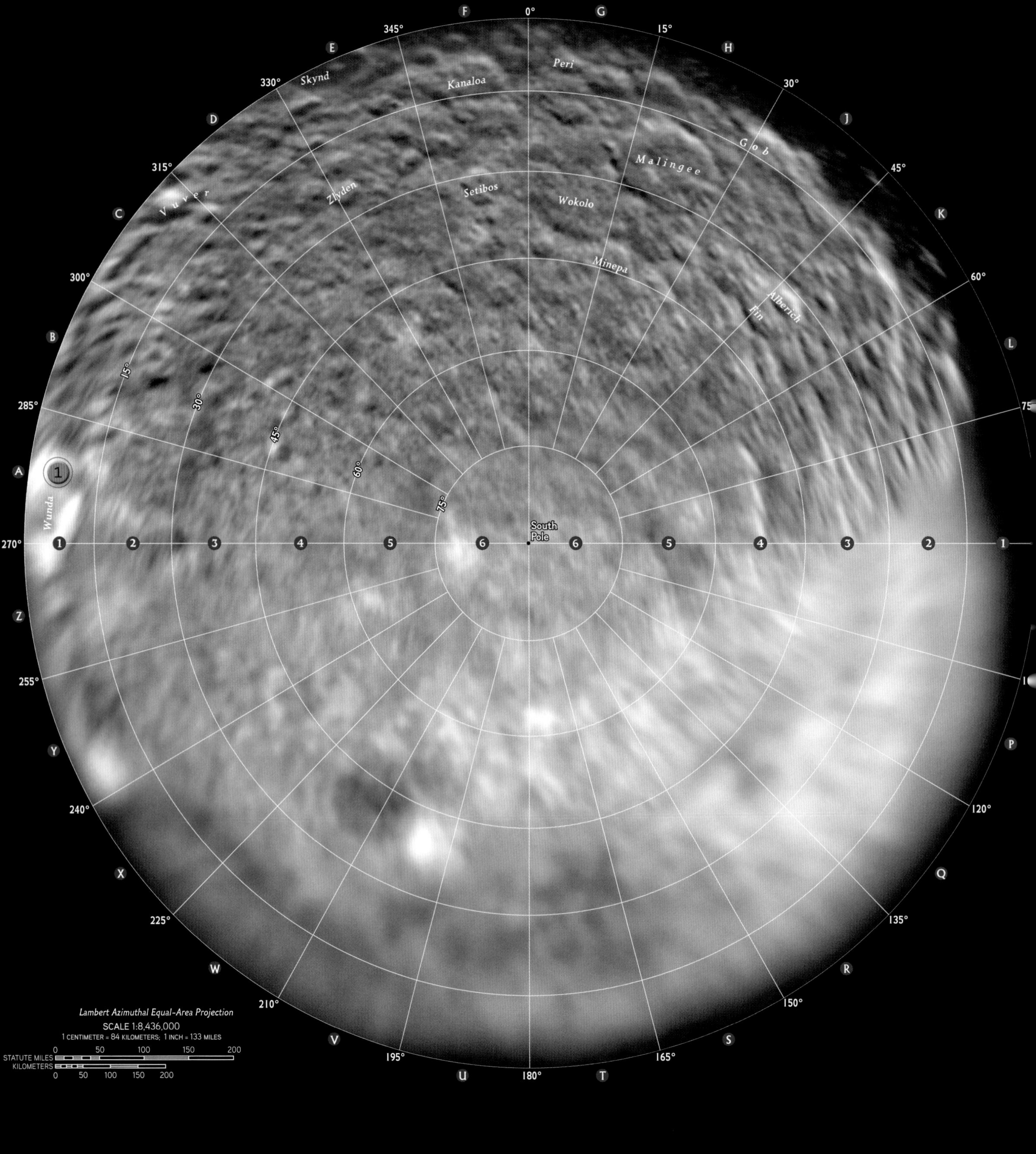

Peri
Skynd
Kanaloa
Gob
Malingee
Vuver
Zlyden
Setibos
Wokolo
Minepa
Alberich
Fin
Wunda
South Pole
0°
15°
30°
45°
60°
75°
120°
135°
150°
165°
180°
195°
210°
225°
240°
255°
270°
285°
300°
315°
330°
345°

Lambert Azimuthal Equal-Area Projection
SCALE 1:8,436,000
1 CENTIMETER = 84 KILOMETERS; 1 INCH = 133 MILES
STATUTE MILES 0 50 100 150 200
KILOMETERS 0 50 100 150 200

1 WUNDA: Crater with oddly bright rim, possibly due to frost or impact deposits

2 MESSINA CHASMATA: Canyon almost 1,500 kilometers (930 mi) long

3 ROUSILLON RUPES: Young scarp (cliff) 400 kilometers (250 mi) long

TITANIA–SOUTHERN HEMISPHERE

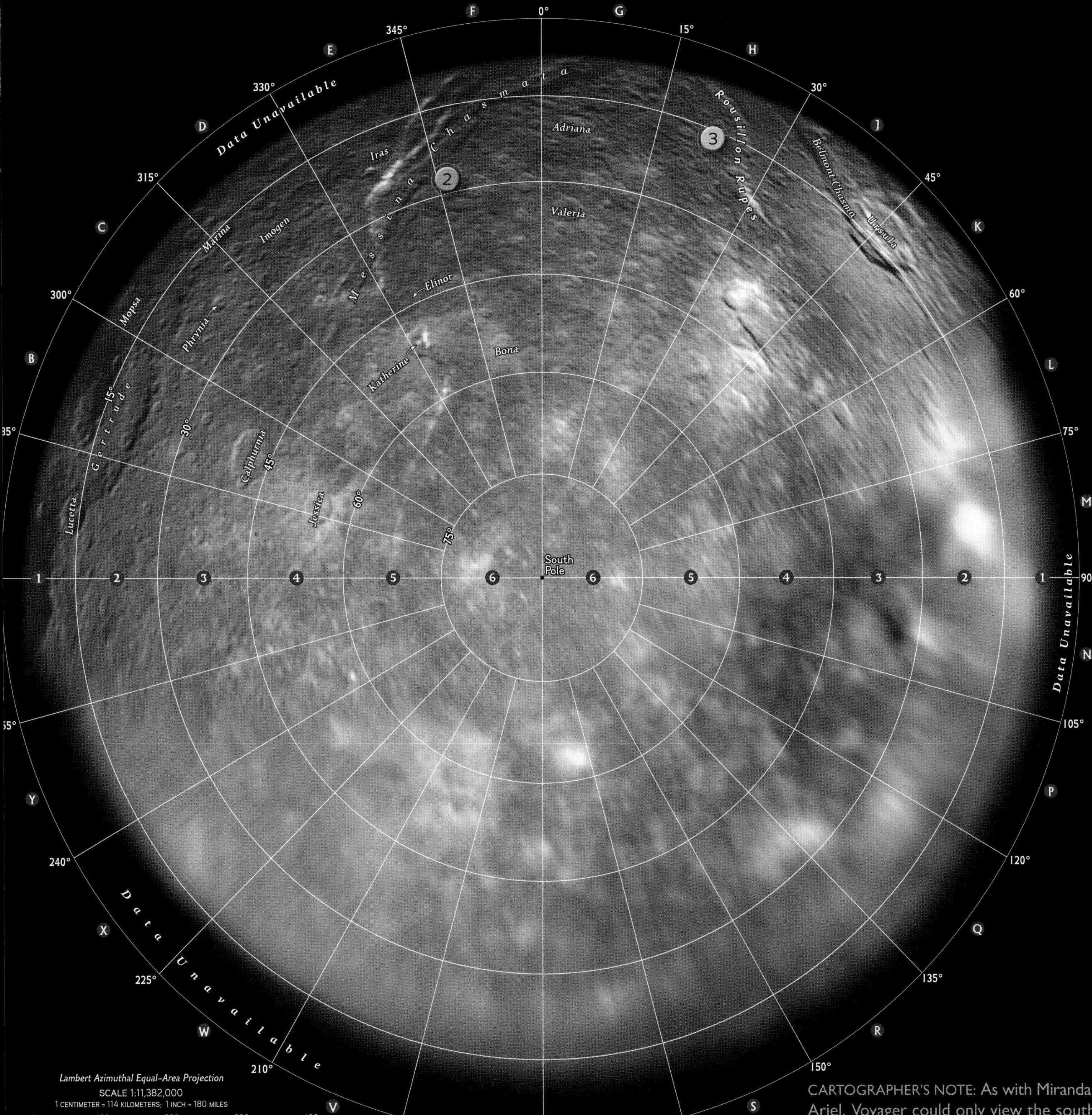

F 0° G
345°
E
330°
D
315°
Data Unavailable
Adriana
15°
H
30°
Rousillon Rupes
J
Iras
Messina Chasmata
Belmont Chasma
45°
C
Imogen
Marina
Valeria
Ursula
K
300°
Mopsa
Phrynia
Elinor
60°
B
Gertrude
Bona
L
Calphurnia
Katherine
15°
30°
45°
75°
M
Lucetta
Jessica
60°
75°
1 2 3 4 5 6 South Pole 6 5 4 3 2 1 90°
N
Data Unavailable
105°
Y
P
240°
120°
X
Data Unavailable
Q
225°
135°
W
R
210°
150°
S
V 195° U 180° T 165°

URANUS & NEPTUNE | SPACE ATLAS | **163**

Lambert Azimuthal Equal-Area Projection
SCALE 1:11,382,000
1 CENTIMETER = 114 KILOMETERS; 1 INCH = 180 MILES
MILES 0 100 200 300 400
METERS 0 100 200 300 400

CARTOGRAPHER'S NOTE: As with Miranda and Ariel, Voyager could only view the southern hemisphere of each moon. Umbriel was a character from Pope's *The Rape of the Lock*, and spirit characters out of the story name its landforms. Titania pulls names from characters and places in Shakespeare's plays.

NEPTUNE

Sometimes orbiting beyond Pluto, icy Neptune shines bright blue.

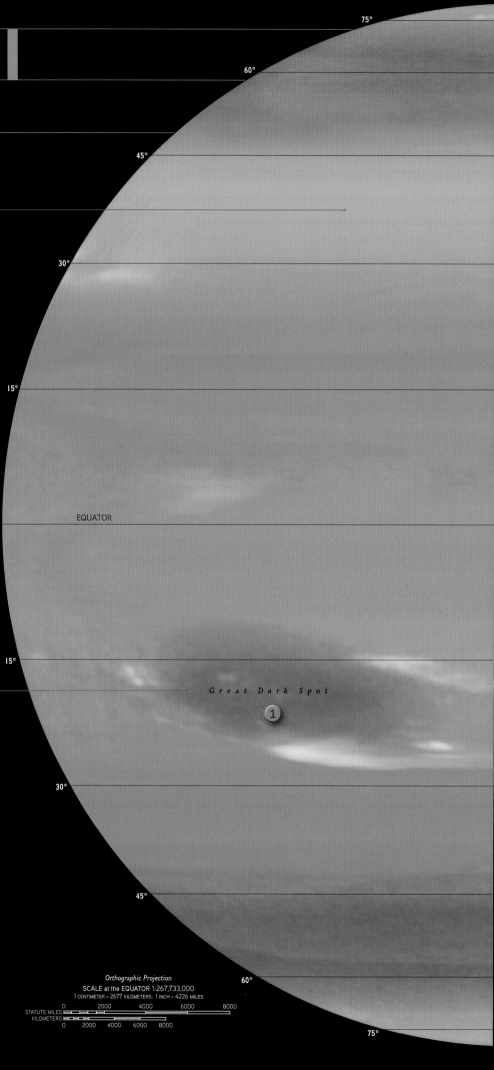

Like the other outer giants, Uranus's atmosphere is divided into belts and zones. Unique to this planet, those winds move easterly, counter to Neptune's rotation.

Similar to Jupiter and Saturn, Neptune radiates more heat than it receives from the sun. Scientists don't know the process by which this occurs.

Seen in many Voyager images of Neptune, some scientists think the dark feature may not be a storm like the Great Red Spot on Jupiter. Variable in size, it could be a vast hole in the atmosphere's upper cloud layers. Once the Hubble Space Telescope began observing the blue planet, the feature had disappeared.

75°
60°
75°
45°
30°
15°
0° EQUATOR
15°
Great Dark Spot
①
30°
45°
60°
75°

Orthographic Projection
SCALE at the EQUATOR 1:267,733,000
1 CENTIMETER = 2677 KILOMETERS; 1 INCH = 4226 MILES

STATUTE MILES
0   2000   4000   6000   8000
KILOMETERS
0   2000   4000   6000   8000

75°

60°

KEY FEATURES

(1) ANTICYCLONE: Huge storms appear and disappear in the atmosphere.

(2) WINDS: High-altitude winds blow at supersonic speeds.

(3) BANDS: Southern cloud bands are increasing during the 40-year summer.

45°

30°

15°

(2)

EQUATOR   0°

15°

(3)

30°

Voyager 2 flew by Neptune,
achieving its closest approach
to the planetary system on
August 24, 1989. Passing
just under 5,000 kilometers
(3,100 miles) from the planet's
cloud surface, nearly all detailed
information on Neptune came
from this sojourn.

45°

60°

CARTOGRAPHER'S NOTE: Visited only by the Voyager probe, Neptune is
mapped here in a construction based on those data. Warmth from inside
the planet stirs active weather in the hydrogen-helium atmosphere; winds
are immensely fast.

75°

TRITON

Largest of Neptune's moons, at minus 240°C
(-400°F) Triton is colder even than Pluto.

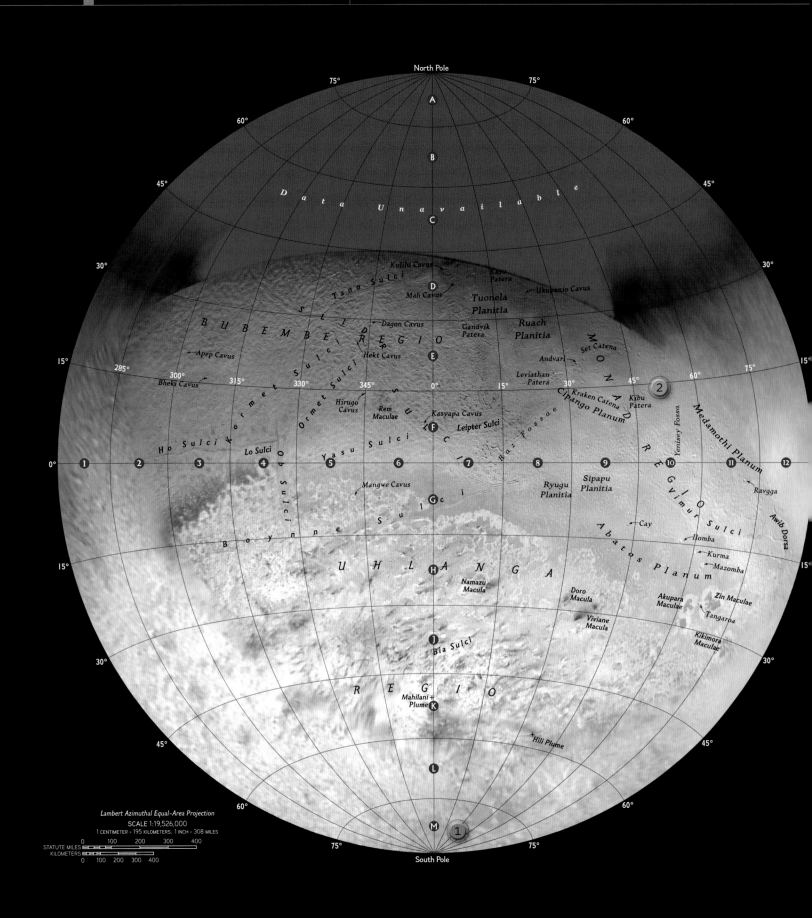

North Pole

Data Unavailable

Kulilu Cavus
Kasu Patera
Ukupanio Cavus
Mah Cavus
Tuonela Planitia
Tano Sulci
Dagon Cavus
Gandvik Patera
Ruach Planitia
BUBEMBE REGIO
Set Catena
Apep Cavus
Hekt Cavus
Andvari
MONAD
Leviathan Patera
Kraken Catena
Kibu Patera
Bheki Cavus
Hirugo Cavus
Cipango Planum
Yenisey Fossa
Medamothi Planum
Ormet Sulci
Rem Maculae
Kasyapa Cavus
Leipter Sulci
Baz Fossae
REGIO
Ho Sulci
Kormet Sulci
Lo Sulci
Yasu Sulci
Gvimur Sulci
Mangwe Cavus
Ryugu Planitia
Sipapu Planitia
Cay
Ravgga
Obo Sulci
Ilomba
Awib Dorsa
Boynne Sulci
Abatos Planum
Kurma
Mazomba
UHLANGA
Namazu Macula
Doro Macula
Zin Maculae
Akupara Maculae
Tangaroa
Viviane Macula
Kikimora Maculae
Bia Sulci
REGIO
Mahilani + Plume
+ Hili Plume

Lambert Azimuthal Equal-Area Projection
SCALE 1:19,526,000
1 CENTIMETER = 195 KILOMETERS; 1 INCH = 308 MILES
0 100 200 300 400
STATUTE MILES
KILOMETERS
0 100 200 300 400

South Pole

KEY FEATURES

① SOUTH POLE: This pole has an ice cap of frozen nitrogen and methane.

② CRYOVOLCANOES: Dark streaks may indicate deposits from ice volcanoes.

③ CANTALOUPE TERRAIN: An image highlights Triton's frozen, buckled surface.

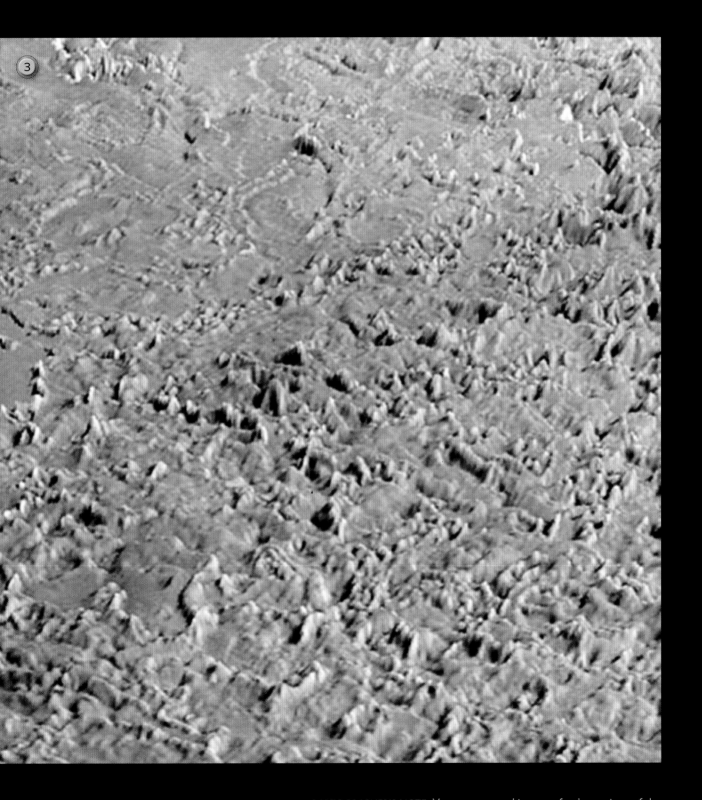

③

CARTOGRAPHER'S NOTE: Voyager returned images of only portions of the moon as the spacecraft streaked past on its way to the outer reaches of the solar system. Aquatic names from Earth are applied to the moon's physiographic features. Cryovolcanoes dot its surface, erupting with what

Beneath their upper cloud layers, Uranus and Neptune are thought to have very similar structures. Each has an atmosphere consisting of hydrogen, helium, and methane gases that gets more and more dense as you descend. At some point there is a transition, without a clearly defined surface, to a hot liquid composed of the same materials. The liquid is hot because of the intense pressure inside the planets—they are much too far from the sun for solar heat to have much of an effect. (It is an idiosyncrasy of astronomical nomenclature that a mixture like this is called an ice, even though it is a dense liquid at a temperature of several thousand degrees.) At the very center of each planet is a small rocky core roughly the size of Earth. Like all the giant planets, both Uranus and Neptune have multiple moons and ring systems.

PLANET SIDEWAYS

In many ways, the most interesting thing about the ice giants is the way they were discovered.

Uranus was actually detected several times before it was officially discovered—it is so faint and so slow-moving that it was mistaken for a star. The realization that this faint, slow-moving object was actually a planet was made by the amateur British astronomer William Herschel. The German-born Herschel worked as chief organist in a chapel in Bath. In his spare time he made telescopes and observed the heavens, and on March 13, 1781, he saw a strange object in his telescope. At first he thought it might be a comet, but as data accumulated on the new object, it became clear that Herschel had become the first human being in recorded history to discover a new planet. (Herschel wanted to name the planet Georgium Sidus in honor of George III. The name didn't stick, but Herschel did get a lifetime pension from the king.)

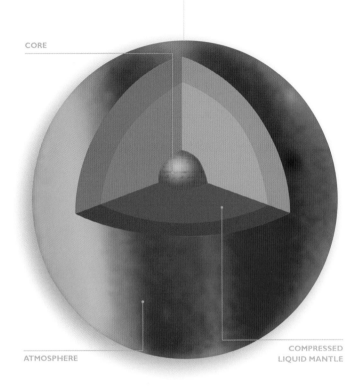

Cutaway view of Uranus. Smaller than Jupiter and Saturn, the ice giant has a rocky core, an icy mantle, and an outer envelope of gaseous hydrogen and helium. This atmosphere makes up 20 percent of the planet's radius.

CORE

ATMOSPHERE

COMPRESSED LIQUID MANTLE

Much of our modern information about the planet was garnered when the spacecraft Voyager 2 flew by Uranus in 1986. Methane (natural gas) is the third most abundant element in the Uranian atmosphere, and it is this gas that gives the planet its aquamarine color. However, the property that catches our attention is the planet's axis of rotation: It lies in the same plane as its orbit, meaning that the planet is tilted 98 degrees, rotating on its side. The skewed rotation is probably the result of a collision shortly after the planet formed. Each pole gets 42 Earth years of sunlight followed by 42 Earth years of darkness as Uranus completes an 84-year orbit around the sun.

Uranus has 27 moons, named after such Shakespearean characters as Ariel and Miranda, and 13 narrow rings.

PLANET OF WINDS

If the discovery of Uranus depended on a bit of observational serendipity, Neptune's discovery was the result of careful calculation. As the newly discovered Uranus was tracked around its orbit, discrepancies started to appear between what was observed and what the law of gravity predicted. Two young astronomers—John Couch Adams in England and Urbain Leverrier in France—independently asked themselves whether these discrepancies could be due to the gravitational tug of another, unseen planet still farther from the sun. Following a series of complex maneuvers and communications, astronomers at the Berlin Observatory trained their telescopes on the place where the unknown planet was predicted to be. On September 23, 1846, the planet we now call Neptune was seen and recorded by Johann Galle.

An illustration depicts Uranus (in the background), its thin rings, and its five largest moons—(left to right) Umbriel, Miranda, Oberon, Titania, and Ariel—shown to scale.

The tilt of Neptune's axis of rotation is similar to that of Earth, so, unlike Uranus, it has seasons and weather. Visited by Voyager 2 in 1989 (after the craft had visited Uranus), Neptune was revealed to have large storms on its surface. Called the Great Dark Spot, the Small Dark Spot, and, incongruously, Scooter, these storms are similar in appearance to Jupiter's Great Red Spot, but they seem to last for only months rather than centuries. Neptune is home to the strongest sustained winds in the solar system—2,100 kilometers an hour (1,300 mph).

Like Uranus, Neptune has thin rings, whose arcs include some with the quintessentially French names of Liberté, Égalité, and Fraternité.

TRITON

Like all the giant planets, Neptune has a large number of moons (13 and counting), but one of them stands out from the others. Triton is a large body—larger than Pluto, for example—and is big enough to be pulled into

NEPTUNE'S GREAT DARK STORM WAS LARGER THAN EARTH

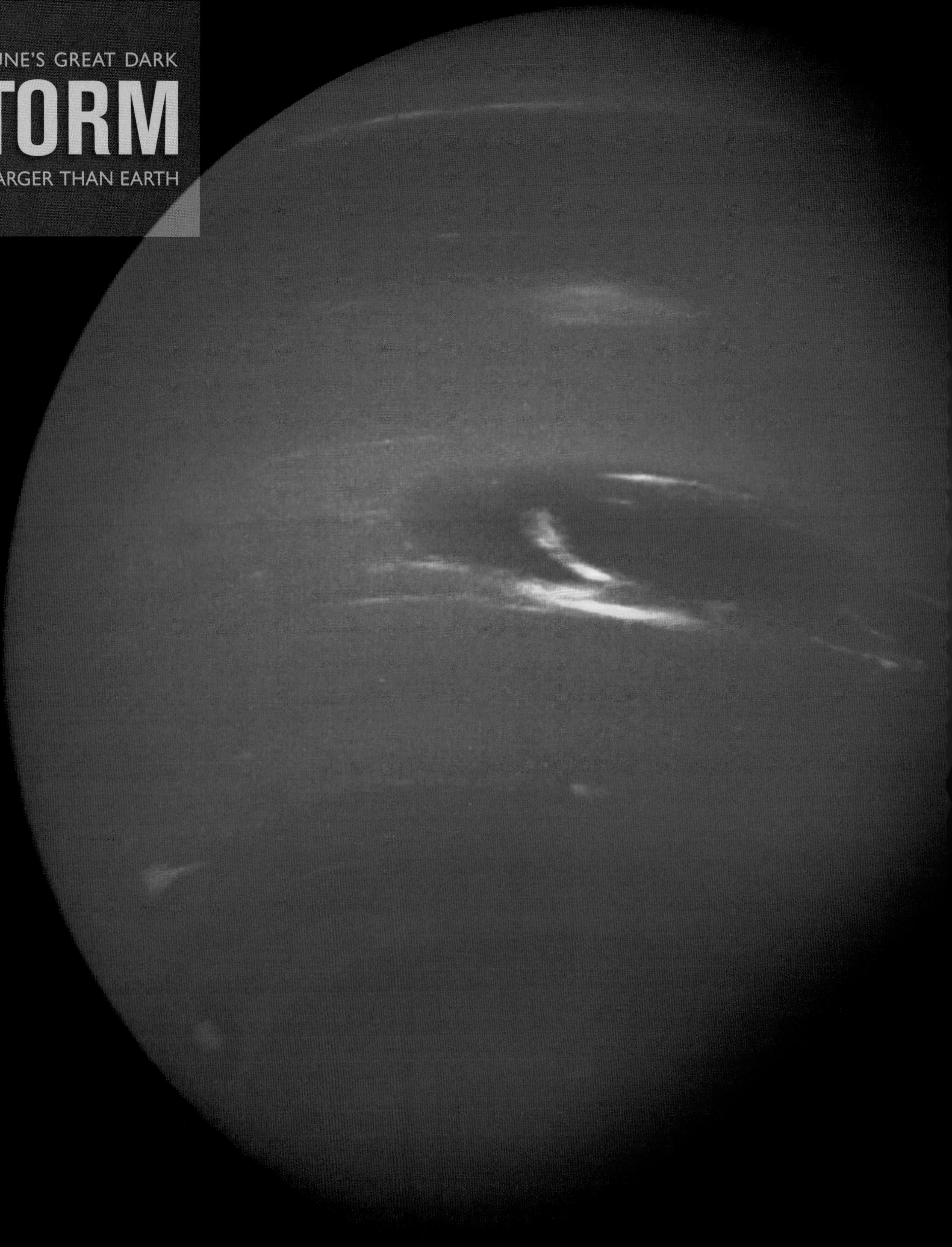

A cutaway view of Neptune shows its rocky, Earth-size core; dense water and ammonia mantle; and gaseous atmosphere, composed of hydrogen, helium, and methane. The outer atmosphere is extremely windy and cold.

CORE

COMPRESSED
LIQUID MANTLE

ATMOSPHERE

A view of the planet Neptune taken by the Voyager spacecraft shows the Great Dark Spot—a titanic storm raging on the planet's surface—and several smaller storms, including one named Scooter (triangular spot at lower left).

a spherical shape by its own gravity. What is interesting about it is that, alone among the solar system's large moons, it moves around Neptune in the opposite direction from the planet's rotation (astronomers call this a retrograde orbit). This fact suggests that the moon did not form at the same time as the planet, but that it took shape elsewhere and was captured. In fact, current thinking is that, like Pluto (which is discussed in the next section), Triton is actually an object that formed in the Kuiper belt (see pages 182–84). It is believed to have a rocky core covered by a layer of frozen nitrogen, and it is one of the coldest objects in the solar system, with a temperature only about 40 degrees above absolute zero. Geysers of frozen nitrogen may even spout from its surface.

Remember Pluto? It used to be a planet. • For generations, in fact, schoolchildren learned that Pluto was the outermost planet in the solar system. Then in 2006 the headlines suddenly shouted that Pluto had been "demoted." What in the world is going on? • Oddly enough, we can begin answering this question by visiting a Kansas farm in the late 1920s. A young farm boy by the name of Clyde Tombaugh, barely past his 20th birthday, had just finished building a small telescope, scavenging parts from old machines to complete his construction project. Using his new instrument, he made some sketches of the surface of Mars and sent them off to Lowell Observatory in Flagstaff with a note asking the professional astronomers there for a critique.

PLUTO

[LAST PLANET, FIRST PLUTOID]

DISCOVERER: CLYDE TOMBAUGH
DISCOVERY DATE: MARCH 13, 1930
NAMED FOR: ROMAN GOD OF THE UNDERWORLD

MASS: 0.002 × EARTH'S
VOLUME: 0.006 × EARTH'S
MEAN RADIUS: 1,151 KM (715 MI)
MIN./MAX. TEMPERATURE: -233/-223°C (-387/-369°F)
LENGTH OF DAY: 6.39 EARTH DAYS (RETROGRADE)
LENGTH OF YEAR: 248 EARTH YEARS
NUMBER OF MOONS: 4 (3 NAMED)
PLANETARY RING SYSTEM: NO

Artist's conception of Pluto and three of its moons.
(Inset) Pluto and Charon flanked by tiny Nix and Hydra

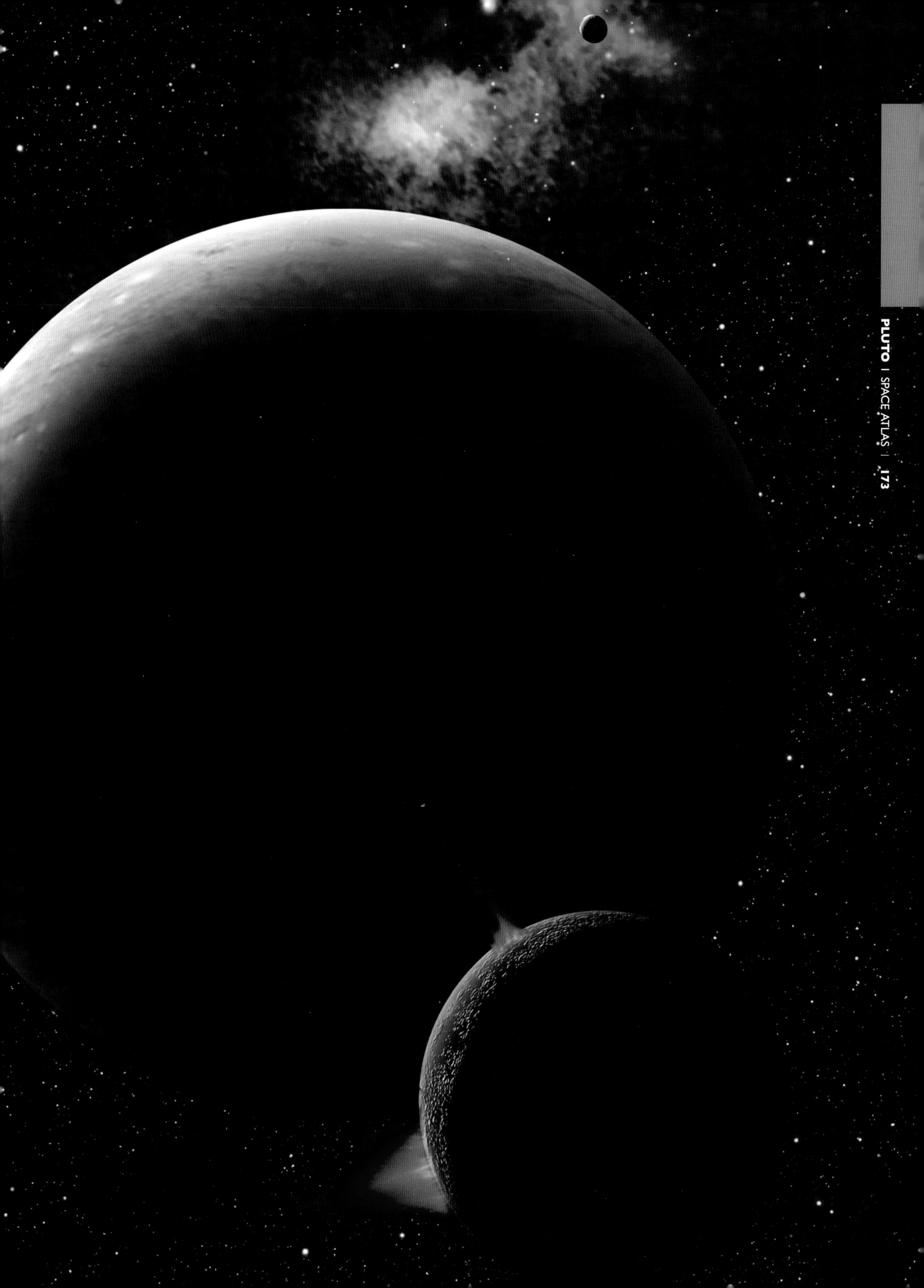

The astronomers at Lowell were so impressed by what they saw in Tombaugh's sketches that they hired him as an assistant and brought him to Arizona, where they put him to work on a project known as the search for Planet X.

A word of explanation: It was thought at the time (incorrectly, as it turned out) that there were discrepancies in the observed orbit of Neptune that indicated the presence of an unknown planet orbiting farther out from the sun, a body astronomers dubbed Planet X. Searching for Planet X was a straightforward, though tedious, process. Astronomers would take photographs of the same section of the sky a couple of weeks apart, then look for an object that had moved just the right amount to be a planet. Boring work—just the thing to foist off on the new hire.

So as Earth swept around its orbit, Tombaugh alternated between photographing the sky (during the dark of the moon) and examining his plates. On February 18, 1930, six months into the project, his hard work paid off. The body we now call Pluto showed up, moving just the right amount between photos to be a new planet. After taking three weeks to verify their results—an astonishingly long interval by today's standards—scientists at Lowell announced that a new planet had been discovered, and Tombaugh

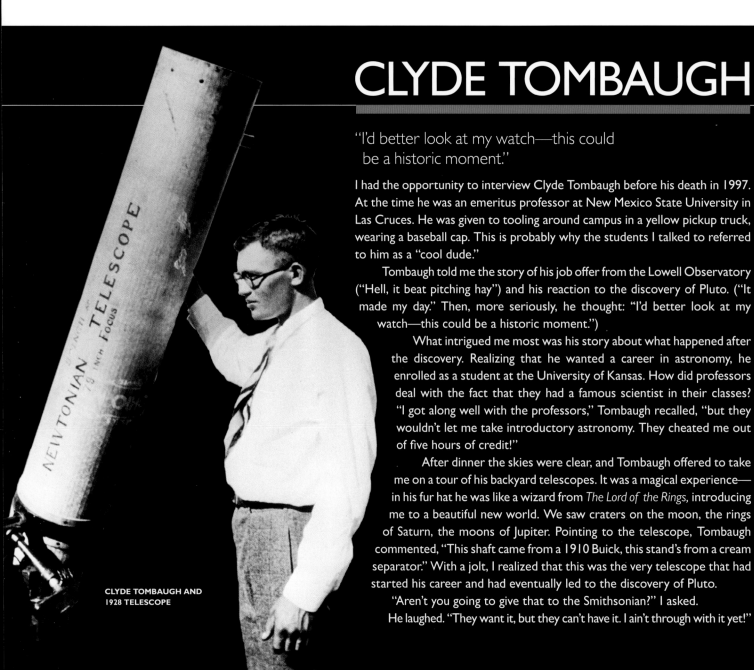

CLYDE TOMBAUGH

"I'd better look at my watch—this could be a historic moment."

I had the opportunity to interview Clyde Tombaugh before his death in 1997. At the time he was an emeritus professor at New Mexico State University in Las Cruces. He was given to tooling around campus in a yellow pickup truck, wearing a baseball cap. This is probably why the students I talked to referred to him as a "cool dude."

Tombaugh told me the story of his job offer from the Lowell Observatory ("Hell, it beat pitching hay") and his reaction to the discovery of Pluto. ("It made my day." Then, more seriously, he thought: "I'd better look at my watch—this could be a historic moment.")

What intrigued me most was his story about what happened after the discovery. Realizing that he wanted a career in astronomy, he enrolled as a student at the University of Kansas. How did professors deal with the fact that they had a famous scientist in their classes? "I got along well with the professors," Tombaugh recalled, "but they wouldn't let me take introductory astronomy. They cheated me out of five hours of credit!"

After dinner the skies were clear, and Tombaugh offered to take me on a tour of his backyard telescopes. It was a magical experience—in his fur hat he was like a wizard from *The Lord of the Rings*, introducing me to a beautiful new world. We saw craters on the moon, the rings of Saturn, the moons of Jupiter. Pointing to the telescope, Tombaugh commented, "This shaft came from a 1910 Buick, this stand's from a cream separator." With a jolt, I realized that this was the very telescope that had started his career and had eventually led to the discovery of Pluto.

"Aren't you going to give that to the Smithsonian?" I asked.

He laughed. "They want it, but they can't have it. I ain't through with it yet!"

CLYDE TOMBAUGH AND
1928 TELESCOPE

(see sidebar opposite) joined the select circle of those who had discovered new worlds. The distant, frigid new planet, fittingly enough, was named for the Greek god of the underworld—a name suggested by 11-year-old British schoolgirl Venetia Burney and passed on to the Lowell Observatory astronomers through a relative connected with Oxford University.

PROBLEMS WITH PLUTO

It wasn't long, though, before problems began to develop with Pluto. For one thing, the plane of its orbit was tilted with respect to the orbits of the other planets. For another, the orbit was somewhat unusual—from 1977 to 1999, for example, Pluto was actually closer to the sun than Neptune. Finally, in 1978 detailed observations revealed that Pluto had a large moon, eventually named Charon after the mythological figure who ferried souls to the underworld. (In addition to Charon, Pluto is now known to have three tiny satellites: Nix, Hydra, and an as-yet-unnamed moonlet discovered in 2011.) The discovery of Charon allowed astronomers to calculate Pluto's mass, which turned out to be smaller than that of our own moon. In fact, if you weigh 100 pounds on Earth, you would weigh only 8 pounds on Pluto. Instead of the gas giant that our theories led us to expect out there, we had found a small, rocky, icy world. Because of all these issues, throughout the latter part of the 20th century Pluto was something of an odd man out—it was there, but nobody really wanted to talk about it.

But mystery continued to follow mystery. Theorists calculated (by means described in the next section) that the solar system didn't end with Pluto, but extended outward in a ring of rocky debris called the Kuiper belt. This is discussed in more detail in the next section, but for the moment we'll simply note that in 2005 astronomer Mike Brown at the Palomar Observatory

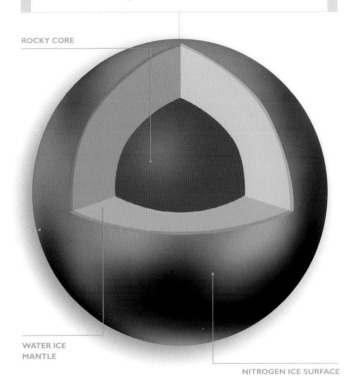

A cutaway shows the probable structure of Pluto. Observations from the Hubble Space Telescope suggest that it is about 50 to 70 percent rocky core, with the remainder ice. At Pluto's temperature, even substances like nitrogen and methane can be found as ices.

ROCKY CORE

WATER ICE MANTLE

NITROGEN ICE SURFACE

discovered a large, planetlike object orbiting out beyond Pluto. Eventually named Eris for the Greek goddess of strife and discord, this object is actually bigger than Pluto. Since that time, a half dozen other planetlike objects have been found in the Kuiper belt, and astronomers expect to find many more.

PLUTOIDS

All of which brings us to the August 2006 meeting of the International Astronomical Union in Prague. On the last day of that conference, with only a little more than 400 of the 2,400 participants voting, Pluto was designated a dwarf planet. Later, all of the newly discovered Kuiper belt planets were put in a class called plutoids. The idea behind this switch is that all of the difficulties that arise when we try to lump Pluto with the inner planets disappear when we realize that Pluto belongs in an entirely

different category. It is, in fact, the first plutoid rather than the last planet, the beginning of the end rather than the end of the beginning.

NEW HORIZONS

The New Horizons spacecraft was launched from Cape Canaveral in 2006; it is scheduled to fly by Pluto on July 15, 2015. After all that travel, the instruments will have about an hour of closest approach as New Horizons speeds by the planet (although, of course, it will be taking data

An artist's conception of the New Horizons spacecraft during its encounter with Pluto and its large moon, Charon. Launched in 2006, the spacecraft will reach Pluto in 2015. Scientists expect to find snow made of frozen methane (natural gas) falling when the spacecraft arrives, since Pluto's orbit is now taking it farther away from the sun.

long before and long after this event). Mission scientists jokingly refer to this period as the "platinum hour" (on the grounds, one explained to me, that NASA deserved more for its money than just gold). It is ironic that the planetary status of Pluto changed after the launch of

New Horizons—it may, in fact, be the only vehicle that started out to explore a planet and wound up exploring a plutoid without ever changing its course.

The planet scientists expect to find when New Horizons reaches its destination is a strange one. It has a thin atmosphere composed of nitrogen, methane (natural gas), and carbon monoxide. As Pluto's orbit takes it nearer to the sun, these substances convert themselves from surface ices into atmospheric gases, and the reverse happens as it moves farther out. Having reached perihelion (its closest point to the sun), in 1989, the plutoid is now moving away again in its 248-year orbit. It is likely, in other words, that it is snowing methane on Pluto as you read this.

Along with its scientific cargo, New Horizons is carrying the usual sort of miscellaneous stuff—an American flag, a Florida state quarter showing a shuttle launch, and other mementos. In addition, as a result of what has to be one of the classiest decisions ever made by a government agency, the spaceship is carrying some of the ashes of Clyde Tombaugh, a Kansas farm boy far from home.

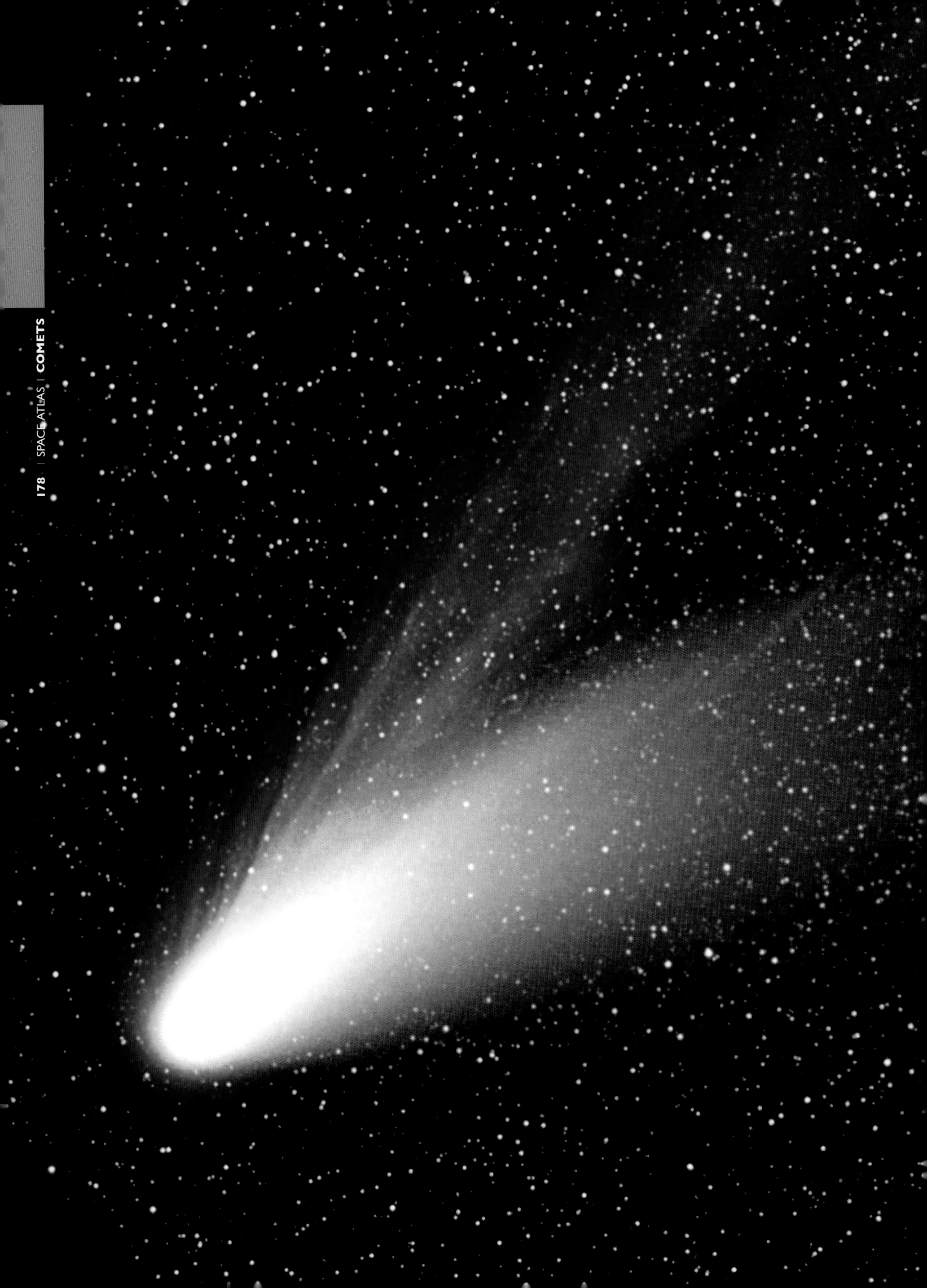

Comets have always been a problem. From ancient times, their appearance in the sky has been taken as an evil omen. There was, for example, a comet in the sky in 1066 when the Normans invaded England. (The comet may have portended disaster for the Saxons, but it certainly brought good luck to the Normans.) • Comets were also a problem in the scientifically minded 17th century, when Isaac Newton described a clockwork universe, with the orderly motion of the planets being the hands of a clock and the rational laws of nature the gears that drove them. There was no room in this universe for bodies that showed up at unpredictable times, stayed in the sky for a while, and then disappeared.

COMETS

[VISITORS FROM THE OUTER LIMITS]

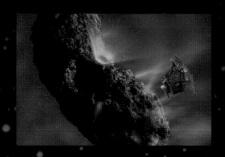

**RECENT APPEARANCES OF
SELECTED BRIGHT COMETS**

CAESAR'S COMET: 44 B.C.

GREAT COMET OF 1577: 1577

GREAT COMET OF 1744: 1744

GREAT MARCH COMET: 1843

GREAT SEPTEMBER COMET: 1882

COMET WEST: 1976

HALLEY'S COMET: 1986

COMET HYAKUTAKE: 1996

COMET HALE-BOPP: 1997

COMET MCNAUGHT: 2007

Comet Hale-Bopp. (Inset) Art of Deep Impact
space probe at comet Tempel 1

ur current understanding of comets began with a dinner between Newton and his friend Edmond Halley (1656–1742), who would eventually become Britain's Astronomer Royal. Halley apparently asked Newton what sorts of orbits comets would have if they were material objects subject to Newton's law of universal gravitation. Newton had actually worked the problem out, but he had not bothered to publish the result. He told Halley that comets would follow elliptical paths through the solar system. Armed with this knowledge, Halley examined the data on 26 comets. To his surprise, he found that three followed the same ellipse. The inference was obvious—these were not three different comets, but one comet coming back three different times. Halley established that some comets are in elongated orbits that bring them near Earth periodically. He predicted that this particular comet, which now bears his name, would return in 1758. When a German amateur astronomer sighted it on Christmas Eve of that year, it was a great triumph for Newton's clockwork universe.

Since then, historians have found sightings of Halley's comet in Chinese and Babylonian records going back to 240 B.C. The comet last visited Earth's neighborhood in 1986 and will come around again in 2061.

THE DIRTY SNOWBALL

The best picture we have now of comets was put forward by the American astronomer Fred Whipple (1906–2004) in the early 1950s. His ideas have acquired the evocative name dirty snowball, which turns out to be a pretty good description.

Astronomer Fred Whipple describes the "dirty snowball" structure of a cometary nucleus with the aid of a model. The diagram on the blackboard shows material boiling off the comet.

The central body of a comet, which can measure anywhere from a few hundred meters to some tens of kilometers across, is called the nucleus. Most of the nuclei that have been analyzed contain dust and mineral grains embedded in water ice, along with a surprising array of trace constituents like methane, ammonia, and even, in some cases, complex molecules like amino acids.

When a comet is far away from the sun, the nucleus is basically frozen solid by the cold of space. As it approaches the sun, however, it heats up and the volatile materials in the nucleus begin to boil off. This gives rise to two structures—a thin atmosphere surrounding the nucleus (called a coma) and a tail—the thing we most associate with a comet. In fact, every comet has two tails. One is made up of the gases that boil off the nucleus; the solar wind—particles streaming out from the sun—exerts a pressure on these gases, blowing the tail so that it always points away from the sun. The other tail is a trail of dust that has drifted off the surface. This dust pretty much stays put along the track of the comet, like the trail a muddy dog leaves on a living room carpet.

Comets are usually divided into long-period and short-period bodies. Short-period comets complete their orbits in less than 200 years. They are thought to originate in the region outside the orbit of Neptune called the Kuiper belt (see pages 182–84). Long-period comets—and we know of some that take thousands of years to complete an orbit—are thought to originate still farther out, in an icy collection of distant bodies known as the Oort cloud (see pages 184–85).

Astronomers have cataloged thousands of comets, and roughly one a year passes close enough to Earth to be visible to the naked eye. Most of these, however, are faint and, frankly, pretty unremarkable. About once a decade, though, a comet will be bright enough to be seen easily. Hale-Bopp in 1997 was a recent example. Unfortunately, the last visit of Halley's comet wasn't spectacular—it just looked like a faint star—and the next one won't be either. To be noticeable, a comet has to pass close to Earth after its tail has grown to its maximum brightness—not an easy set of requirements to satisfy.

SPACECRAFT AND COMETS

In the late 20th century spacecraft visited comets for the first time, either flying by or, in one case, bringing material from the comet's tail back to Earth. There have been close to a dozen such encounters, but let me just describe two, the Deep Impact and Stardust missions.

In 2005 the Deep Impact spacecraft (later renamed EPOXI) released a probe that blasted a crater in the surface of comet Tempel 1. By monitoring the material that came off, scientists were able to establish that most of the water ice in a comet is beneath the dust layer at the surface.

The Stardust spacecraft was launched in 1999 and flew through the tail of comet Wild 2 (pronounced vilt) in 2004. Material from the tail was absorbed in special materials in the spacecraft and returned to Earth in a capsule in 2006. The results from this mission caused a stir in the astronomical community, because they showed that the comet contained grains of materials that could only have formed at high temperatures, lending support to the idea that the formation of the solar system (see page 45) must have been more turbulent than had been thought.

Finally, we should point out that some scientists think that comets, with their high water content and brew of organic molecules, may have played an important role in shaping the early Earth. Some have suggested that comets provided the water for Earth's oceans and perhaps even brought in the molecules from which the first life developed.

f this book had been written 20 years ago, our discussion of the solar system would end at this point, with Pluto as the outermost planet. Today, however, we understand that the planets are just the beginning, and that the actual solar system extends far out into space—farther than we ever imagined. • To understand what is meant by this, we're going to have to change our scale of perspective. Scientists usually use something called the astronomical unit, or AU, to talk about distances in the solar system. The AU is the distance between Earth and the sun—roughly 150 million kilometers (93 million mi), or about 8 light-minutes. In terms of this unit, Mars is about 1.5 AU from the sun, Jupiter about 5.2 AU, and Neptune about 30 AU.

KUIPER BELT &
OORT CLOUD

[THE ICY OUTSKIRTS]

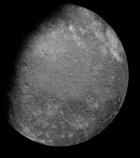

KUIPER BELT DISCOVERERS: DAVID JEWITT AND JANE LUU
CONFIRMED DISCOVERY DATE: 1992
DISTANCE FROM SUN: 30–55 AU
OORT CLOUD PROPONENTS: JAN OORT AND ERNST ÖPIK
DATE FIRST PROPOSED: 1932
DISTANCE FROM SUN: 5,000–100,000 AU

KNOWN PLUTOIDS AND AVERAGE DISTANCE FROM SUN:
PLUTO: 39 AU
HAUMEA: 43 AU
MAKEMAKE: 46 AU
ERIS: 68 AU

Artwork shows the disk and shell of the Oort cloud around
the sun. (Inset) Artist's conception of the plutoid Eris

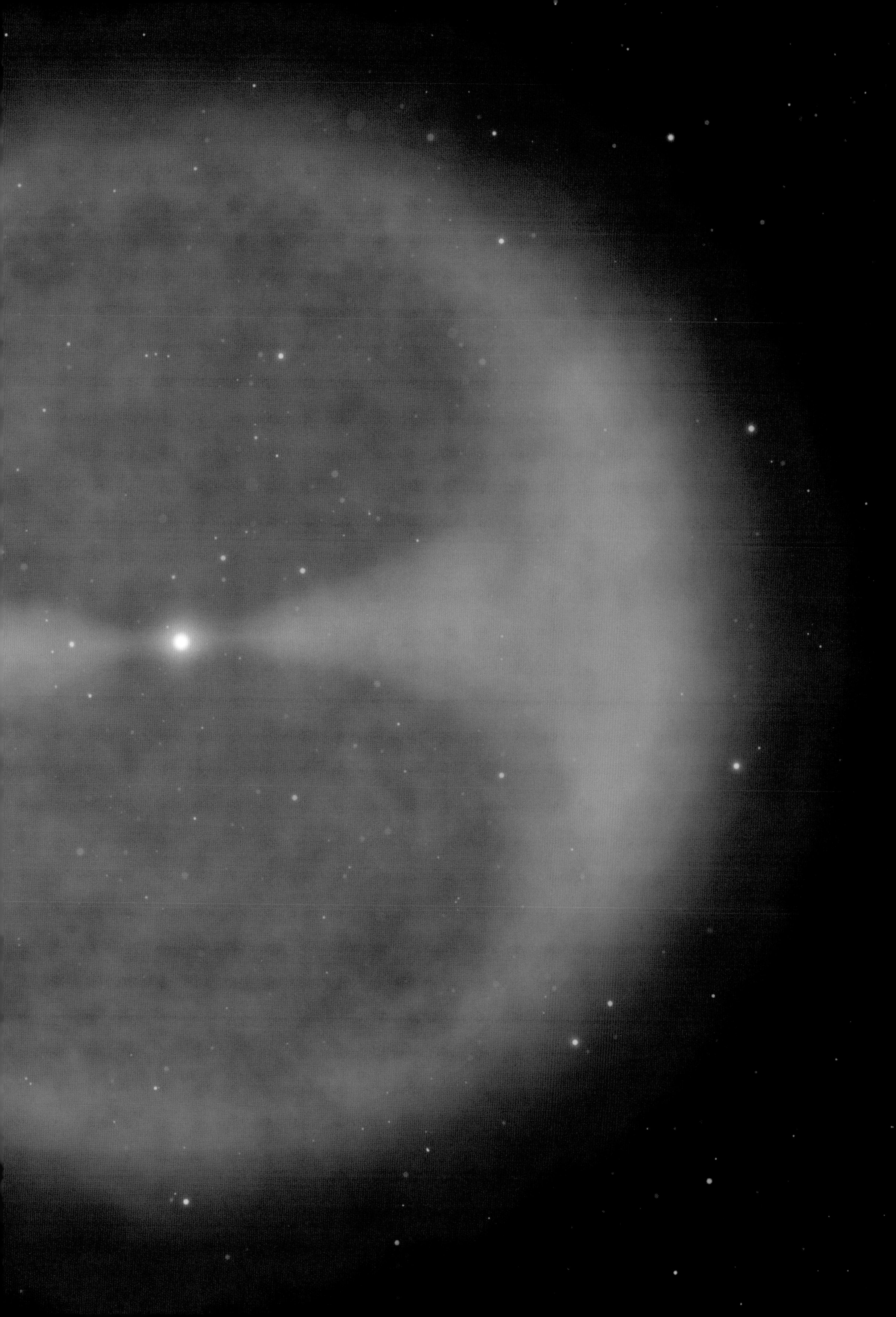

From Neptune's orbit to about 55 AU, the solar system extends outward in a giant doughnut-shaped structure known as the Kuiper belt. (Think of the planets as fitting into the hole in the doughnut.) The structure is named after the Dutch astronomer Gerard Kuiper (rhymes with "viper"), one of the scientists who did an early calculation of the belt's properties in 1951.

The Kuiper belt consists primarily of material that appears to be planetesimals left over from the formation of the solar system—indeed, one author referred to it as a "reservoir of rejects." Most likely the belt represents the remnants of the protoplanetary disk from which the solar system formed—remnants that survived the migration of Uranus and Neptune to their present orbits (see page 47). Today it is just a shadow of its former self, comprising no more, in aggregate, than 10 percent of the mass of Earth. Since 1992, telescopic exploration of the Kuiper belt has uncovered over 1,000 Kuiper belt objects (KBOs), and astronomers expect to find many more. Pluto is now seen as the first KBO rather than as the last planet, and Neptune's moon Triton is considered a captured KBO.

Telescopic exploration will continue with a survey called Pan-STARRS (Panoramic Survey Telescope and Rapid Response System). Started in 2008 with a telescope located on Mount Haleakala in Hawaii, its primary mission is to locate asteroids and comets that might impact Earth. As a side benefit, however, it will produce a detailed map of faint objects in the sky, including KBOs.

Outside of the Kuiper belt "doughnut" is a sparsely populated region of objects in highly eccentric orbits, called the scattered disk, extending out to about 100 AU. Unlike objects in the Kuiper belt proper, which are in stable orbits, objects in the scattered disk have orbits that bring them in to 30 AU, where they can be affected by the gravitational pull of Neptune. It is believed that most short-period comets (see page 181) had their ultimate origin in the scattered disk.

DWARF PLANETS

We can't leave the Kuiper belt without mentioning a couple of objects found out beyond the orbit of Neptune. One is Eris, larger than Pluto and actually part of the scattered disk. Discovered in 2005 by astronomer Mike Brown and his team at Palomar Observatory, the dwarf planet was first nicknamed Xena after the TV character. Right now Eris is about 97 AU from the sun, well outside of the Kuiper belt proper, and is the farthest known member of our solar system.

Another strange discovery is the dwarf planet Sedna. Found in 2003 by Brown's team, it is smaller than Pluto but has a very unusual orbit. Circling 88 AU from the sun right now, Sedna's orbit never takes it closer than 76 AU. Thus, it orbits well outside both the planets and the Kuiper belt. Astronomers calculate that its farthest distance from the sun will be an astonishing 975 AU, far beyond anything we've discussed so far. This fact has led Brown to suggest that Sedna may not by a scattered disk object at all, but the first member of the Oort cloud—which brings us to the last part of our story.

OORT CLOUD

In 1950 the Dutch astronomer Jan Oort (rhymes with "sort") suggested that somewhere in the outskirts of the solar system there had to be a reservoir of comets. His argument was simple: Comets can't last forever—they lose some of their mass every time they get near the sun and are subject to the gravitational effects of the planets. Since we still see comets today, Oort argued, there has to be a process by which new comets can be formed, some reservoir of comets out beyond Pluto.

An artist's imagining of the plutoid Sedna in the Kuiper belt shows the distant sun as a bright star. The reddish color of Sedna's surface is what we see in our telescopes, but the moon the artist has put into orbit has not yet been confirmed.

Today this reservoir is thought to consist of a huge cloud at the very edge of the solar system. Named the Oort cloud, it extends from a few thousand AU out to at least 50,000 AU, and perhaps farther. It has two parts—an inner doughnut-shaped section that extends out to about 20,000 AU and a thinly populated outer sphere. The Oort cloud is thought to be the remains of the original protoplanetary disk. Objects in the disk probably formed closer to the sun but moved out to their present location following the great reshuffling of the solar system four billion years ago.

The Oort cloud may be the point of origin for what are known as long-period comets—comets that have a period of more than about 200 years. The comet Hale-Bopp that visited Earth in 1997 is a recent example of a long-period comet. Oddly enough, so is Halley's comet. Even though it now has a period of roughly 72 years, it is believed to have originated in the Oort cloud and been pulled into its present short orbit by the gravitational attraction of the planets. Most comets with a period of less than 200 years—the so-called short-period comets—are thought to originate in the scattered disk.

As befits a region so remote and mysterious, there are many imaginative explanations for the origin of larger Oort cloud candidates, including the idea that they were captured from the Oort cloud of a passing star. Thus, our exploration of our own cosmic back-yard—our first "universe"—brings us to the realm of the stars.

THE MILKY WAY

Seen from Earth, the central plane of our galaxy, the Milky Way, forms a dense stream of stars across the sky.

Our planet, as magnificent as it is, is just one planet circling an ordinary star in a low-rent section of our galaxy. The Milky Way isn't just a passive collection of stars, however, but a roiling, dynamic place. Stars are born in the collapse of gigantic clouds of dust and debris. They keep themselves alive by consuming the primordial hydrogen of the universe. They use nuclear processes to create the heavier elements, including the carbon that makes up much of your body. Eventually stars run out of fuel and die, returning their heavy elements to the interstellar clouds from which new stars and planets are made.

THE GA

When stars die they leave behind a strange menagerie of objects. Some leave white dwarfs, the dying embers of stars like the sun. Others become pulsars, incredibly dense objects tens of kilometers across. A few collapse into black holes, representing the ultimate triumph of gravity.

In recent decades astronomers discovered that, just as the inner planets are actually a small part of the solar system, the mighty pinwheel of the Milky Way is just a small part of the galaxy. In fact, the spiral arms of the Milky Way are enclosed in a mysterious substance known as dark matter. Puzzling out the nature of dark matter remains a major research project today.

ALAXY

THE MILKY WAY

NGC 5272

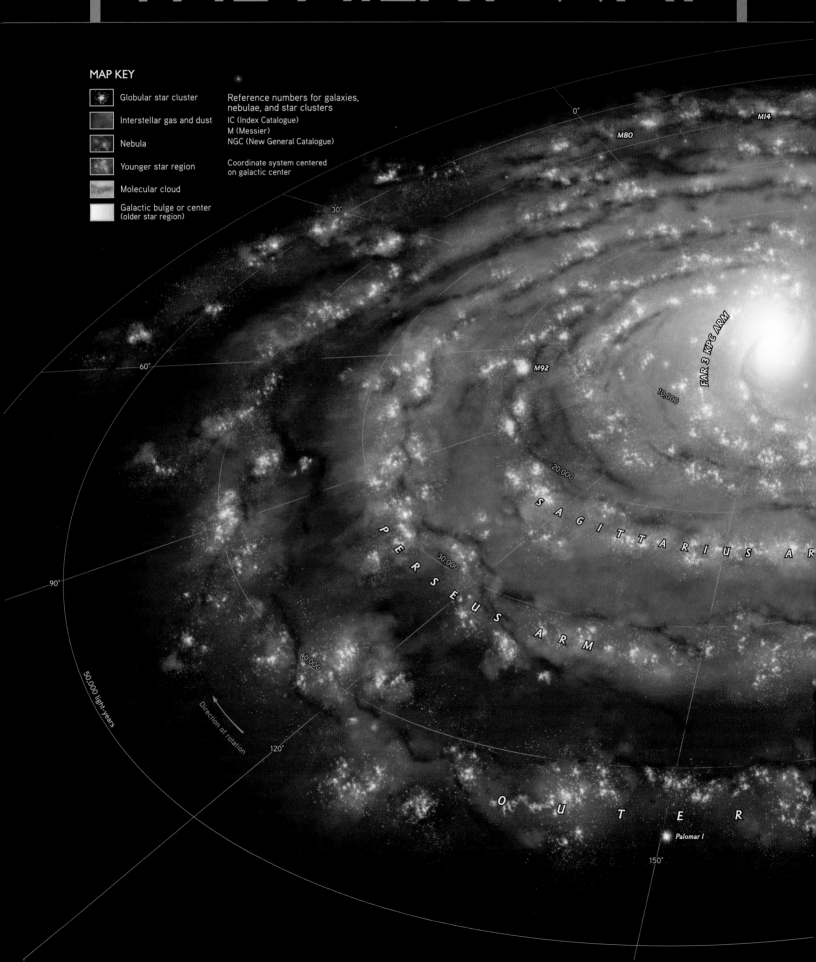

MAP KEY

- **Globular star cluster**
- **Interstellar gas and dust**
- **Nebula**
- **Younger star region**
- **Molecular cloud**
- **Galactic bulge or center** (older star region)

Reference numbers for galaxies, nebulae, and star clusters
IC (Index Catalogue)
M (Messier)
NGC (New General Catalogue)

Coordinate system centered on galactic center

0°

M14

M80

30°

FAR 3 KPC ARM

10,000

M92

60°

20,000

S A G I T T A R I U S   A R

P E R S E U S   A R M

30,000

90°

40,000

Direction of rotation

50,000 light-years

120°

O U T E R

Palomar I

150°

An artist's conception of the Milky Way galaxy shows its dense central core and flat disk of spiral arms, whose brightest regions contain young stars. Our own solar system is located about 25,000 light-years out on one of these arms and rotates with the entire structure, making a complete circuit every 250 million years or so. The galaxy's crowded center may contain a supermassive black hole.

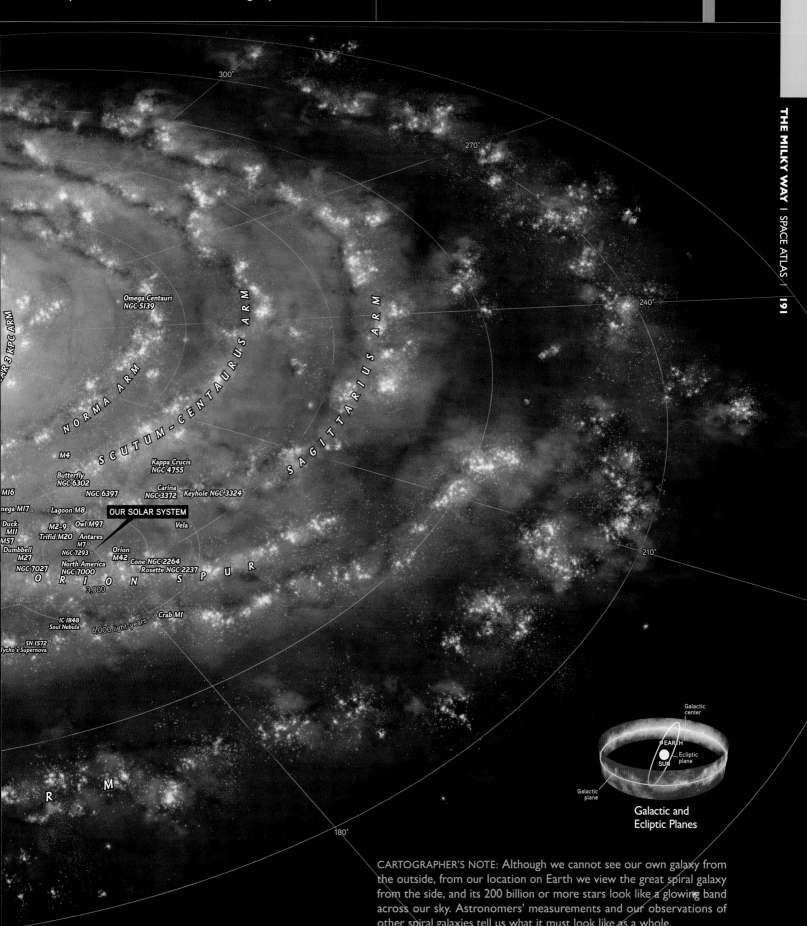

300°

270°

240°

Omega Centauri
NGC 5139

AR 3 KPC ARM

NORMA ARM

SCUTUM-CENTAURUS ARM

SAGITTARIUS ARM

M4

Kappa Crucis
NGC 4755

Butterfly
NGC 6302

MI6

NGC 6397

Carina
NGC 3372   Keyhole NGC 3324

mega MI7

Lagoon M8

OUR SOLAR SYSTEM

Vela

Duck
MII

M2-9   Owl M97

M57

Trifid M20   Antares

Dumbbell

M7

Orion
M42

210°

M27

NGC 1293

NGC 7027

North America
NGC 7000

Cone NGC 2264
Rosette NGC 2237

O   R   I   O   N   S   P   U   R

3,000

IC 1848
Soul Nebula

Crab MI

6,000 light-years

SN 1572

Tycho's Supernova

R   M

Galactic
center

EARTH

SUN   Ecliptic
plane

Galactic
plane

Galactic and
Ecliptic Planes

180°

CARTOGRAPHER'S NOTE: Although we cannot see our own galaxy from the outside, from our location on Earth we view the great spiral galaxy from the side, and its 200 billion or more stars look like a glowing band across our sky. Astronomers' measurements and our observations of other spiral galaxies tell us what it must look like as a whole.

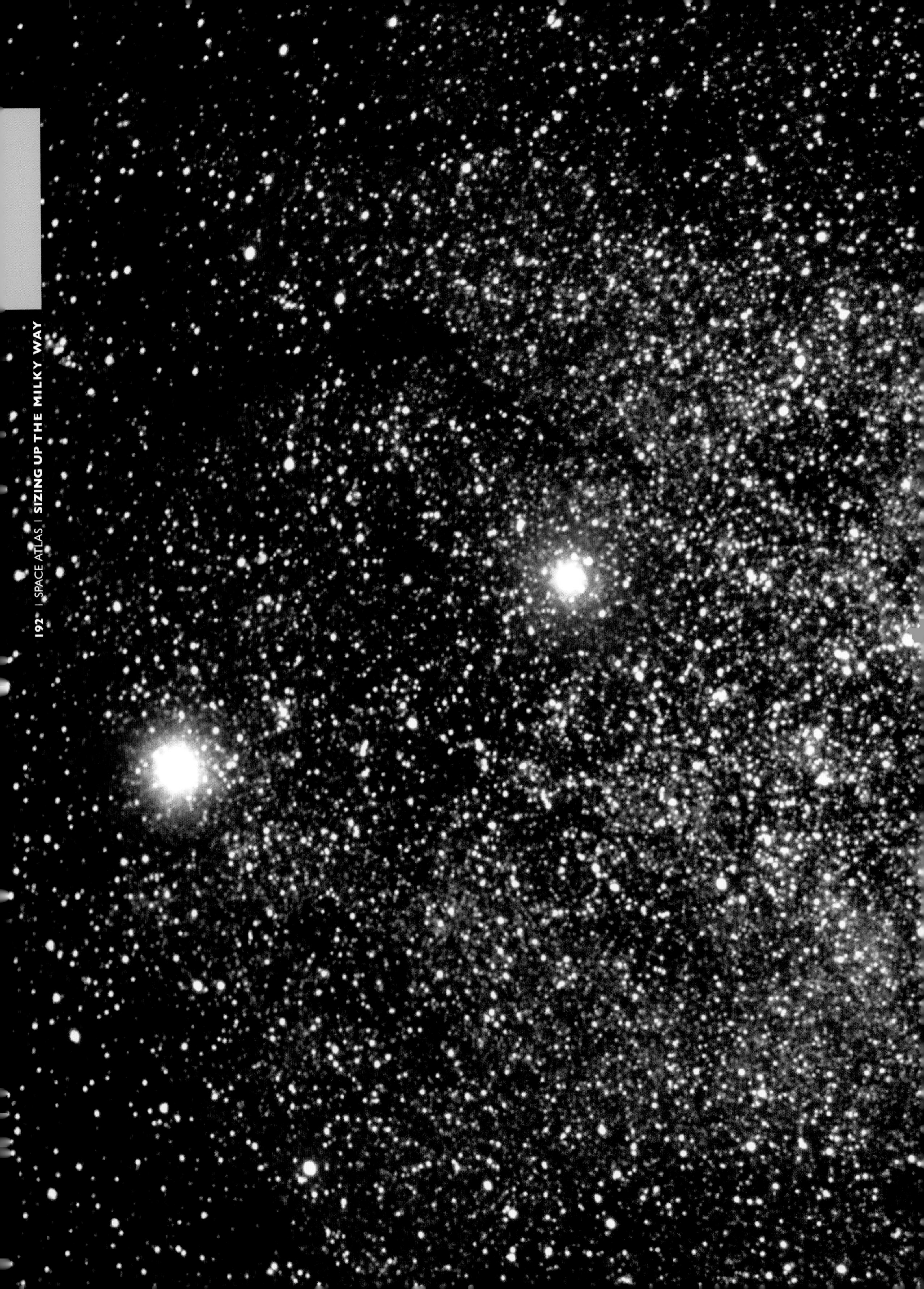

When we venture out of the confines of our own solar system to explore the Milky Way galaxy, the first thing we have to do is reset our thinking about distances. Consider this analogy: If the sun were a bowling ball in the center of a city on the U.S. East Coast—Washington or New York, for example—then all of the planets (including Pluto) would be found within a dozen or so city blocks and the very outermost reaches of the Oort cloud would be somewhere around Saint Louis. You would then have to travel to Hawaii to get to the nearest star, and the rest of this giant city of stars we call our home galaxy would take you right off the planet Earth.

SIZING UP THE MILKY WAY

[HOW FAR ARE THE STARS?]

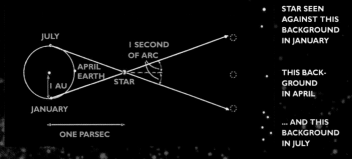

DISCOVERER: UNKNOWN
DISCOVERY DATE: PREHISTORIC
DISTANCE TO MILKY WAY CENTER: 28,000 LIGHT-YEARS

DIAMETER: 100,000 TO 120,000 LIGHT-YEARS
THICKNESS: 1,000 LIGHT-YEARS
ROTATION: 250 MILLION YEARS AT SUN'S DISTANCE
MASS: ABOUT 1 TRILLION (10^{12}) SOLAR MASSES
NUMBER OF STARS: 300 BILLION ± 100 BILLION
AGE: 13.2 BILLION YEARS
GALAXY TYPE: BARRED-SPIRAL
MAIN SATELLITES: LARGE AND SMALL MAGELLANIC CLOUDS

Bright stars of the southern sky include Alpha Centauri and Beta Centauri (at left, left and right). (Inset) Diagram of measurement by parallax

We introduced the astronomical unit (AU)—the distance from Earth to the sun—as a convenient unit for dealing with our solar system, since it is comparable to the distance between planets. In the same way, we need a new distance measure in our new universe. After all, measuring the distances between stars in terms of the AU is a little like measuring the distance between cities in inches—you could do it, but it would be inconvenient at best. Consequently, astronomers use a new yardstick in this universe—the light-year. This measurement is defined to be the distance light travels in one year: 9.5×10^{12} kilometers, 6×10^{12} miles, or 63,000 AU. (For technical reasons, astronomers also use a unit called the parsec, which is 3.3 light-years.) Roughly speaking, stars in the Milky Way are separated by a few light-years—the nearest star to us, for example, is a little over 4 light-years away. The Milky Way itself is about 100,000 light-years across and about 10,000 light-years thick at the center. How do astronomers determine these distances? After all, when you look at the sky, what you see is a two-dimensional display, lights on an inverted bowl. The third dimension—the distance to each object—is not immediately apparent. A star might appear faint either because it really is faint and close, or because it is bright but far away. Over the millennia a lot of effort has been devoted to putting the third dimension into our picture of the heavens. Astronomers have developed a variety of techniques to measure distances to celestial objects, depending on how far away they are. In this section we'll look at the two most important of these, parallax (or triangulation) and standard candles.

PARALLAX

Hold out your finger and look at it with one eye closed, then the other. Notice how your finger appears to move

HENRIETTA LEAVITT

"A comparison [of the plates] led immediately to the discovery of an extraordinary number of new variable stars."

Educated at Oberlin College and what is now Radcliffe, Leavitt joined the staff of the Harvard College Observatory in 1893 as a "computer." (In those days, computers were human beings who carried out long, tedious calculations with pencil and paper.) Analyzing countless photographs of the night sky on glass plates, she noticed the connection between a variable star's brightness and its cycle time. In 1908 Leavitt published the results of her painstaking study of some 1,777 variable stars, establishing the basis for the Cepheid distance scale. Eventually she was appointed head of photometry at the observatory, a position she held until her death in 1921. Because her work proved to be the foundation of so many important later advances in astronomy, her name has been give to an asteroid and a crater on the moon.

against the background? This, basically, is what parallax is all about. Your finger appears to move because you are looking at it from two different positions, separated by the distance between your eyes.

Here's how you can use this effect to measure the distance to an object out of reach: Suppose you want to find out how far away a flagpole is, but you can't actually get to the flagpole itself. (Imagine it being on the other side of a river, for example.) You could find the distance by looking at the flagpole from two different spots on your side of the river and measuring the angles between your two lines of sight and the line connecting the two observation points. If you then measure the distance between the two observation points—what is called the baseline of the measurement—you have one side and two angles of a triangle, a triangle whose apex is the flagpole. Some simple geometry will then give you the distance you want.

This technique, also called triangulation, gives us a way of adding that third dimension to our picture of the heavens, *provided* that we can actually measure the difference between the two angles. And there's the rub, because the farther away the object is, the harder it is to use this method.

Imagine that our flagpole is being moved farther and farther away from us. Depending on the kind of instrument we are using to measure the angles, there will eventually come a point where the flagpole is so far away that we won't be able to tell the difference between the two angles we are measuring. As far as we are concerned, the two lines are parallel. At this point, the parallax measurement has run out of oomph and we can no longer measure the distance we want. We now have two alternatives:

1. Increase the baseline, thereby increasing the difference between the two angles so that it becomes detectable with the instruments we have.

HENRIETTA LEAVITT (OPPOSITE PAGE). VARIABLE STAR V1 (INSETS) IS FOUND IN THE ANDROMEDA GALAXY (BACKGROUND).

DEC. 17, 2010

DEC. 21, 2010

DEC. 30, 2010

JAN. 26, 2011

LOCATION

2. Get better instruments, so that we can detect the difference in the angles while keeping the same baseline.

Being confined to the planet Earth, we are somewhat limited in our ability to increase our baseline. Nevertheless, the most obvious baseline is the diameter of Earth itself. We could take simultaneous measurements from opposite sides of Earth, or we could take a measurement from one place, wait 12 hours until the rotation of Earth has carried us around a half turn, then take the second measurement. In either case, the baseline of the measurement is limited to about 13,000 kilometers (8,000 mi), the diameter of our planet.

WITHIN THE MILKY WAY

After Eratosthenes estimated the circumference of Earth in about 240 B.C. (see sidebar page 78), celestial distance measurements ran out of steam for almost two millennia. The reason is simple: The next available longer baseline is the diameter of Earth's orbit, which you can use by taking angle measurements six months apart. The problem is that to use this baseline you need to know the distance from Earth to the sun, and you can't get that if you are restricted to using Earth's diameter as a baseline and don't have telescopes. It wasn't, in fact, until 1672 that French astronomers, using the best telescopes available, were able to get a reasonably accurate measurement of the distance between Earth and Mars, and from this, using some simple math, the distance between Earth and the sun. Even with this much larger baseline, it still took more than a century for telescopes to get good enough to measure the distance to a star.

This feat was accomplished in 1838 by the German scientist Frederick Bessel, who measured the way the star 61 Cygni appeared to move against the background of more distant stars and determined that its distance was 10.9 light-years. This discovery established once and for all that the universe was a much bigger place than had ever been imagined.

As telescopes got better, astronomers used the triangulation method to measure the distance to stars some tens of light-years away. This process ran into a roadblock, however, because fluctuations in Earth's atmosphere set a limit on our ability to measure angles. Then, in 1989, the European Space Agency launched the satellite Hipparcos. From above the atmosphere, it accumulated a massive amount of data and pushed the limit on parallax measurements out to over 130 light-years. In this century, astronomers using the extreme accuracy available to radio telescopes have been able to determine the parallax of objects called pulsars (see pages 235–37) and have pushed this limit past 500 light-years—an impressive achievement, but still not enough to get us out of the Milky Way galaxy. To go farther, we need a new measurement technique.

STANDARD CANDLES

A standard candle is an object for which we know the total energy output. A good example of a standard candle is a 100-watt lightbulb, since all we have to do to find out how much energy it is putting out is to read the label on the bulb. With this information, we can measure how much energy actually reaches us from the bulb (for example, by using the light meter on our camera). Then, knowing the way that distance dilutes energy, we can find out how far away the bulb is. For example, if we knew we were looking at a 100-watt bulb and were getting 10 watts in our detector, we could use standard equations to find the distance to the bulb. If we had a way to do that with a distant star, then, we could find out how far away it is.

If we know how bright an object is intrinsically, then we can tell how far away it is by measuring how much light we receive. Such an object is called a standard candle and is used to measure the distance to the stars.

The trick, of course, is to "read the label" on a star—to find out how much energy it is actually sending into space. This was first done by a rather remarkable person: Henrietta Leavitt (1868–1921), America's first woman astronomer (see sidebar page 194). While working at the Harvard College Observatory, she noticed something interesting about a certain class of stars known as Cepheid variables, and she exploited this insight to produce what is now known as the Cepheid distance scale.

Most stars shine with a constant light—their brightness doesn't vary except on astronomical timescales. Some stars, however, do not share this property. Watch them for weeks or months and you will see them get brighter or dimmer, sometimes on a regular cycle. These are called, appropriately enough, variable stars. The type of star that Leavitt studied was first seen in the constellation Cepheus (visible only in the Southern Hemisphere), so now all stars of this type are called Cepheid variables.

The point about these stars is that their brightness varies regularly, first brightening, then dimming, then brightening again. We now understand that this behavior is due to processes in the outer atmospheres of certain kinds of stars as they reach the end of their life. What Leavitt found was that the time it took this type of star to go through its cycle depended on how much energy the star was pouring into space—the longer the cycle took, the more energy was being emitted. Watch a Cepheid variable go through its bright-dim-bright cycle, in other words, and you have, in effect, read the label on the lightbulb. It is then a simple matter to measure how much light you are actually receiving in your telescope and figure out how far away the star is. This means that as long as we can see the variable star, we can get its distance. As we shall see in chapter 3, it was Leavitt's work that allowed Edwin Hubble, several years later, to establish both the existence of other galaxies and the expansion of the universe.

Every day on Earth begins when one representative of the Milky Way galaxy, one very ordinary star, pokes its nose above the eastern horizon, and every day ends when that same star disappears in the west. The sun plays such a central role in life on Earth that it is easy for us to forget that it is, after all, just one more star in a galaxy full of them. The proximity of the sun does have an advantage, though—it allows us to study a star up close. In fact, everyday experience can lead you to one of the great scientific questions that occupied astronomers when they began thinking seriously about the sun. Stand outside on a summer day and you will feel warmth on your face. Energy—in the form of infrared radiation—is coming to you from the sun.

THE SUN

[THE STAR NEXT DOOR]

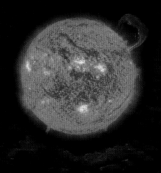

AGE: 4.567 BILLION YEARS
BECOMES RED GIANT: 5.5 BILLION YEARS FROM NOW
MASS (EARTH = 1): 333,000

DIAMETER: 1,392,000 KILOMETERS (865,000 MI)
ROTATION: 25.1 DAYS (EQUATOR), 34.4 DAYS (POLES)
CORE TEMPERATURE: 16,000,000 KELVINS (16,000,000°C/29,000,000°F)
SURFACE TEMPERATURE: 5,800 KELVINS (5500°C/10,400°F)
COMPOSITION: HYDROGEN 74.9 PERCENT, HELIUM 23.8 PERCENT
ENERGY: TURNS 400 MILLION TONS HYDROGEN
INTO HELIUM PER SECOND
SURFACE GRAVITY (EARTH = 1): 28
VISIBLE DEPTH: ABOUT 160 KILOMETERS (100 MI)

Sunset in Indonesia. (Inset) The sun, with a large,
handle-shaped prominence

THE SUN

A cutaway view shows the dynamic structure of the sun, our nearest star.

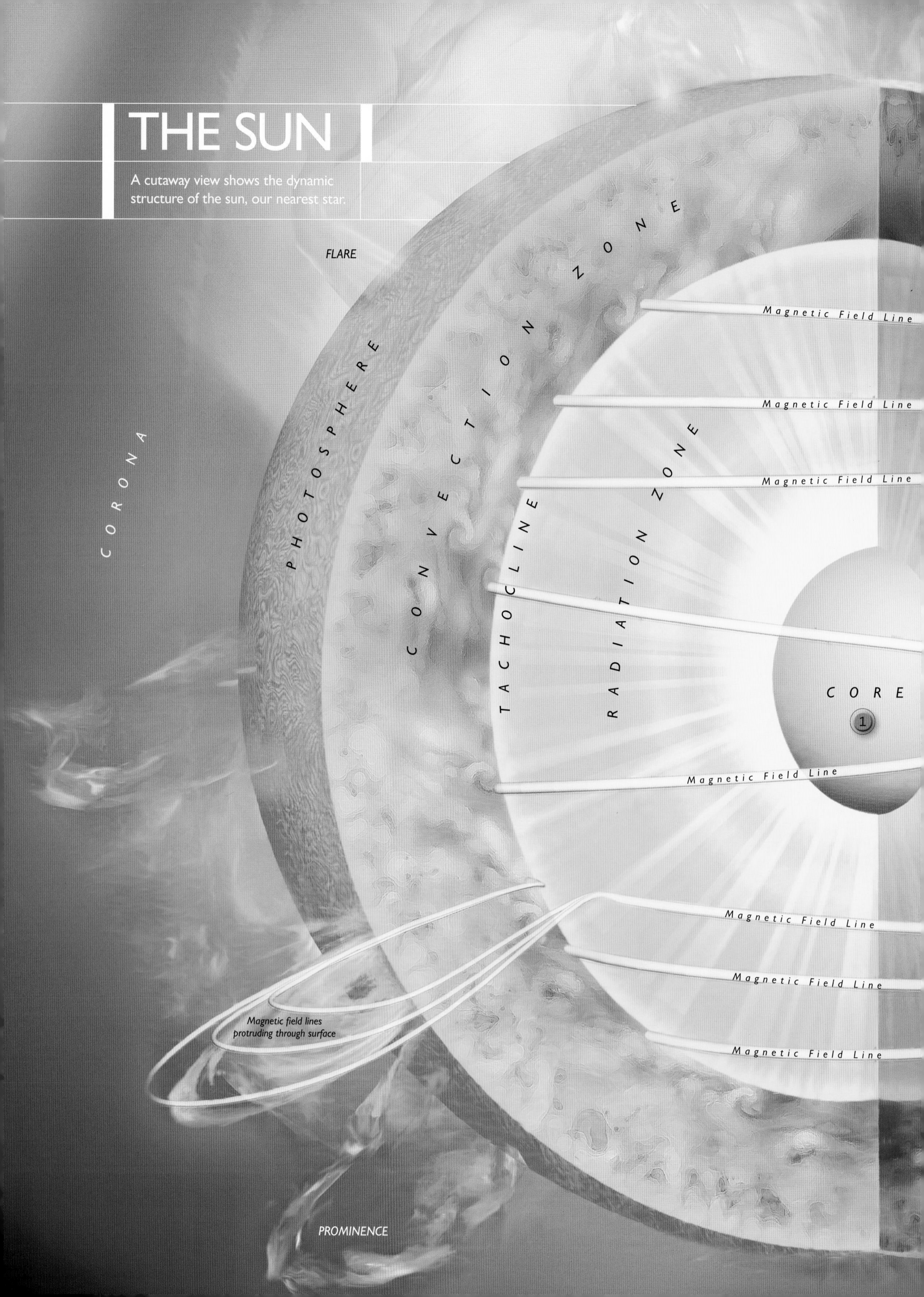

FLARE

CORONA

PHOTOSPHERE

CONVECTION ZONE

TACHOCLINE

RADIATION ZONE

CORE

1

Magnetic Field Line

Magnetic Field Line

Magnetic Field Line

Magnetic Field Line

Magnetic Field Line

Magnetic Field Line

Magnetic Field Line

Magnetic field lines protruding through surface

PROMINENCE

KEY FEATURES

(1) SOLAR CORE: Center of fusion and the sun's energy

(2) SOLAR CORONA: The sun's superheated outer atmosphere

(3) SUNSPOTS: Cool spots on the sun's surface where magnetic lines break through

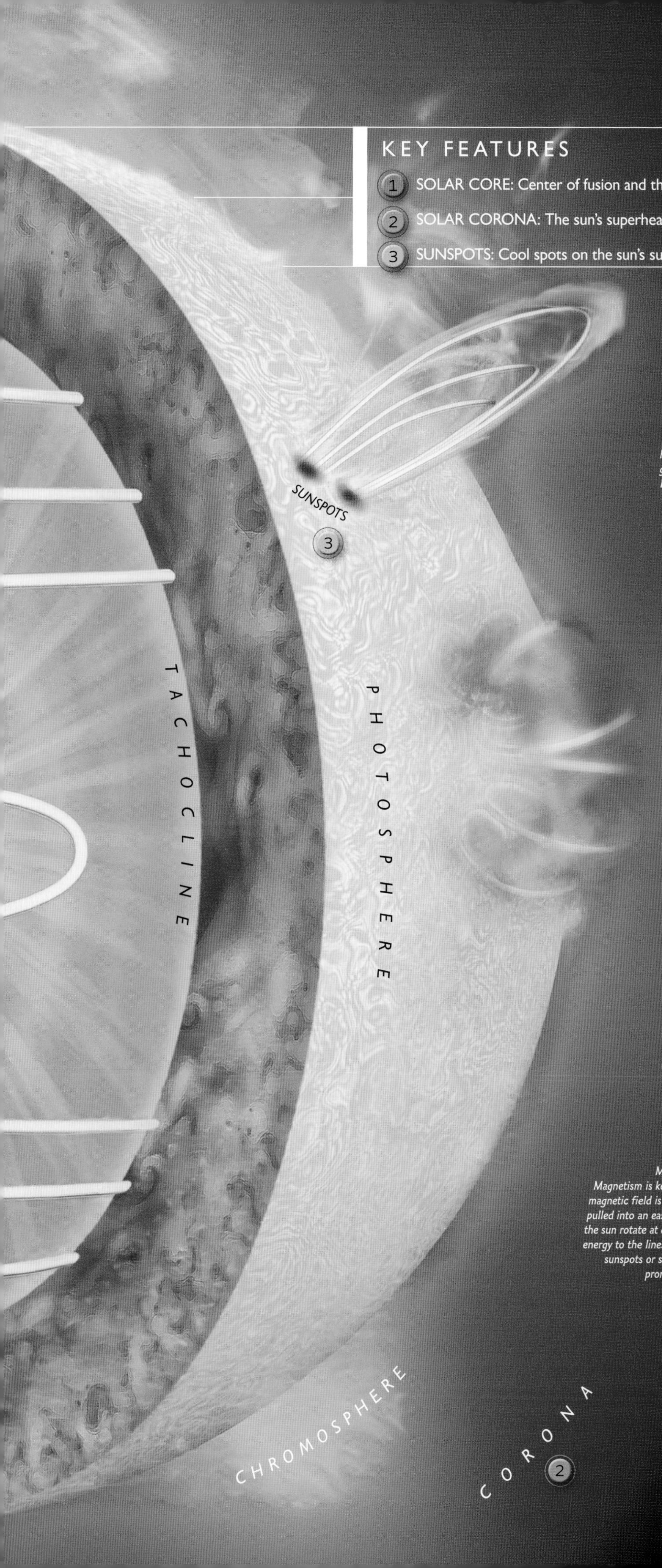

SUNSPOTS
(3)

TACHOCLINE

PHOTOSPHERE

CHROMOSPHERE

CORONA
(2)

STAR POWER

The sun's core is a thermonuclear reactor, fusing hydrogen into helium. Because of the intense heat, these gases exist in an electrified state of matter called plasma. It takes hundreds of thousands of years for light to cross the dense interior to the convection zone, where plasma then bubbles to the surface the way water boils in a pot.

MASSIVE STORMS

A flare explodes when magnetic field lines become overloaded with electrical current, not unlike a fuse blowing. More threatening to Earth, a coronal mass ejection (CME) erupts when field lines snap in a way that lets billions of tons of plasma lift off. Moving as fast as five million miles an hour, the plasma cloud can expand to a width of tens of millions of miles.

DAILY SPACE WEATHER

The halo-like corona is the sun's outer atmosphere. Its plasma particles flow nonstop into space as the solar wind. The shifting magnetic field can open areas where solar wind escapes at a higher speed. When the sun's rotation brings those areas toward Earth, auroras increase and radio transmissions can falter.

MAGNETIC DYNAMO

Magnetism is key to solar behavior. A north-south magnetic field is generated in the tachocline, then is pulled into an east-west pattern as different layers of the sun rotate at different speeds. The stretching adds energy to the lines, which break through the surface as sunspots or soar into the corona as loops and prominences (bottom left).

Until the middle of the 19th century, the fact that the sun gives off energy would have been an unremarkable observation. It was about that time, however, that what we now call the law of conservation of energy was discovered. That law tells us that energy is neither created nor destroyed, but just shifted from one form and place to another. The energy that warms your face, in other words, has to come from some source inside the sun, and the fact that you feel the warmth means that that energy has left the sun forever. The warmth of a summer day contains within it one of the most important messages about the Milky Way—every star must sooner or later run out of energy. Stars are not forever, but are born and die like everything else.

Once this fact sank in, a whole constellation of theories as to the source of the sun's energy appeared. One late 19th-century astronomy textbook, for example, devoted several pages to calculating how long the sun could burn if it was made of the best fuel known at the time, anthracite coal. (The answer is about ten million years.) Infalling meteorites and a gradual shrinking due to gravity both had their day in the sun, as it were, as possible energy sources before they were dropped. (The idea in the latter case being that material moving toward the center of a shrinking sun would give up energy just as a ball does when it falls from a height.) In fact, it wasn't until the early 1930s that the young German-American physicist Hans Bethe showed that the energy source of stars was the process of nuclear fusion.

THE POWER AT THE CORE OF THE SUN

To understand the basis of Bethe's breakthrough, we have to go back to page 44, where we talked about how the sun and the solar system formed from the gravitational collapse of an interstellar dust cloud. In that discussion we concentrated on the small amount of material that went into the formation of the planets, but the fact of the matter is that more than 99 percent of that dust

LITTLE NEUTRAL ONES

When we look at the sun, we can see down only 160 kilometers (100 mi) or so. No light reaches us directly from the interior because the solar material absorbs it. There is, however, an elusive particle called a neutrino that can travel to Earth from the sun's center without being absorbed. Produced in nuclear reactions in the sun's core, neutrinos give us a way of seeing into the heart of our star.

Neutrinos (the name means "little neutral one" in a combination of Latin and Italian) have no electric charge, almost no mass, and interact very weakly with matter. Billions of them have passed through your body since you started reading this sentence, for example, without disturbing a single atom. The only way to detect neutrinos is to put a large target in their way and measure the occasional interaction.

The first attempt to make such a measurement took place a mile underground in a South Dakota gold mine in the late 1960s (the overlying rock shielded the apparatus from cosmic rays). The detector was basically a tank full of carbon tetrachloride, or cleaning fluid. Once a day or so a neutrino from the sun would convert one of the chlorine atoms to a detectible atom of argon. At first scientists were mystified— too few neutrinos were being detected— but eventually they realized that during their transit from the sun, some of the neutrinos changed form and could no longer produce argon atoms in the tank. Today there are neutrino detectors in many places around the world, and they are producing results consistent with our notions of the power at the core of the sun.

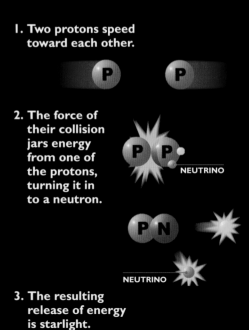

1. Two protons speed toward each other.

2. The force of their collision jars energy from one of the protons, turning it in to a neutron.

NEUTRINO

NEUTRINO

3. The resulting release of energy is starlight.

cloud was incorporated into the body we call the sun. Let's look at what happened as that part of the dust cloud contracted.

One of the basic rules of nature is that as objects contract, they heat up. The rule applies to the cloud that became the sun, so as the great gas cloud contracted, it got warmer. The particles and atoms that made up the cloud moved faster and faster, and the collisions between them became more and more violent. Eventually they became so violent that electrons started to be torn loose from their atoms (this actually doesn't take all that much energy—it happens all the time in fluorescent lightbulbs, for example). The material in the sun became what physicists call a plasma—a collection of loose, negatively charged electrons and their positively charged nuclei moving around independently of each other.

When the temperature in the core of the sun got up into the range of millions of degrees, particles in the center of the newly forming star began moving very fast indeed. Eventually, protons (the positively charged nuclei of hydrogen atoms) started moving so rapidly that they overcame their mutual electrical repulsion and got close enough to initiate a nuclear reaction. In a series of reactions, four hydrogen nuclei fused together to make the nucleus of a helium atom (two protons and two neutrons) and a miscellaneous spray of fast-moving particles. The mass of the final particles was less than the mass of the four initial particles, and the difference was converted, via Einstein's famous formula $E = mc^2$, to energy. This energy streamed outward, creating a pressure that balanced the inward force of gravity, and the sun stabilized.

THE SUN'S STRUCTURE

Since that time about 4.5 billion years ago, the sun has been burning hydrogen at the rate of more than 400 million *tons* per second, and will continue to do so for another 5.5 billion years. The core of the sun, where it's hot enough for nuclear reactions to occur, extends outward about a

NEUTRINO DETECTOR IN HOMESTAKE GOLD MINE, SOUTH DAKOTA

SOLAR ACTIVITY PEAKS EVERY

11
YEARS

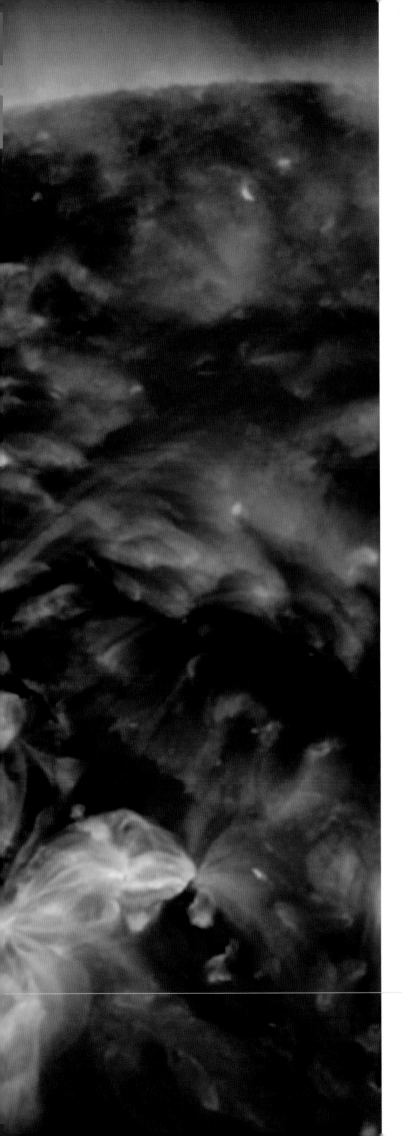

quarter of the way to the surface. Moving outward from the core about 70 percent of the distance to the surface, we encounter what is called the radiative zone. In this region, where matter is still very dense, the fast-moving particles streaming out from the core suffer a series of collisions—think of the process as being like a giant pinball machine—as the energy streams through. Beyond that, the density of matter becomes too low for these kinds of collisions to impede the energy flow, and the sun actually boils, like water on the stove. This convective zone extends almost all the way to the surface.

Looking at the sun from the outside is a little like looking down into murky water. We can see only about 160 kilometers (100 mi) down into the sun, and this thin outer layer, called the photosphere, is what presents itself to us. Above the photosphere are tenuous layers of atmosphere such as the corona (visible during eclipses) and the heliosphere, which actually extends out past the orbit of Pluto.

Because the sun is not solid, different parts of it rotate at different speeds. The poles, for example, complete a revolution in about 35 days, while the equator takes about 25 days to do the same. This difference, together with the constantly churning convection beneath the photosphere, causes the magnetic field of the sun to be constantly twisted, distorted, and pulled around. This gives rise to phenomena like sunspots (dark spots that move across the face of the sun) and solar flares (in which huge numbers of particles are thrown into space). Sunspots go through an 11-year cycle, in which the number of observed spots increases and drops in a regular pattern. The solar cycle, and particularly solar flares, can have an effect on things like satellite operations and radio transmissions on Earth.

A false-color image shows the sun emitting a large solar flare (white spot at the center). The flare was associated with a coronal mass ejection, a fast-moving cloud of electrical particles.

ne of the grand questions we ask about our galaxy is, Are there other life-forms out there—or are we alone in the universe? Science fiction is full of encounters with other intelligent beings—but there are lots of possible life-forms besides Klingons. For most of its history, for example, life on our planet consisted of little more than, essentially, green pond scum. Much of the motivation behind the search for the life-nurturing Goldilocks planet comes from the fact that we want to know if life developed on those planets as it did on ours. • The problem is this: Every living thing we know about works the same way, through a DNA-based chemical code. In effect, all life on Earth is the result of one experiment.

THE ORIGINS OF LIFE

[ARE WE ALONE IN THE UNIVERSE?   ]

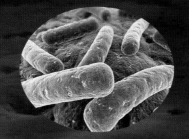

SOLAR SYSTEM'S AGE: 4.6 BILLION YEARS
EARTH'S AGE: 4.56 BILLION YEARS
OLDEST MINERAL (ZIRCON): 4.404 BILLION YEARS OLD
FIRST OCEANS: 4.4 TO 4.2 BILLION YEARS AGO
OLDEST DATED ROCKS: 4.031 BILLION YEARS OLD
OLDEST GEOCHEMICAL SIGNS OF LIFE: 3.8 BILLION YEARS OLD
OLDEST FOSSIL STROMATOLITES: 3.45 BILLION YEARS OLD
ATMOSPHERE TURNS OXYGEN-RICH: 2.4 BILLION YEARS AGO
"SNOWBALL EARTH" GLACIAL ERA: 2.3 BILLION YEARS AGO
FIRST MULTICELLULAR FOSSILS: 2.1 BILLION YEARS AGO
FIRST ANIMAL FOSSILS: 580 MILLION YEARS AGO

Algae color a hot basin at Yellowstone National Park.
(Inset) Bacteria are an ancient form of life.

With only one data point, there is no way we can tell whether we are the result of some wildly improbable cosmic coincidence (as some have argued) or the result of normal chemical processes that will occur many times in the galaxy. Obviously, one important step in solving this puzzle would be to understand how life developed on our own planet.

The watershed experiment in this field took place in the basement of the chemistry building at the University of Chicago in 1952. Harold Urey, a Nobel Prize–winning chemist, and his then student Stanley Miller decided to try a rather unusual experiment. They attempted to set up a miniature model of what they thought Earth was like when it was very young. In a closed system they had water (to simulate the ocean), heat (to simulate the action of the sun), an electric spark (to simulate lightning), and a collection of gases—hydrogen, ammonia, methane, carbon dioxide—that they thought were present on the early Earth. They sealed the system off, turned on the heat and the sparks, and watched to see what happened. After a few weeks, the water turned a brownish maroon color, and on analysis they found that it contained amino acids, one of the basic building blocks of living systems.

The philosophical impact of the Miller-Urey experiment was more important than the chemical details. What the two men showed was that it is possible to start with quite ordinary substances and, by normal chemical processes, produce the molecules that make up living systems. In effect, they moved origin-of-life studies from philosophy into serious science. The consensus today is that they had the wrong mixture of materials in their simulated atmosphere, but it really doesn't matter. Organic molecules of the type they produced have been found in meteorites, in comets, and even in interstellar dust clouds. It seems to be easy, in other words, to produce the molecules of

life by normal chemical means. With this understanding, origin-of-life research moved from the question of how you make the basic building blocks of life to the question of how those blocks were assembled into a living cell.

PRIMORDIAL SOUP

The picture that developed after the Miller-Urey experiment was one in which the chemical process they discovered (or, alternatively, the impact of organic-rich meteorites or comets) turned the oceans into a thin broth of the molecules present in living systems, a broth that was given the wonderful name of primordial soup. Once this broth formed, the argument went, a chance collection of just the right molecules would eventually produce a primitive cell that could take in energy and reproduce. Locales for this chance assembly varied according to the theory, but tidal pools, oily drops in the ocean, clays on the ocean floor, and, more recently, deep-sea vents have all been suggested.

Theories like this, in which life arises through a random occurrence, go by the name of frozen accident theories, since they rely on the idea that a set of molecules, assembled first by random interactions, lock in the chemistry of all of their descendants. A sophisticated modern version of this type of theory is called RNA world. In this scenario, a molecule called RNA, which plays a role in the mechanics of modern cells, appeared first, more or less by chance, and then played the role of facilitator of the chemical reactions that led to life.

METABOLISM FIRST

There is, however, another way that life could have developed, and that is to imagine that there are normal chemical processes, which we are still in the process of discovering, that will drive certain kinds of reactions in the chemical stew of the early ocean—reactions that

Stromatolite reefs, like these in Shark's Bay, Australia, were some of the first living communities that appeared on our planet. They contributed oxygen to the early atmosphere.

will lead directly to a primitive living system without the need of complex molecules like RNA. This is called the metabolism-first school of thought, one that the author finds particularly appealing.

Here's an analogy that may help in visualizing the metabolism-first approach: Consider the modern interstate highway system. If we wanted to explain this system in all of its complexity, we would not start with a highway system and try to find a way that it could have produced cars. Instead, we would look at the most primitive transportation system, such as that provided by Native American foot trails. We would talk about how these developed into unpaved wagon roads, how the first primitive cars appeared, and we would eventually wind up with the present system in all its complexity without

needing recourse to highly improbable chance events. In just the same way, life could have started with simple chemical reactions and evolved its present complexity over billions of years.

Whichever path life took on our planet, we know that it developed quickly, at least on a geological time-scale. A half billion years after the end of the Late Heavy Bombardment (see page 47) made it possible for life to develop without interruption, we find fossils of a complex bacterial ecosystem—the green pond scum mentioned above. The complexity of these organisms suggests that the earliest, simplest cells must have appeared fairly quickly after the end of the bombardment, and this, in turn, suggests that the chemical reactions that produced living systems on our planet could operate with equal speed. However, the question of whether life has developed elsewhere won't be answered until we actually find an exoplanet covered with green pond scum. Or Klingons.

he idea that planetary systems circle other stars is an old one. Only in the last few decades, though, has the search for extrasolar planets, or exoplanets, become a major feature of galactic astronomy. The reason for the delay is simple: Planets shine by reflected light, and are thus much dimmer than stars. Furthermore, they are located close to stars, so whatever light they send out is swamped by light from their parent star. One astronomer likened the problem of seeing an exoplanet directly to detecting a birthday candle next to a searchlight in Boston by using a telescope in Washington, D.C.! Consequently, the discovery of exoplanets had to wait for new kinds of detection techniques.

EXOPLANETS

[OTHER WORLDS, OTHER EARTHS]

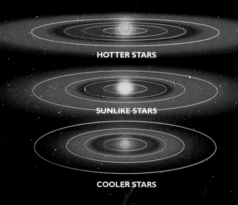

HOTTER STARS

SUNLIKE STARS

COOLER STARS

YEAR FIRST EXOPLANET DISCOVERED: 1992 (PSR 1257+12B)
NUMBER OF EXOPLANETS KNOWN: 760 (AND COUNTING)
SMALLEST: 0.0022 EARTH MASS (PSR 1257+12B)
LARGEST: 9,852 EARTH MASSES (CD-35 2722B)

SHORTEST PERIOD: 2.2 HOURS (PSR 1719-14B)
LONGEST PERIOD: 1,999 YEARS (OPH 11B)
DISCOVERED THROUGH GRAVITY EFFECTS: 699 PLANETS
DISCOVERED BY MICROLENSING: 14 PLANETS
DISCOVERED BY IMAGING: 31 PLANETS
DISCOVERED BY TIMING: 16 PLANETS
NUMBER IN STELLAR HABITABLE ZONES: 534 PLANETS
NUMBER OF KEPLER CANDIDATE EXOPLANETS: 2,321

Artist's conception of ringed exoplanet and its moon.
(Inset) Habitable zones (green) around different kinds of stars

TWO PLANETARY SYSTEMS

The habitable zone of our solar system compared to that of the
newly discovered Kepler-22 system, which has a near-Earth-size planet

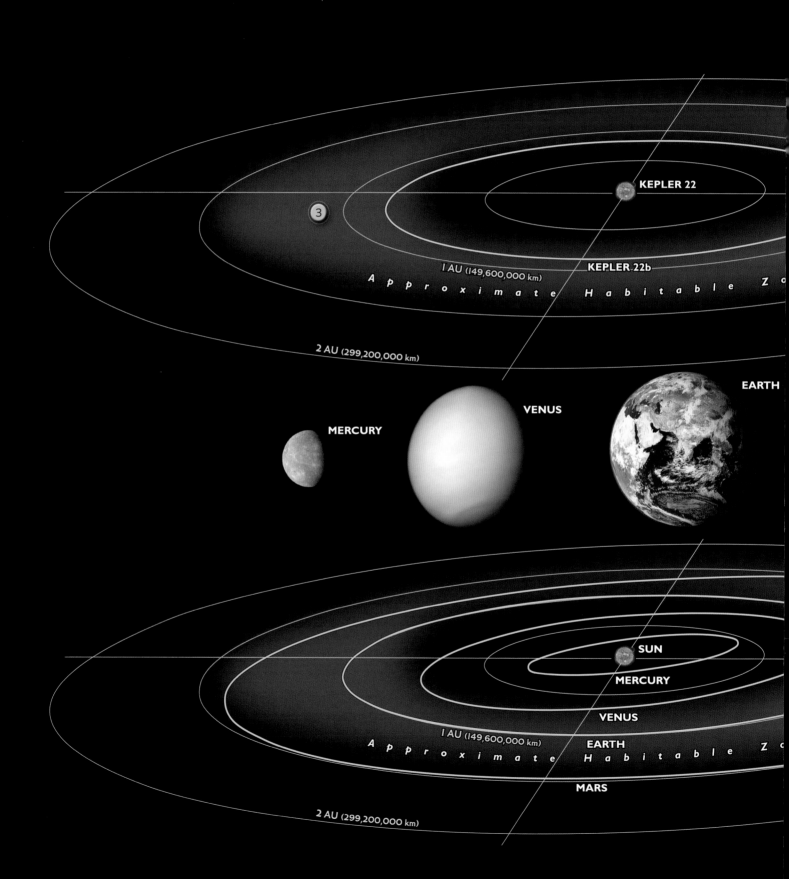

KEPLER 22

1 AU (149,600,000 km)   KEPLER 22b

A p p r o x i m a t e   H a b i t a b l e   Z

2 AU (299,200,000 km)

EARTH

VENUS

MERCURY

SUN

MERCURY

VENUS

1 AU (149,600,000 km)   EARTH

A p p r o x i m a t e   H a b i t a b l e   Z

MARS

2 AU (299,200,000 km)

KEY FEATURES

1. Earth's orbit in the sun's habitable zone
2. A planetary orbit in Kepler-22's habitable zone
3. Kepler-22's habitable zone

KEPLER 22b

MARS

CARTOGRAPHER'S NOTE: Because the Kepler-22 star is smaller than the sun, its habitable zone is closer to the star. Although its planet Kepler-22b orbits within that zone, its orbit alone does not guarantee the presence of life, as can be seen with Mars in our own system.

The first successful detection technique is known in astronomical jargon as radial velocity measurement. To see how it works, imagine that you are an observer looking at our own solar system from a distance of many light-years. We are accustomed to thinking of the sun as being stationary while planets circle in their orbits, but in fact the sun moves around in response to the planets' gravitational pull. For example, if Jupiter lay between you and the sun when you were making your observation, the sun would be pulled slightly in your direction. On the other hand, if Jupiter were behind the sun, the sun would be pulled slightly away from you. Over a ten-year period, then, you would see the sun moving toward you for a while, then away from you. This motion can be detected by observing the Doppler shift in the light the sun emits—blue as it moves toward you, red as it moves away. So, although you couldn't see Jupiter directly, you would know that it's there because of its effect on the sun.

Oddly enough, the first detection of an exoplanet, in 1992, involved a rare case, a planet orbiting a pulsar (see pages 235–37). This planet must have formed after its parent star became a supernova, and the discovery was certainly unexpected. It was followed quickly in 1995, however, by the more conventional discovery of a planet circling a star (in this case in the constellation Pegasus), and it was this that ushered in the modern era of exoplanet

An illustration depicts the Kepler space observatory, launched in 2009, which has detected more than a thousand possible exoplanets. Kepler's discoveries are changing our view of how planetary systems can be organized.

detection. At first, the discoveries came in slowly, at the rate of a few a year, but as techniques improved, the pace picked up. We now know of more than a thousand possible planetary systems around other stars, and some astronomers predict that the number will climb into the tens of thousands once data from the Kepler probe, launched by NASA in 2009, is analyzed.

KEPLER MISSION

Kepler is a spacecraft weighing in at a little over a ton. It is equipped to provide continuous monitoring of the brightness of over 150,000 stars in our immediate galactic neighborhood. Because a satellite in low earth orbit can have up to half the sky blocked by Earth's disk, and because Kepler needs to watch the sky continuously, the spacecraft is actually maintained in an orbit around the sun, not the Earth. You can think of it as trailing along behind Earth like a miniature planet.

The basic planet hunting technique used by Kepler is simple to describe, but it requires sophisticated equipment to make it work. The central point is that if a planet passes in front of a star, the brightness of that star will drop while the planet is making its transit, then pick up again as the planet moves around its orbit. Of course, this so-called transit method of detection works only if the orbit of the planet lies in the line of sight to Earth. (For example, if someone were observing our solar system, Jupiter could be detected by the transit method if the observer were located on the same plane as the planetary orbits, but not if he or she was above or below it.) This means that only a fraction of planetary systems can be detected by transit techniques.

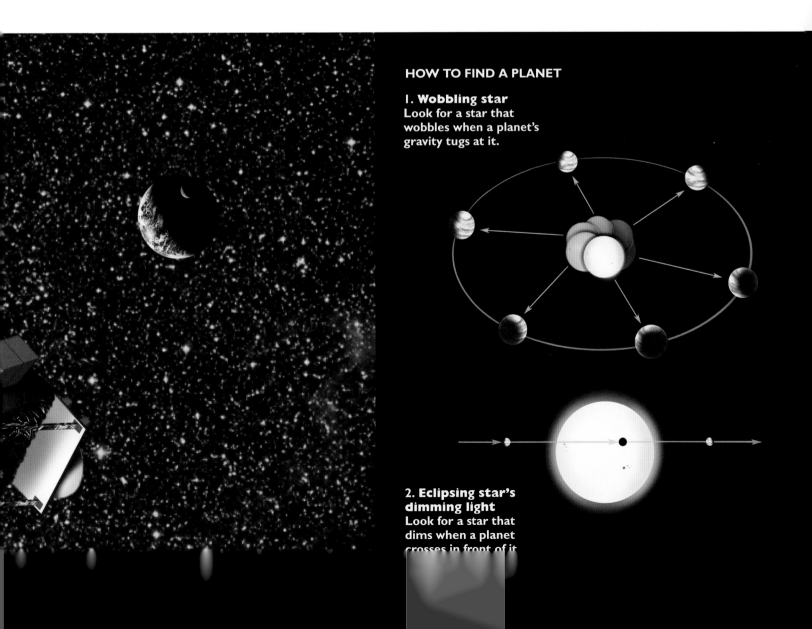

HOW TO FIND A PLANET

1. Wobbling star
Look for a star that wobbles when a planet's gravity tugs at it.

2. Eclipsing star's dimming light
Look for a star that dims when a planet crosses in front of it

On the other hand, Kepler scientists are quick to point out that the transit method has one important advantage over the older radial velocity technique. In order for a planet to jerk a star around enough to produce a detectable Doppler shift, the planet has to exert a large gravitational attraction on the star. This means that the radial velocity method is most likely to detect large planets orbiting close to their stars—the so-called hot Jupiters. Indeed, most of the planets detected before the Kepler launch were of this type. The transit method, on the other hand, detects any planet that affects its star's brightness, regardless of its size or distance from the star. In fact, one of Kepler's most important results is the realization that the hot Jupiters are not all that common—they were just the first exoplanets to be seen. This came as a great relief to scientists who study planet formation, because as we saw on page 43, theory tells us that large planets should form far from their stars.

A TREASURE TROVE OF EXOPLANETS

By early 2012 the Kepler team had identified more than 2,300 possible exoplanets. William Borucki, the principal investigator for the spacecraft, is careful to refer to these objects as candidates. This is not a case of excessive caution, but simply a recognition of the fact that events other than planetary transits can cause a star's brightness to flicker. A large collection of sunspots could do this, for example, as could an eclipsing binary star so close to the line of sight that the spacecraft can't tell that it's a separate object. A minimum requirement, then, is that the transit repeat at the regular intervals appropriate for a planet in orbit.

In fact, Kepler scientists describe two hurdles that a candidate must overcome before it is granted the rank of planet: confirmation and validation. Confirmation requires that enough data be gathered to determine the object's mass—usually by having ground-based telescopes measure the star's radial velocities after Kepler

LIFE AT THE EXTREMES

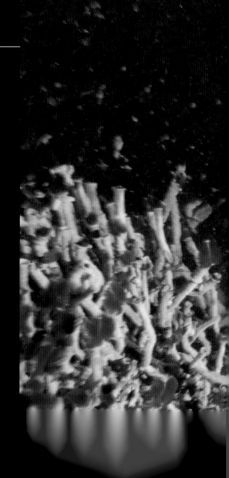

No discussion of exoplanets or life elsewhere in the universe would be complete without a discussion of strange life right here on planet Earth. Over the past 50 years scientists have discovered life in unexpected places. These newly discovered forms of life are called extremophiles, from the Latin *extremus* and the Greek *philia,* and the word can be translated roughly as "loving extremes."

The first extremophiles were discovered in the 1960s in the hot springs of Yellowstone National Park. The temperature of the water—at or above the boiling point—would have killed ordinary bacteria, but the extremophiles thrived. In fact, they are responsible for some of the spectacular coloring of those pools. Since that time, life has been found in more

highly acidic and highly salty environs and at the unimaginable pressure and temperatures of deep-sea vents, to name a few. Experimenters in Japan even found microbes that could thrive in centrifuges producing gravity 400 times that of Earth!

These discoveries have had a profound effect on science. Some biologists have proposed, for example, that life on Earth began with extremophiles at deep-sea trenches and only later migrated to the surface. Astrobiologists who think about life on other planets have been cautioned not to be too restrictive in their definitions of where life might develop. And, of course, there is always the possibility that surprises might be waiting for us here, on our home planet. In the words of physicist Paul Davies, "Life might be under our

has identified a likely system. In some cases, monitoring the transits of several planets in a system allows scientists to work out the complex dynamics of motion and get the masses that way. (I should point out that finding the mass of the many small planets Kepler has found is an extremely difficult job.)

Validation is the process of going through the data carefully and ruling out all the false positives that could fool you into thinking you had found a planetary system. "In the end," says Borucki, "we may never be able to do this for many of the small objects." The team most likely will publish a list of candidates with their best estimate of the probability that each is actually a planet.

GOLDILOCKS PLANETS

But what about the holy grail of planet hunting—the Goldilocks planet (see page 78)? To zero in on these possibly life-sustaining worlds, scientists define something

BLACK SMOKER NEAR HYDROTHERMAL VENTS

called a habitable zone around each star. This is a region in which the temperature is (roughly) between the freezing and boiling points of water, so that there could be liquid water on the planetary surface. Candidates for Goldilocks planets would have to be Earth-size objects in the habitable zones of their stars.

The problem of finding an Earthlike planet is compounded by the fact that Kepler scientists demand that they see three properly timed transits before they will even consider whether the object is really a planet. "A planet like Earth may already be in our data," Borucki says, "but we won't be able to say anything about it for another two years." Moreover, as Sara Seager of MIT is fond of pointing out, there is an important distinction to be made between Earthlike and Earth-size planets. "Venus is Earth-size," she says, "but it is clearly not Earthlike."

So far, the Kepler scientists have found 207 Earth-size planet candidates. Perhaps more exciting is the first confirmed planet in the Goldilocks zone: Kepler 22b, about 2.4 times the size of Earth. Temperatures on the planet should be balmy, but there is no indication yet if the planet possesses an atmosphere or signs of life.

And this point, in turn, illustrates what is perhaps the most important aspect of the Kepler mission. It has already revolutionized the field of planetary science, giving theorists more data to chew on than they ever believed possible. In the view of Kepler scientists, however, their mission is more in the nature of a reconnaissance survey—a kind of Lewis and Clark expedition into the cosmos. Its purpose is to map the main features of what's out there and to guide future explorations to the points of greatest interest. The real holy grail of planetary science—an Earthlike planet that either could support life or already does—will be found by spacecraft following the trail that Kepler is blazing now.

Its not often that the birth of a scientific field can be dated precisely, but SETI—the Search for Extraterrestrial Intelligence—is an exception. It began with a paper by physicists Giuseppe Cocconi and Philip Morrison in 1959 and came to full fruition at a conference in the mountains of West Virginia in 1961. In their paper, the physicists pointed out that with the new availability of radio telescopes, it had become possible to scan the radio bands and see if anyone out there was trying to contact us. "The chance of success if we try is small," they argued, "but the chance of success if we don't try is zero."

SETI

[ARE WE ALONE?]

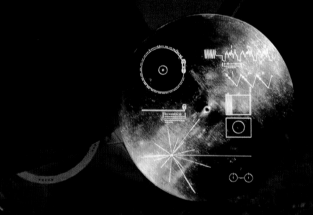

1895: Percival Lowell claims Martians dug canals

1896: Nikola Tesla suggests radio to find extraterrestrials

1924: U.S. Naval Observatory listens for Mars radio signals

1959: Philip Morrison and Giuseppe Cocconi suggest using radio microwaves

1960: Frank Drake heads up Project Ozma, first radio search

1961: R. Schwartz and C. Townes suggest optical masers for interstellar communication

1971: Project Cyclops study uses 1,500 antenna arrays

1981: Paul Horowitz builds multichannel spectrum analyzer

1981: NASA drops SETI from budget after congressional criticism

1999: SETI@home begins

2007: 42-antenna Allen Telescope Array begins, funded by Paul Allen

Allen Telescope Array, California. (Inset)
Diagrams on Voyager golden record cover

The conference at the National Radio Astronomy Observatory in Green Bank, West Virginia, gathered 11 scientists together to talk about this new possibility. The conferees eventually summarized their estimates of the number of extraterrestrial civilizations in a compact notation that has come to be called the Drake equation, after Cornell astronomer Frank Drake, one of the organizers of the conference. The equation estimates the value of N, the number of extraterrestrial civilizations trying to communicate with us right now, as

$$N = R \, f_p \, n_e \, f_l \, f_i \, f_c \, L$$

where the symbols represent, from left to right, the rate at which new stars are forming in the galaxy, the probability that the new star will have planets, the number of those planets capable of supporting life, the probability that those planets will actually develop life, the probability that life will develop intelligence, the probability that intelligent life will develop a technology capable of interstellar communication, and the length of time that that communication, once started, will continue. Obviously, as we move from left to right through these terms, we go from fairly well known astronomy to sheer guesswork. Nevertheless, the Drake equation is a useful way to organize our knowledge (and our ignorance) on the subject of extraterrestrial intelligence.

In 1961 the Green Bank attendees argued that N could be as high as 200 million or as low as 4, but came up with a most probable estimate of around a million. This was a time, remember, when scientists still harbored hopes of being able to find life on Mars and other places in our solar system. The idea of a vast, intercommunicating Galactic Club made up of thousands of intelligent

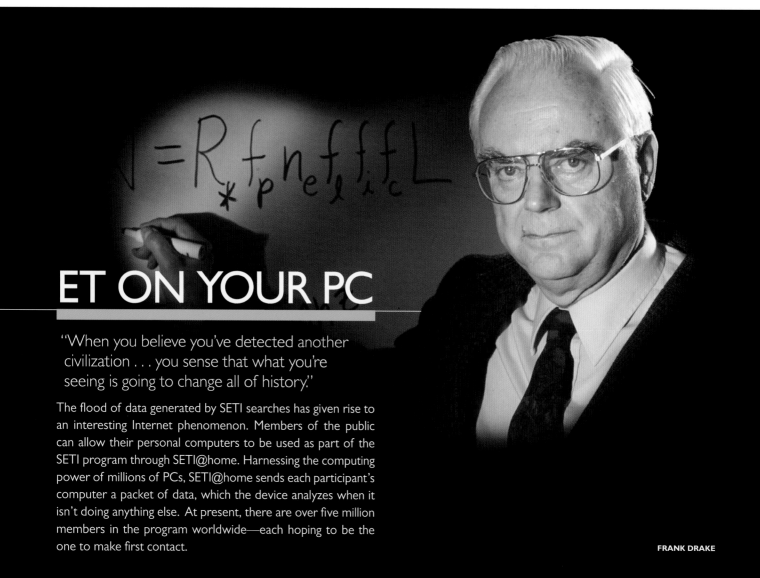

ET ON YOUR PC

"When you believe you've detected another civilization . . . you sense that what you're seeing is going to change all of history."

The flood of data generated by SETI searches has given rise to an interesting Internet phenomenon. Members of the public can allow their personal computers to be used as part of the SETI program through SETI@home. Harnessing the computing power of millions of PCs, SETI@home sends each participant's computer a packet of data, which the device analyzes when it isn't doing anything else. At present, there are over five million members in the program worldwide—each hoping to be the one to make first contact.

FRANK DRAKE

species entered the public consciousness, fueling countless science-fiction scenarios.

Unfortunately, this is one case where science didn't keep up with science fiction. In the beginning, there was modest federal support for SETI programs, but when the searches failed to turn up anything interesting, that source of funding waned. Today, the SETI program is being carried forward using private funding.

When you think about the problems involved in the search, you can see why it is taking so long. There are billions of stars out there, and, as we saw on page 215, a good proportion of them probably have planetary systems. With each star, you have to follow a slow process: You don't know what frequency the aliens might be using for their broadcasts, so, like someone listening to radio programs in a strange city, you slowly move through the dial, pausing long enough at each position to see if something interesting is coming through. Today SETI searches make use of modern fast electronics to monitor thousands of stars and thousands of frequencies at once as they sort through a mountain of data.

WHERE IS EVERYBODY?

But despite these efforts and widespread public interest, the unfortunate fact is that no contact has yet been made with extraterrestrials. One of the engaging things about SETI is that it is tailor-made for "armchair science"—amiable speculation that doesn't require rigorous evidence to support it. Here, for example, are a few of the explanations that have been given for the lack of a SETI signal:

The zoo hypothesis—our solar system has been declared a kind of galactic wilderness area, off-limits to ET.

The gloom and doom hypothesis—any species aggressive enough to win the evolutionary battle and develop technology will wipe itself out with nuclear weapons, thereby producing a low value of L in the Drake equation.

The magic frequency hypothesis—we're looking at the wrong frequencies in our searches, although there is some other magic frequency, recently discovered by the theory's proponent, that we should be monitoring instead.

And so on. You can undoubtedly add to this list yourself.

The most interesting SETI argument, though, was made early on by the Italian-American physicist Enrico Fermi. When presented with the kind of arguments that eventually led to the idea of the Galactic Club, he is supposed to have thought for a while, then asked, "So where is everybody?" Fermi, a genius at seeing through to the heart of complex problems, argued as follows: Modern science is only a few hundred years old—less than the blink of an eye on astronomical timescales. It's pretty likely that in another few hundred years—another eyeblink—we will have solved the problem of interstellar travel and be colonizing the stars ourselves. If there are really millions of other civilizations out there, some must have passed this stage already, so they should be here already. So, he asked, "Where is everybody?"

The point of the argument, of course, is that we shouldn't be looking for ET out there, we should be looking right here. Many authors (myself included) have used the Fermi argument to propose that N is quite small, possibly even 1, and that humans might well be alone in the universe.

The point about SETI, though, is that no matter how the search turns out, it's worth doing. Are there ETs out there? Fantastic! Are we alone in the galaxy? Even more fantastic! There aren't many other kinds of scientific activities with this kind of payoff.

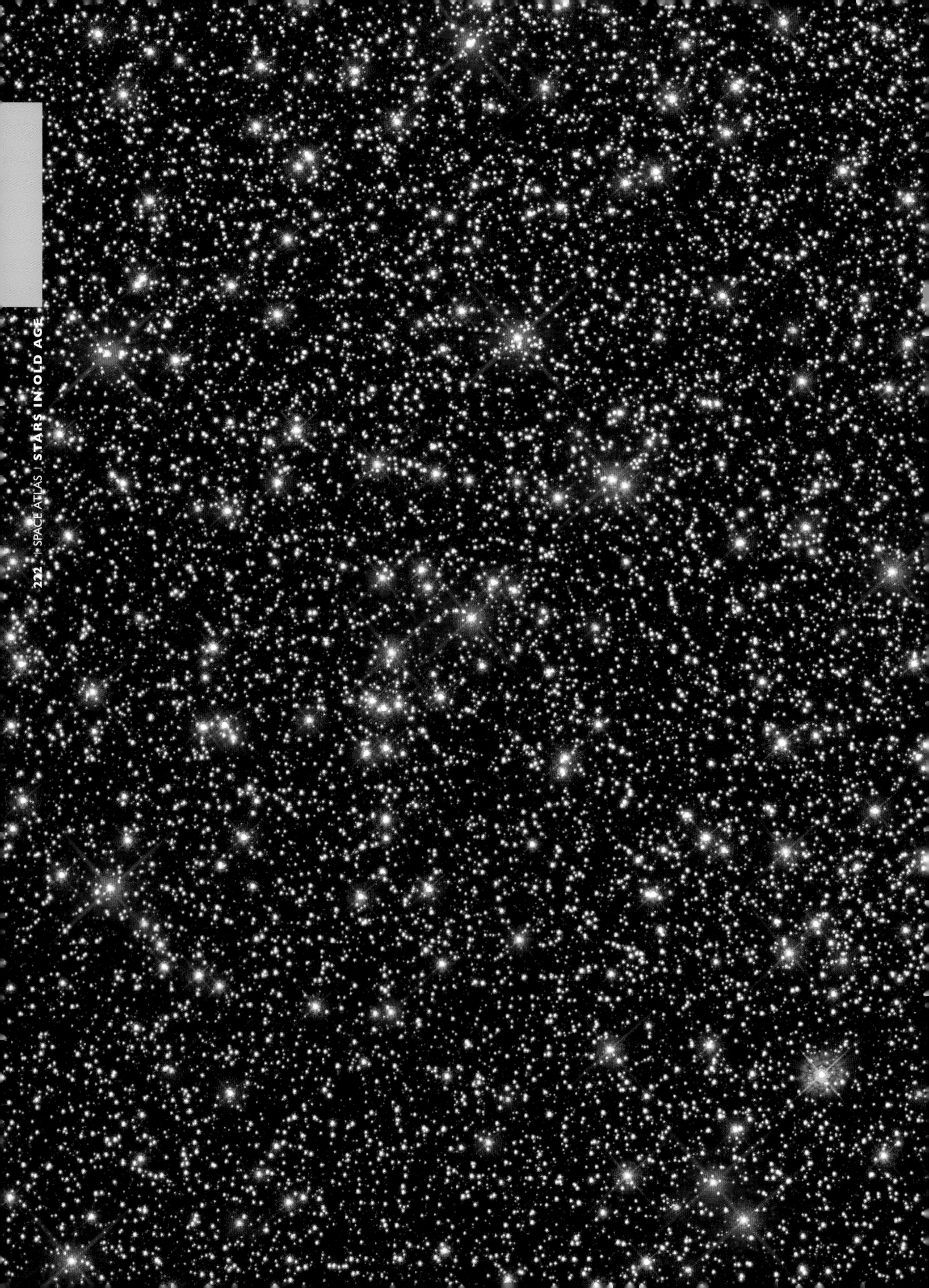

All stars begin life the way our sun did, as a condensing cloud of interstellar dust. From then on, the star is essentially devising strategies to ward off the eternal inward pull of gravity. We discussed the first of these strategies in the birth of our own sun—the initiation of fusion reactions in the core, which sets up an outward pressure to stop the cloud from collapsing. Almost every star you see is in this hydrogen-burning phase—astronomers call them main-sequence stars. Our sun has been in this phase for roughly 4.5 billion years.

STARS IN OLD AGE

[WHAT HAPPENS WHEN THE FUEL RUNS LOW?]

ANTARES

RIGEL   SIRIUS A   SUN

LIFETIMES OF STARS OF DIFFERENT MASSES

0.1 SOLAR MASS: 6 TO 12 TRILLION YEARS
1 SOLAR MASS: 10 BILLION YEARS
10 SOLAR MASSES: 32 MILLION YEARS
100 SOLAR MASSES: 100,000 YEARS

FATES OF STARS OF DIFFERENT MASSES

0.1 SOLAR MASS: RED DWARF
1 SOLAR MASS: RED GIANT, THEN WHITE DWARF
10 SOLAR MASSES: SUPERNOVA, THEN BLACK HOLE
100 SOLAR MASSES: SUPERNOVA, THEN BLACK HOLE

Stars in the globular cluster Omega Centauri. (Inset) Typical stars Sirius A and the sun, hot blue star Rigel, and red supergiant Antares (background)

bviously, hydrogen burning can't go on forever. Sooner or later the hydrogen fuel in the star's core will run out and the star will have to develop a new way of countering gravity. How long that takes depends on how big the star is. There are two competing effects here. On one hand, bigger stars have more fuel to start with. On the other hand, bigger stars also generate a stronger gravitational force and have to burn fuel faster to overcome it. It turns out that the second effect wins, and that the bigger a star is, the shorter the time it can spend burning hydrogen. If you imagine the Milky Way galaxy being a year old, for example, a star like the sun might have fuel enough for ten months or so, while very large stars may last as little as half an hour.

RED GIANTS

So what will happen to a typical star, like our sun, when the hydrogen in its core runs out? Obviously, the nuclear fires will start to fade and the pressure that has held off the forces of gravity for billions of years will grow weaker. Gravity will take over again, the star will start to contract, and once again the contraction will cause the star's interior to heat up. This will have two effects: First, the temperature of the region just outside the core, which still has a lot of unburned hydrogen in it, will increase to the point where that hydrogen can start fusing into helium. Second, the temperature of the core itself goes up until the helium nuclei that were the end product of the original fusion reactions are moving fast enough to start a new cycle of fusion. In the end, three helium nuclei, each with two protons and two neutrons, come together in the core to produce a carbon nucleus (six protons and six neutrons) along with some extra energy. This sort of sequence, in which the ashes of one nuclear fire become the fuel for the next, is one of the major features of energy generation in aging stars.

In the end, for stars up to about six times as massive as the sun, the renewed fusion reactions will increase the star's energy output as well as cause its outer atmosphere to expand. The edge of our sun, for example, will eventually extend outside the current orbit of Earth. Because the star's energy is being sent out through a much larger surface, the color of that surface changes from white-hot (as in the sun today) to a cooler red. A star like this is called a red giant.

What will happen to Earth when the sun turns into a red giant 5.5 billion years from now? Clearly Mercury and probably Venus will be swallowed up. During the run-up to the red giant phase, the sun will throw an appreciable fraction of its mass into space, however, weakening its gravitational hold on the planets. Consequently, Earth's orbit will move outward. If this were the only thing happening, Earth would narrowly escape being engulfed, but recent calculations indicate that tidal effects may pull the orbit in enough to ensure destruction. Even if our planet is not engulfed, however, its oceans would evaporate and surface rocks would melt, ending any life that has survived to that time.

WHITE DWARFS

What's next? The kinds of pressures that stars like the sun can generate are not high enough to initiate any new nuclear fusion reactions with the carbon created earlier in the star's core, so that avenue of escape from gravity is closed off. Gravity takes over again and the collapse resumes.

We know that early in the collapse of the interstellar gas cloud that led to the sun, collisions between atoms became so violent that electrons were torn from their nuclei. Throughout the whole stellar life cycle, from main sequence to red giant, these electrons were basically spectators while nuclear events held center stage. Now it becomes their turn to shine.

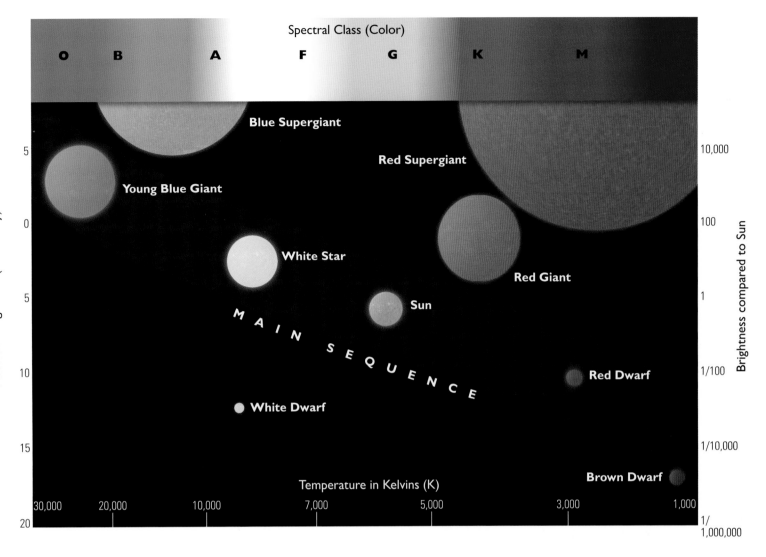

The connection between a star's temperature and its luminosity can be seen on a Hertzsprung-Russell diagram (above). From left to right, stars run from hottest to coolest; from bottom to top, dimmer to brighter. Our sun is a typical star on the central main sequence.

One fact about electrons becomes very important at this stage of the stellar life cycle. It's called the exclusion principle, and it says that no two electrons can be in the same state. Think of a crowd of people: You can push them close together, but eventually you reach a point where each person requires some minimum amount of elbow room, at which point you can't make the crowd any smaller. In the same way, the loose electrons in the sun are pulled in as the final collapse starts, but they eventually reach a point where they just can't be pushed together anymore. At this point gravity is pushing in while the electrons push out, and the star reaches a final point of stability. It will last forever.

For the sun, this new equilibrium will be reached when the star has shrunk down to about the size of Earth. The star is white-hot, like the ember of a recent fire. Astronomers call this sort of object a white dwarf. It is still radiating leftover energy produced during its long life cycle, but like a dying ember it will keep cooling and dimming. This is how our own star will end its life, along with many of the stars we now see in the sky.

This is not the only way for a star to die, however. Some life stories have a much more spectacular endings, as we shall see in the next section.

W
hen you look out into the Milky Way, you see all kinds and sizes of stars. If the sun were the size of a bowling ball, for example, the galaxy would be littered with BB-size stars, interspersed with every other conceivable size ball up to giant beach balls. As you might expect from a collection with this much diversity, not all stars have the same life story, although all of them, like the sun, start by burning hydrogen. Stars like the sun (roughly from golf ball to basketball size) go through the red giant–white dwarf sequence described in the last chapter. Larger stars have a much more spectacular endgame and play a much more important part in the history of life on Earth.

SUPERNOVAE

[GOING OUT WITH A BANG]

MOST FAMOUS SUPERNOVA: 1054 (CRAB NEBULA)

DISTANCE TO 1054 SUPERNOVA: 6,500 LIGHT-YEARS

LAST SUPERNOVA SEEN IN MILKY WAY: 1604

DISTANCE TO 1604 SUPERNOVA: 14,000 LIGHT-YEARS

SUPERNOVA CONCEPT BORN: 1931, BY WALTER BAADE AND FRITZ ZWICKY

MOST RECENT BRIGHT SUPERNOVA: 1987 IN LARGE MAGELLANIC CLOUD

DISTANCE TO 1987 SUPERNOVA: 160,000 LIGHT-YEARS

SUPERNOVA BLAST-WAVE SPEED: 30,000 KM/SEC (19,000 MI/SEC)

NUMBER OF SUPERNOVAE DISCOVERED EACH YEAR: HUNDREDS

MILKY WAY SUPERNOVA FREQUENCY: ABOUT EVERY 50 YEARS

SUPERNOVA TYPES: IA, IB, IC, IIP, IIL

False-color image of supernova remnant Cassiopeia A. (Inset) Shock waves heat material around Supernova 1987A.

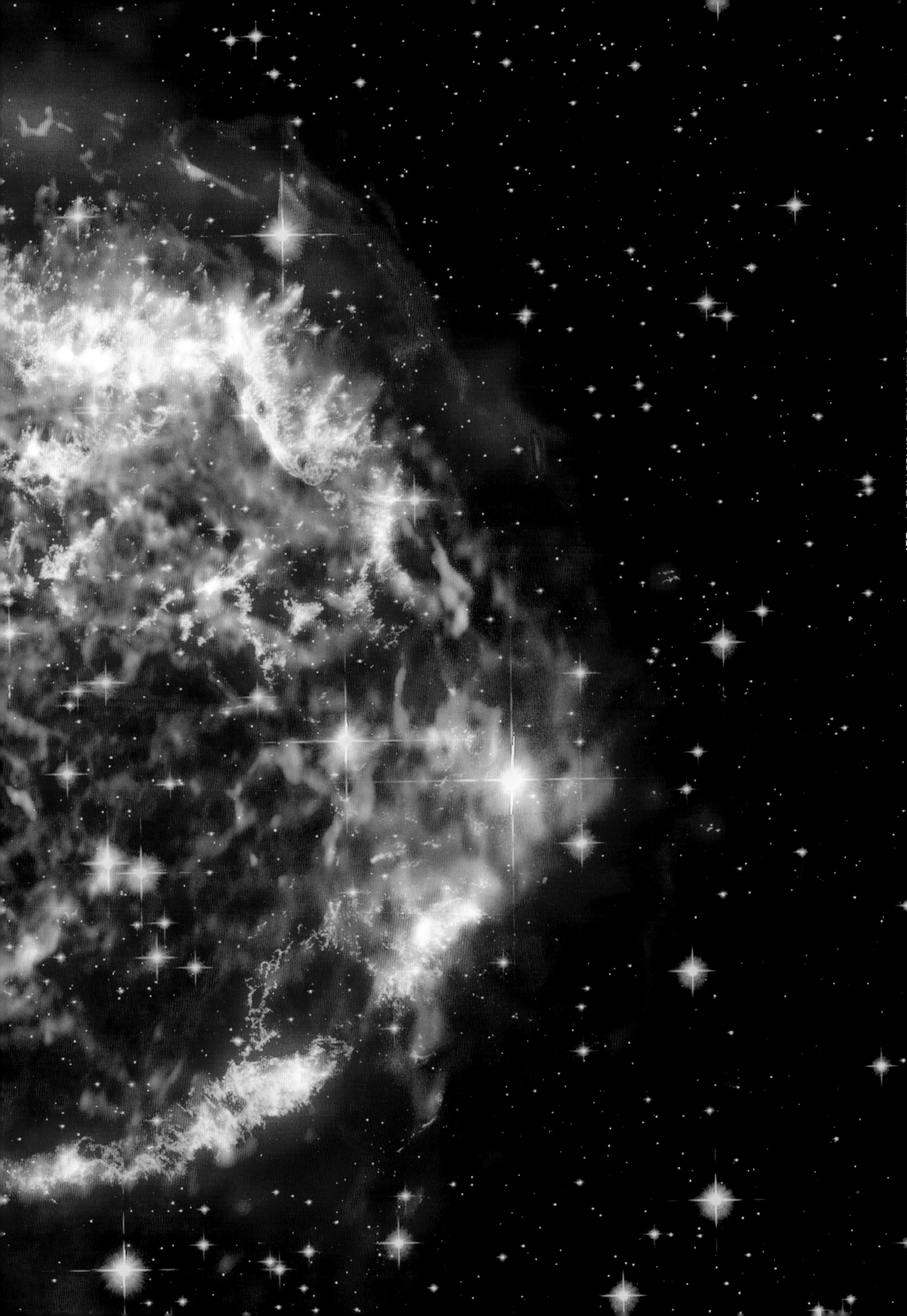

A quick review: Every star begins as a collapsing cloud of interstellar dust. It reaches a first point of stabilization when the temperature in its core gets high enough to begin nuclear fusion, turning hydrogen into helium. When the hydrogen in the core runs out, the collapse continues, the interior heats up, and the helium ash is burned along with some previously unused hydrogen. For stars up to six times the size of the sun, the nuclear story ends here—there just isn't enough mass to push the temperature high enough to ignite any new nuclear fires. The star becomes a white dwarf.

For stars nine or ten times more massive than the sun, though, the story is different. These stars do have enough mass to push both the compression and the temperature high enough to keep the nuclear fires burning. As in the sun, the first nuclear step for these stars once the hydrogen in the core runs out is to burn the resulting helium, ultimately combining three helium nuclei to create a carbon nucleus. This process is accompanied by the burning of previously unburned hydrogen into helium in a shell surrounding the core. The star now has a carbon core surrounded by a helium shell that, in turn, is surrounded by unburned hydrogen. When the collapse starts again, the temperature in all of these regions rises. In the innermost core the carbon (six protons, six neutrons) combines with other nuclei to produce oxygen (eight protons, eight neutrons) and other, heavier products. The helium in the first shell burns to carbon, and the hydrogen in the next shell burns to helium.

This process, with the ashes of one fire becoming the fuel for the next, continues through the periodic table of the elements, with the star developing an onionlike shell structure as successively heavier elements are created by each collapse. Each new reaction adds another

NOVAE AND SUPERNOVAE

The word "nova" means "new" in Latin, and the term has long been applied to a specific type of astronomical event: A star suddenly appears in the sky where no star had been before. Today we realize that there are many different processes that can produce new stars, and we give them different names if they have different causes.

A nova occurs in double-star systems when one of the stars has gone through its life cycle and become a white dwarf (see pages 224–25). If the two stars are close enough, the dwarf will pull material, mainly hydrogen, from its partner. When this hydrogen accumulates to a depth of several feet on the dwarf's surface, nuclear reactions ignite. In essence, the layer goes off like a huge hydrogen bomb, temporarily brightening the sky in a nova. This process can be repeated as the layer builds up again.

Although people often confuse novae and supernovae, only those stars that undergo this temporary surface brightening are properly called novae. Supernovae are stellar explosions that have their own dramatic processes (see page 230).

SUPERNOVA SEQUENCE

layer to the onion. As this process goes on, however, the star gets less bang for the buck with each new cycle. Indeed, some calculations suggest that the last stage of the process, which creates nuclei of iron in the core, buys the star only a few days' respite from the relentless pull of gravity.

CRITICAL MASS

Iron is the ultimate nuclear ash. You can't get energy from iron by splitting it, and you can't get energy from iron by adding to it. It just builds up in the core of the star like ashes in the grate of a woodstove, setting the stage for one of the most spectacular events in the universe. As the iron accumulates in the heart of the star, nuclear forces cannot keep it from collapsing under the influence of its own gravity. Electrons in the iron core for a time provide the pressure to counteract gravity. In effect, the core of the massive star becomes a kind of white dwarf, except that it is made of iron.

But as more and more iron "ash" falls into the star's center, the mass of the core approaches a critical value. When the mass gets to be about 40 percent more than that of the sun, the electrons begin to combine with the protons in the iron nuclei to produce neutrons. As each electron disappears, the ability of the remaining electrons to counteract gravity decreases, and in a very short time the core turns into a mass of neutrons that collapses catastrophically.

Depending on the mass of the star, the collapse will continue until the neutrons can't be pushed together any further, or it will go on to form a black hole. We'll talk about both of these possibilities in the next sections. For the moment, though, let's think about the rest of the star—that huge, onionlike envelope of heavy elements that the nuclear reactions have built.

RS OPHIUCHI, A RECURRENT NOVA SYSTEM

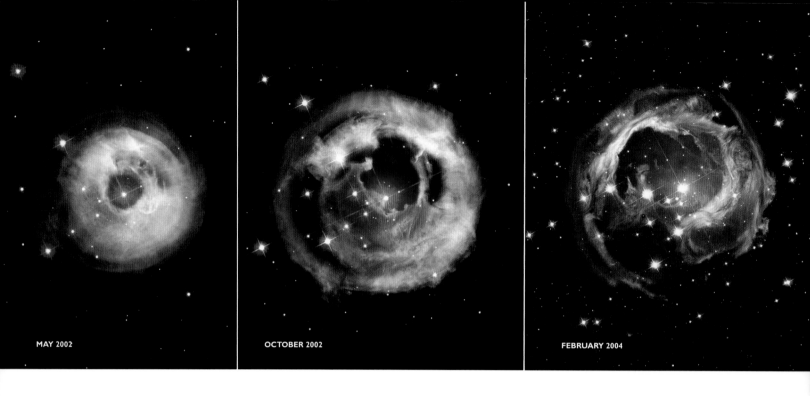

EXPLOSION

From the point of view of that envelope, the rug has just been pulled out from under its feet. The iron core that had supported the weight of the rest of the star has suddenly disappeared. In response, the envelope starts to collapse, falling inward until it hits the new neutron core, at which point it rebounds, producing massive shock waves—and the entire star explodes. In the resulting maelstrom, nuclear reactions produce all of the elements up to uranium. It is this titanic event that gives rise to a new star in the sky: a supernova.

The remnants of the envelope speed outward, carrying with them all of the star's heavy elements. Over the next few thousand years these clouds of materials will cool and mix with the interstellar medium, becoming part of the dust clouds from which new stars and new planetary systems form. Our own solar system formed late in the history of the Milky Way, incorporating into its planets and even its life-forms the heavy elements made by long-dead stars.

TYPE I SUPERNOVAE

For historical reasons, the kind of event we've just described is called a Type II supernova. A Type I supernova operates differently, although it produces the same kind of massive explosion. Type I supernovae occur in double-star systems in which one star has gone through its life cycle and is a white dwarf. It can happen that this dwarf starts to pull hydrogen off of its partner, increasing its mass until it reaches that critical point, 40 percent higher than the mass of the sun. At this point the entire star explodes in a massive nuclear inferno.

Let's think about the dynamic picture the supernova story gives us of our home galaxy. It started off, billions of years ago, as a cloud made up primarily of hydrogen and helium left over from the big bang. As large, short-lived stars were born and became supernovae, heavier elements began to appear—a few at first, then more and more as time went on. One way to think about the Milky Way, then, is to picture it as a giant machine constantly taking in primordial hydrogen and churning out the rest of the chemical elements.

Oh, and in case you're interested, astronomers expect the star IK Pegasi (in the constellation Pegasus), 150 light-years from Earth, to become a Type I supernova some time in the next few million years or so.

Beginning in 2002, the Hubble Space Telescope captured a remarkable sequence of images as the star V838 Monocerotis suddenly swelled, heating and illuminating a shell of surrounding dust clouds.

When we described supernovae in the last section, we concentrated on events in the exploding stellar envelope. Now it's time to go back and see what happened to the collapsing core. You may recall that the envelope suddenly collapsed because electrons were being forced to combine with the protons in the iron nuclei of the core, turning most of the core into a mass of neutrons. Because neutrons have no electrical charge and therefore do not repel each other, there was nothing to counter the force of gravity in the heart of the star, so the material in the core went into free fall. This fall continued until some force could be found to overcome the inward pull.

NEUTRON STARS & PULSARS

[MAGNETIC SPINNERS]

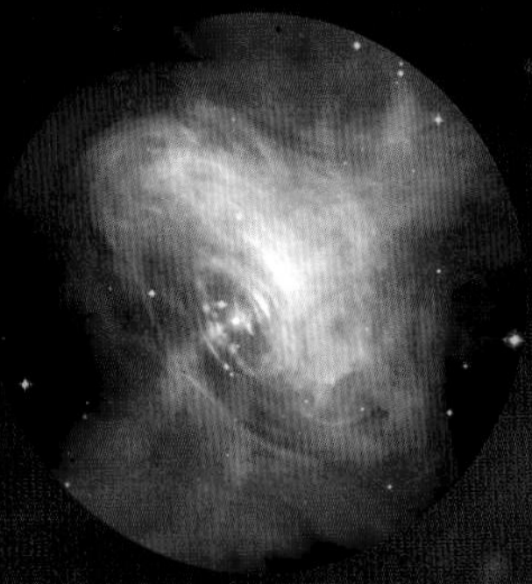

FIRST PULSAR DISCOVERED: NOVEMBER 28, 1967
DISCOVERERS: JOCELYN BELL BURNELL AND ANTONY HEWISH
IDENTIFIED AS NEUTRON STARS BY: THOMAS GOLD AND FRANCO PACINI

PERIOD OF FIRST PULSAR: 1.33 SECONDS
LONGEST PERIOD PULSAR: 8.51 SECONDS (PSR J2144-3933)
FIRST BINARY PULSAR: PSR 1913+16
FIRST DOUBLE-PULSAR BINARY SYSTEM: PSR J0737−3039
FIRST X-RAY PULSAR: CENTAURUS X-3
FIRST PULSAR WITH PLANETS: PSR B1257+12
FASTEST ROTATION: 716 TIMES A SECOND (PSR J1748-2446AD)
CLOSEST TO EARTH: 510 LIGHT-YEARS (PSR J0437-4715)

Illustration of pulsar in supernova remnant.
(Inset) A pulsar in the Crab Nebula

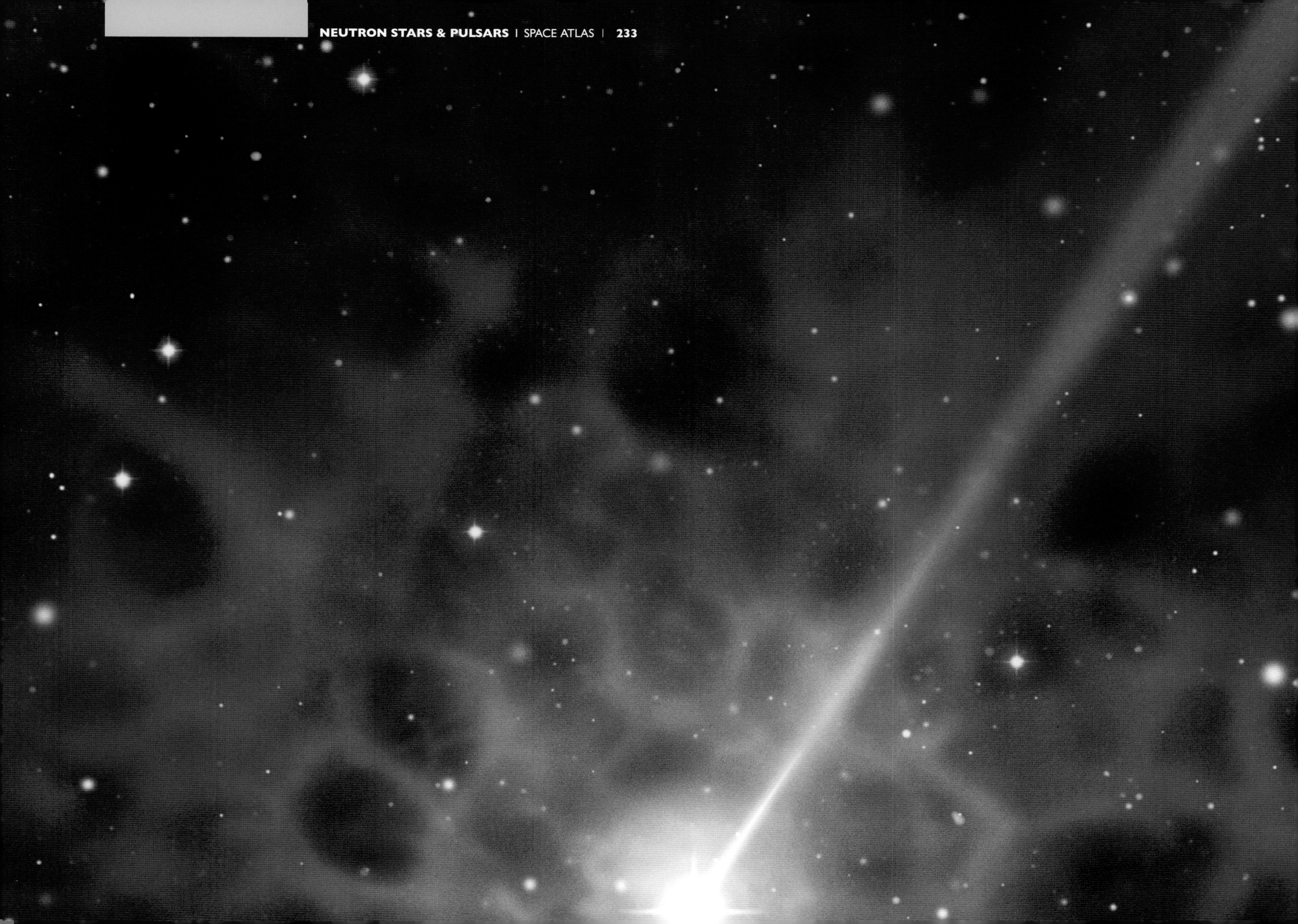

Neutrons, like electrons, can't be crowded too closely together, so if the star isn't too big, eventually the pressure of the crowded neutrons will balance gravity. The laws of quantum mechanics tell us, however, that the heavier a particle is, the less elbow room it needs and the more closely it can be packed. Since neutrons are almost 2,000 times heavier than electrons, this means that the object formed from the collapsing core will be much smaller than the white dwarfs discussed on pages 224–25. In fact, most neutron stars are thought to be less than about 16 kilometers (10 mi) across—small enough to fit inside the city limits of many urban areas. When I want to make this point to my students in the suburbs of Washington, D.C., I point out that a neutron star would fit comfortably inside the Capital Beltway.

PROPERTIES OF A NEUTRON STAR

An object like this has several amazing properties. In the first place, a supernova's iron core doesn't start to collapse until it is significantly more massive than the sun (typically, between 40 and 200 percent more massive). If you cram that much material into something the size of a small city, the material is going to be incredibly dense. In fact, a tiny drop of neutron star stuff would easily outweigh the Great Pyramid at Giza. Because the mass is so concentrated, the force of gravity at the surface is huge—perhaps a hundred billion times stronger than on the surface of Earth.

The second amazing feature has to do with rotation. All stars rotate—we saw, for example, that the sun spins on its axis about once a month. The iron core of the supernova would have this sort of rotation rate as well. But just as an ice-skater's rate of spin increases when she pulls in her arms, the rate of rotation of the core will increase during the collapse. Sometimes, the spin can get very fast indeed. Some neutron stars rotate at nearly a thousand times a second!

Finally, the collapse will produce an extremely strong magnetic field in the neutron star. Stars normally have moderate magnetic fields—the field of the sun, for example, is about half that of Earth. This field is locked in to the matter of the star, however, so it is concentrated by the collapse of the core. The magnetic fields in some neutron stars may be a million billion times greater than that on Earth.

We can't study neutron stars close up, of course, but there are fairly solid theoretical models of what they must be like. A neutron star 16 kilometers (10 mi) in diameter is thought to have a solid crust—nuclei crushed into a kind of lattice—about 1.5 kilometers thick. Because of the intense gravity, the atmosphere (composed of atomic nuclei and electrons) is under a meter high, and the surface is extremely smooth, with a maximum irregularity less than the thickness of a dime. In this model, the interior of the star would be a kind of nuclear liquid made up mostly of neutrons.

Putting all of this together, we get a compelling picture of the star. It is a compact, rapidly spinning object with a very strong magnetic field. In general, we would expect the star's north magnetic pole to be different from the geographic North Pole, just as it is on Earth. (Remember that the north magnetic pole of our planet is in Canada, not at the geographic North Pole). This means that as the star rotates, the magnetic field is swept around in a circle. Because of the field's intensity, the neutron star emits a beam of radio waves that travels outward along the direction of the star's magnetic axis. Like a lighthouse beam sweeping around in a circle, then, the rotating neutron star sends out a radio beam that sweeps through space. It's what happens if Earth happens to be in the path of that beam that is interesting.

THE STELLAR LIGHTHOUSE

Imagine standing on a shore near a lighthouse. You will see a flash of light when the beam is in your direction, followed by a period of darkness as the beam travels around, and then another flash. In the same way, if you have a radio receiver in the path of the beam from a rotating neutron star, you will see a pulse of radio waves when the magnetic axis points in our direction, a period of no signal, then another pulse, and so on.

In 1967 scientists at a radio telescope in England first observed these sorts of regular pulses. No one had anticipated anything like this, and at first the scientists referred to them, jokingly, as the LGM signals—LGM standing for "little green men." Once the signal was understood to be from a rotating neutron star, the word "pulsar" was coined for these objects.

Since that first discovery, a couple of thousand pulsars have been discovered in the Milky Way. Their periods of rotation vary from somewhat less than ten seconds down to a few milliseconds. Pulsars have also been discovered that emit pulses in the x-ray and gamma-ray regions of the spectrum; a few even have planets. The closest pulsar to Earth is 280 light-years away in the constellation Cetus.

PULSAR SCIENCE

Pulsars come in many varieties and have figured prominently in a number of different investigations. We will talk about only a few of them here.

JOCELYN BELL BURNELL

"Finding the first [pulsar] was disturbing—scary— because we weren't sure what it was."

Every year when the Nobel Prizes are announced, there is a mini-debate on the subject of who was left out. In any scientific discovery, there are always many people whose work was important but who do not get the ultimate recognition, and therefore they join the ranks of the "also-rans." These debates are usually low-key, but in a few cases they continue for decades. The story of Jocelyn Bell and the discovery of pulsars is one of those.

Bell was a graduate student at Cambridge University. In the late 1960s she participated in building one of the first radio telescopes designed to use a technique suggested by her thesis adviser, Antony Hewish, to detect compact sources of radio waves in the sky. Bell was in charge of op erating the telescope and doing the first analysis of the data. In 1967 she began to notice signals she called "scruff"; in the face of hostility from senior astronomers, she showed that these signals—regularly timed radio pulses—were real, and not caused by man-made interference. The signals were originally dubbed LGM, for "little green men," because of the possibility that they came from extraterrestrials. It was quickly realized, however, that they were what we now call pulsars (see above).

For this discovery, Hewish and radio astronomer Martin Ryle shared the Nobel Prize in physics in 1974, the first physics prize awarded in astronomy. The exclusion of Bell triggered protests from many prominent astronomers—though not from Bell (now Jocelyn Bell Burnell) herself, who is still active as an astronomer and has won many other awards in her career.

JOCELYN BELL BURNELL

THE CRAB NEBULA SUPERNOVA

EXPLODED

ALMOST 1,000 YEARS AGO

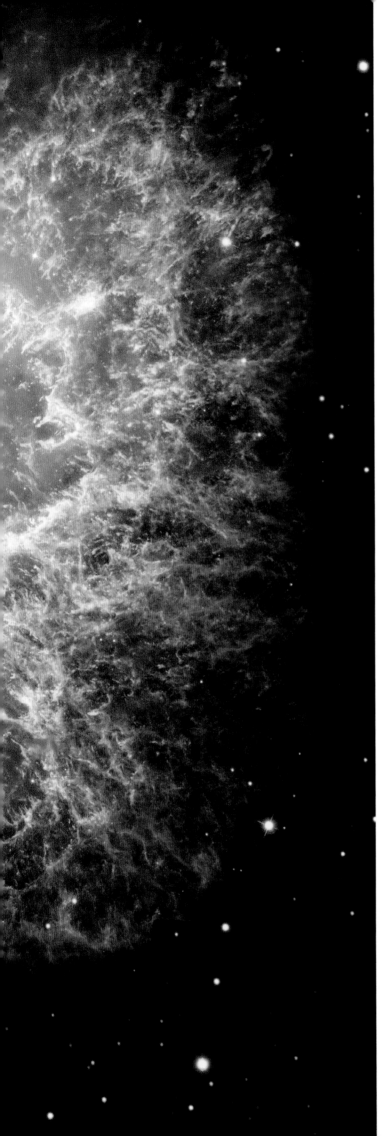

The Crab Nebula is the remains of a supernova whose light was first recorded by Japanese and Chinese astronomers in 1054. The remnants now form a gaseous cloud six light-years across. Its bluish glow is powered by the magnetic field of a rapidly spinning neutron star in the center, the collapsed core of the exploded star.

Sometimes the pulsar's rotation will speed up. Remembering the analogy of the ice-skater, we can conclude that the size of the neutron star must have decreased slightly. This is interpreted as a "starquake" — a rupturing and crunching of the star's crust.

In 1974 Russell Hulse and Joseph Taylor at Princeton University discovered a pulsar in orbit around another star. Using the precise timing of the pulses, they were able to document a slow decay in the pulsar's orbit. The energy loss turned out to be precisely that predicted by general relativity—basically, the system is giving off a type of radiation known as gravitational waves. For this discovery, Hulse and Taylor were awarded the Nobel Prize in physics in 1993.

When neutron stars are part of binary systems, in fact, they provide scientists with a fascinating natural laboratory. For instance, while studying a neutron star in orbit around a larger companion in the constellation Volans, astronomers were able to observe how light from the companion star was bent and redshifted through the extreme gravity of the neutron star's centimeter-high atmosphere.

Some astronomers have suggested using the precise rotation rate of pulsars to define a new time standard to improve on the current atomic clocks. Atomic clocks are accurate to "only" 13 decimal places, and pulsar timing is one system being proposed to push that limit to 15 decimal places. Given that time standards began with observations of heavenly bodies, there would be a certain philosophical rightness to returning those standards to the sky.

There is probably no object in the sky that has been taken up so wholeheartedly in fiction and common usage as the black hole. It is certainly the most exotic object we know. • The short definition of a black hole is that it is an object so massive, so compact, that nothing can escape the gravitational pull at its surface. Light that falls into it never comes out. We expect that stars that begin life with 30 times the mass of the sun and go through the supernova process will wind up as black holes. The gravitational force exerted by the core of such a star is so huge that it will overcome the attempts of the neutrons to oppose it.

BLACK HOLES

[THE TIMELESS END OF THE BIGGEST STARS]

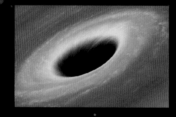

FIRST PREDICTED: 1783, BY JOHN MICHELL
FIRST MODERN THEORY: 1939, BY J. ROBERT OPPENHEIMER

MICRO BLACK HOLE SIZE: UP TO ABOUT 0.1 MILLIMETER
MICRO BLACK HOLE MASS: UP TO ABOUT THE MASS OF THE MOON
STELLAR BLACK HOLE SIZE: ABOUT 30 KM (19 MI)
STELLAR BLACK HOLE MASS: ABOUT 10 SUNS
INTERMEDIATE BLACK HOLE SIZE: ABOUT 1,000 KM (600 MI)
INTERMEDIATE BLACK HOLE MASS: ABOUT 1,000 SUNS
SUPERMASSIVE BLACK HOLE SIZE: 150,000 TO 1.5 BILLION KM
(93,200 TO 932 MILLION MI)
SUPERMASSIVE BLACK HOLE MASS: 100,000 TO A BILLION SUNS

Gas jets out from the edges of a black hole in galaxy Centaurus A.
(Inset) Illustration of matter sucked into a black hole

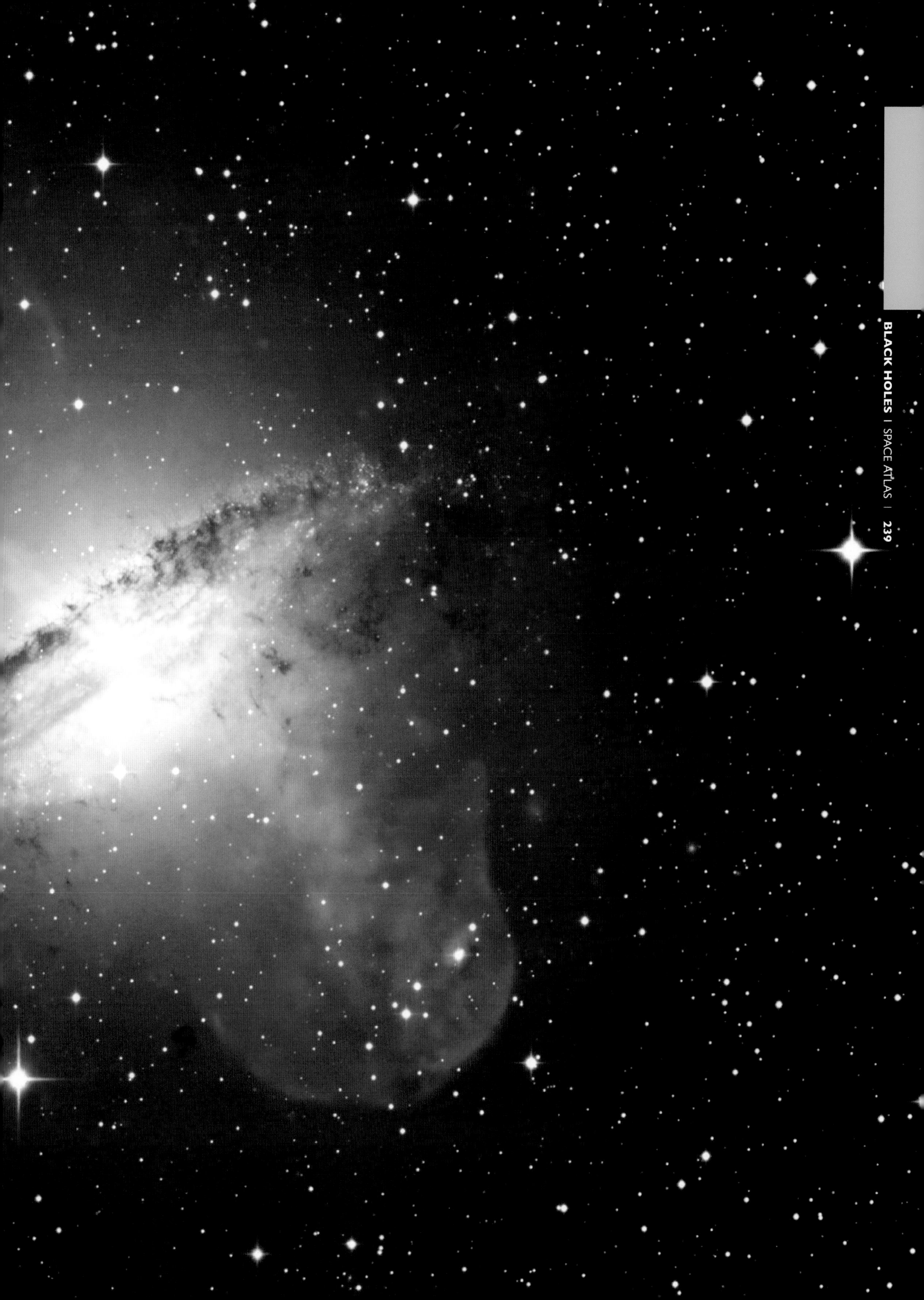

nce these supermassive stars begin to collapse, the process takes a remarkable turn. Instead of forming a neutron star, the core of the dying star will continue its collapse until a stellar black hole is formed. There are other kinds of black holes, as we shall see below, but before we get into that subject, let's consider what a black hole is in terms of Einstein's theory of relativity.

The best way to visualize the force of gravity as Einstein saw it is to picture a flexible plastic sheet marked off with a coordinate grid and stretched tightly over a frame. Roll a lightweight marble across the sheet and it will travel in a straight line. Now imagine putting a heavy object, like a bowling ball, on the sheet. The ball will drag down and distort the sheet; if you roll the marble again, its path will be warped around the slope created by the bowling ball. In Einstein's language, the mass of the bowling ball distorts the grid (he would say "space-time grid"), and what we interpret as the force of gravity is actually the result of this distortion.

Now imagine making the bowling ball heavier and heavier, pressing the distortion deeper and deeper into the sheet. Eventually, you might get to the point where the plastic just wraps itself around the ball and snaps off, isolating itself from the grid. In essence, you now have a black hole—a region of space that has cut itself off from the rest of the universe.

ACROSS THE EVENT HORIZON

German physicist Karl Schwarzchild predicted the existence of black holes in 1916, shortly after Einstein published his theory of relativity. For a long time, Schwarzchild's solution of the Einstein equations was regarded as an oddity—the duck-billed platypus of the astrophysical world. In fact, I can remember that when it was discussed in my general relativity class at Stanford in the 1960s, we were told that although black holes were possible in theory, they could never form in the real world. This was the prevailing orthodoxy for much of the 20th century.

What makes the Schwarzchild solution so strange is the existence of what is called the event horizon, or Schwarzchild radius. This marks a point of no return, a boundary in space that separates the interior of the black hole from the rest of the universe. Event horizons surround an incredibly small volume—you would have to pack all of the mass of an object like the sun into a sphere a little over a mile across to create one, for example. At the event horizon of a black hole, though, really strange things happen.

Here's an analogy that may help you understand the event horizon. Suppose you and a friend get into a couple canoes and start down a river. You communicate with people at your starting point by shouting regularly (for example, by sending sounds waves every minute by your clock) so that they can keep track of where you are. Suppose further that there is a waterfall downstream, and that the speed of the water increases as you approach it. Finally, suppose that at some point the speed of the water exceeds the speed of sound. This will be a kind of event horizon. How will your trip downstream look to you and to your friends on the bank?

To them, your shouts will get farther and farther apart as you speed up. To be technical, they would see your clock (as measured by the arrival of your shouts) slowing down until, when you pass the event horizon, the shouts (and hence your clock) stop. You and your partner in the neighboring canoe, however, don't notice anything strange, and can go on communicating normally. As far as you are concerned, nothing particular happens as you cross the event horizon.

In the same way, a distant observer watching an object falling into a black hole will see time on the object slow down and stop at the event horizon. An observer on the infalling object, however, sees no change in the clock that is traveling with her. Such is the nature of the event horizon.

FINDING BLACK HOLES

From our description of black holes, it's obvious that they cannot be detected in the normal way, by bouncing light off them, since any light sent into a black hole cannot, by definition, return. Black holes, however, do exert a gravitational force depending on their masses, and that can be used to find them.

In the 1980s scientists, observing the motion of objects in the gravitational field of an unseen body, found the first evidence for black holes in our universe. Researchers monitoring bodies near the center of our galaxy (the center being in the constellation Sagittarius) realized that objects there were in orbit around something extremely massive—what we call today a galactic black hole. The black hole at the center of the Milky Way has a mass several million times that of the sun.

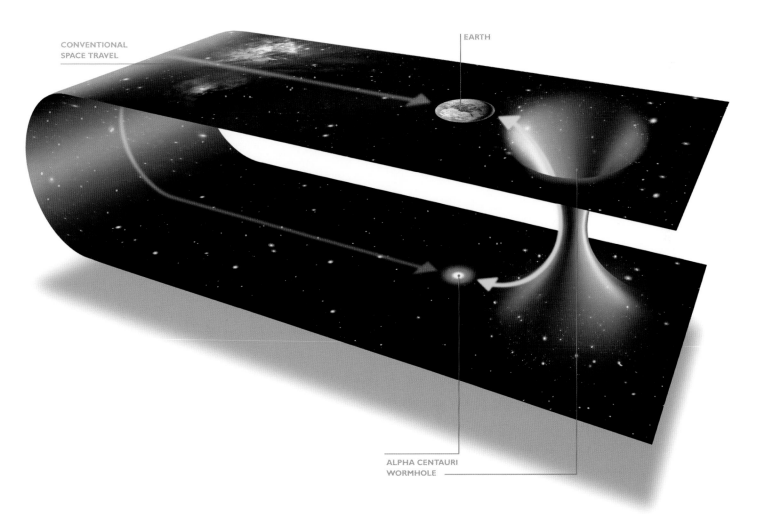

CONVENTIONAL
SPACE TRAVEL

EARTH

ALPHA CENTAURI
WORMHOLE

A diagram shows how a wormhole might lead from Earth (upper center) to the nearby star Alpha Centauri, allowing travel far faster than a more conventional trip (red arrow) at or below the speed of light. Wormholes, a theoretical possibility, are a favorite device in science fiction.

As impressive as this is, you have to remember that the galaxy contains many billions of stars, so the black hole at the center represents considerably less than one-tenth of one percent of the galaxy's mass. Astronomers think that it is likely that almost all (and perhaps all) galaxies have black holes at their centers.

Another way to spot black holes is through radiation from their outskirts. Matter falling into a black hole tends to bunch up, forming what is called an accretion disk. Collisions in the disk heat it up, so that it emits energetic radiation. It is this effect that allows us to see the smaller black holes that result from the collapse of stars—the so-called stellar black holes. The best candidates we have for stellar black holes are double-star systems in which one of the partners has gone through its life cycle and evolved into a black hole. In this case, the black hole can pull material away from the other star and form an accretion disk, which would then heat up and emit radiation such as x-rays. The star Cygnus X-1, a strong x-ray source, is one of our best candidates for a stellar black hole.

The Cygnus X-1 system (below left) is found near star-forming regions of the Milky Way. It is thought that this double star contains a black hole of about 15 solar masses in orbit around a blue giant star (illustration, below right). Scientists believe the black hole pulls gas from its companion and ejects some of it in high-speed jets.

OPTICAL IMAGE OF CYGNUS X-1

INSIDE A BLACK HOLE

Up to this point we have talked only about observing black holes from the outside. We can have no direct knowledge of the interior of a black hole, for the simple reason that there is no way for information to get out. Our mathematical models, however, suggest that the interior may be very strange. The Schwarzchild solution, for example, predicts that at the center of the black hole may be a singularity—a place where the curvature of space-time becomes infinite. The known laws of physics would break down at a singularity. In the case of electrically charged or rotating black holes, it is theoretically possible that someone entering a black hole at one point could exit in a different space-time continuum. Such a path, beloved of science-fiction writers, is called a wormhole. The same models suggest that moving through a black hole could produce time travel.

Before you get too intrigued by these exotic results, though, you have to realize that the gravitational fields near a black hole are so strong that the difference in the force between your head and feet would be enough to stretch you out and tear you apart—a process astrophysicists call "spaghettification." Nothing, not even the sturdiest starship imaginable, would survive a trip through a singularity. Best to avoid black holes if you can!

ILLUSTRATION OF CYGNUS X-1 BLACK HOLE, CLOSE UP

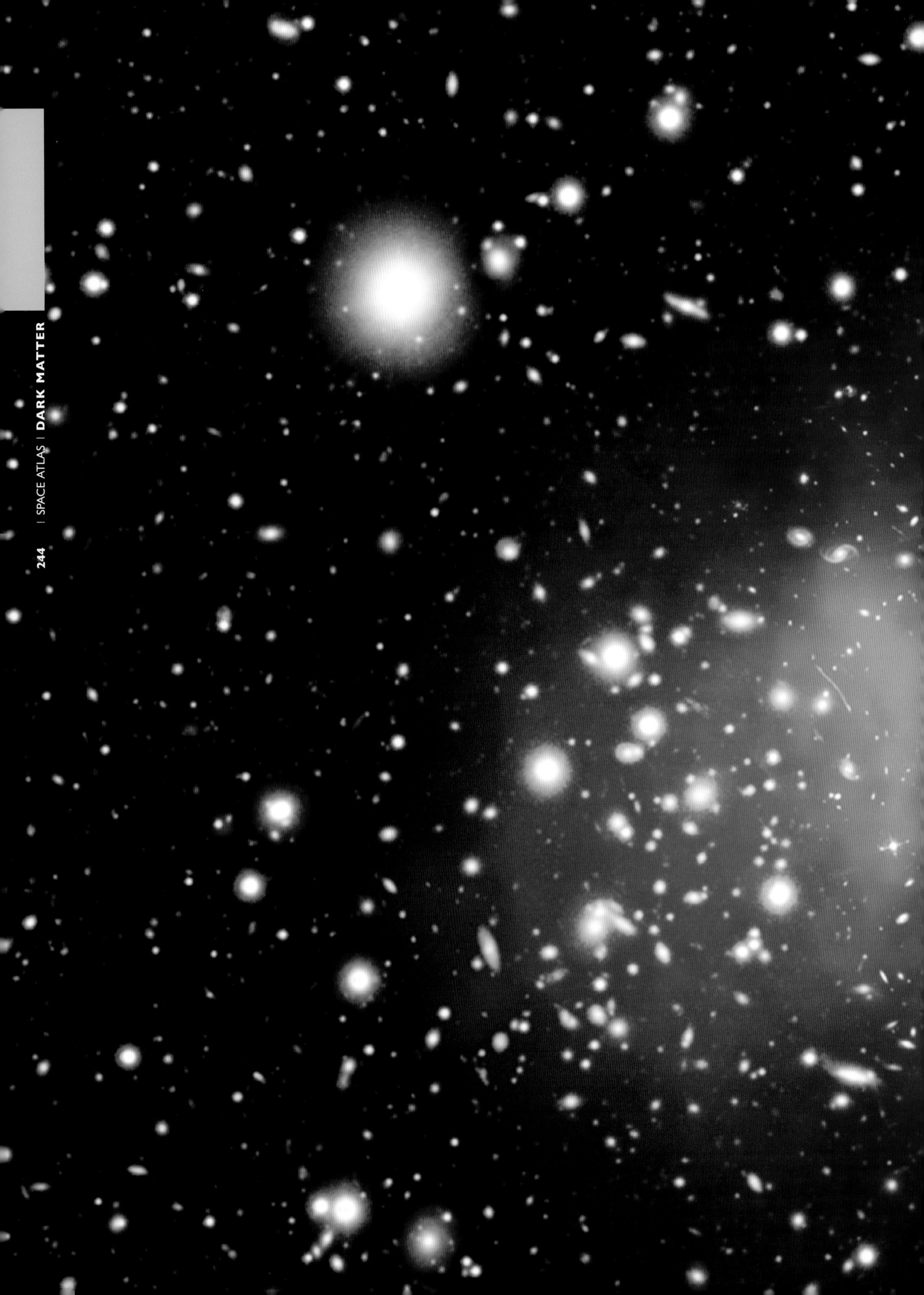

One of the most startling discoveries of the late 20th century was that ordinary matter, the stuff we're made of, is just a small part of our universe. We'll come back to this subject when we discuss the expansion of the universe, but here we will talk about the discovery of dark matter, another insight into our own insignificance. • To do this, we'll have to think about how galaxies rotate. Our galaxy, for instance, rotates like a giant pinwheel over periods of hundreds of millions of years, with our sun making a grand circuit every 220 million years or so.

DARK MATTER

[AN INVISIBLE GALACTIC HALO]

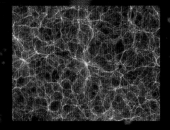

FIRST PROPOSED BY: 1932 (JAN OORT) AND 1933 (FRITZ ZWICKY)
DISCOVERED BY: VERA RUBIN, EARLY 1970S
HOW DISCOVERED: ROTATION SPEEDS OF GALAXIES

PROPORTION OF ORDINARY MATTER IN TODAY'S UNIVERSE: 5 PERCENT
PROPORTION OF DARK MATTER IN TODAY'S UNIVERSE: 23 PERCENT
PROPORTION OF DARK ENERGY IN TODAY'S UNIVERSE: 73 PERCENT
TYPES OF DARK MATTER: COLD, WARM, AND HOT
COLD DARK MATTER: OBJECTS MOVING AT ORDINARY SPEEDS
WARM DARK MATTTER: PARTICLES MOVING RELATIVISTICALLY
HOT DARK MATTER: PARTICLES MOVING NEARLY AT LIGHT SPEED
GALAXY MADE LARGELY OF DARK MATTER: VIRGOHI21

Blue regions in the Bullet galaxy cluster indicate dark matter.
(Inset) Simulation of dark matter distribution in local universe

One way that astronomers study the structure of galaxies is to look at the details of galactic rotation. The main tool in this study is something called a rotation curve, in which astronomers plot how fast a star is moving as a function of how far away from the galactic center it lies.

To get the concept of a rotation curve firmly in mind, we can look at a simple example of a rotating object—a merry-go-round. If you are standing near the inside of the platform, you will be moving fairly slowly, but as you move outward to the periphery, you will be whirling around faster and faster. The rotation curve for this situation shows the speed increasing steadily as you get farther away from the center. In the jargon of astronomers, this kind of rotation is called wheel flow, since it is characteristic of any solid rotating object.

In the central regions of a galaxy like the Milky Way, you would find something similar. Stars in the densely populated galactic center, locked together by gravity, exhibit wheel flow. But as you move outward, at some point (exactly where depends on the details of the structure of the galaxy), wheel flow stops. Beyond this point, we find that all the stars are moving at the same speed, regardless of how far they are from the center. They are like runners who have to stay in their lanes on a curving track. Those on the outside, because they have farther

VERA RUBIN

"It's not really good for an optical astronomer to learn that most of the universe is dark."

Vera Rubin is a pleasant, friendly woman who doesn't look at all like someone who could turn the world of astronomy upside down. About the only way you could guess that she is an astronomer is to notice that she occasionally wears a necklace made of stones that progress in color from red to blue, like the spectrum of light that members of her tribe love to study.

Rubin's love of the night sky started when, at the age of ten, her family moved to Washington, D.C. "From my bedroom window I could see the northern sky," she remembers, "and I suppose that's when my love of astronomy started." She attended Vassar College, an institution known for training astronomers, and then went on to graduate school at Cornell. It was work on her master's thesis that brought her into contact with the study of galaxies that was to dominate her professional career. Three weeks after she and her husband celebrated the birth of their first child, she presented the results of her thesis work

which involved a study of the motion of other galaxies, at a conference in Haverford, Pennsylvania. She still laughs at a headline in the local paper the next day. "It read, 'Young Mother Figures Center of Creation by Star Motion.' "

When the young Rubin family moved back to Washington, D.C., she enrolled in the doctoral program in astronomy at Georgetown University. Her adviser was the well-known physicist George Gamow, who taught at neighboring George Washington University. The two met at the Carnegie Institution of Washington. "I decided the first time I walked into the building that this was where I wanted to be," she recalls. She eventually joined the Carnegie staff in 1965.

So how does she feel about the discovery of dark matter? "Well, it's not really good for an optical astronomer to learn that most of the universe is dark," she says, laughing, then adds in a more serious tone: "I've been looking at the sky all my life, and I'm still looking."

to go and yet must move at about the same speed as those on the inside, will start to fall behind. This is why the spiral arms of our galaxy bend as they do.

What happens to the galactic rotation curve when we move still farther out? Here's a little thought experiment: Imagine getting so far away from the galaxy that the whole pinwheel shrinks down to a single faint point of light in the distance. The galaxy will still exert a gravitational force on you, so you will still be orbiting that distant point. Your situation, though, will be analogous to the planets circling the sun in our solar system. Just as Jupiter moves more slowly in its orbit than Mars, once you are this far away from a galaxy, you expect the speed

of the stars and dust clouds out there to get slower the farther away they are. This is called Kepler rotation, after Johannes Kepler (1571–1630), who discovered the law for planets in the solar system. It is represented by the downward-trending tail of the rotation curve.

A GRAVITATIONAL MYSTERY

As astronomers traced the rotation curve into the far outskirts of other galaxies, they fully expected to see that downturn. The problem was that they didn't. In the early 1970s Vera Rubin, then a young astronomer at the Carnegie Institution of Washington, began using advanced imaging instruments to measure the rotation

VERA RUBIN

curves of galaxies. Starting with the nearby Andromeda galaxy, she was amazed to observe that the curve stayed flat out to the limits of what she could measure—the stars kept moving at the same speed no matter how far they were from the galactic center. As she extended her search outward, galaxy after galaxy gave the same result, and by 1978 astronomers realized that their expectations about the rotation of galaxies were simply wrong.

In fact, scientists quickly realized that the only way to explain the observed rotation of the galaxies was to say that the visible part of a galaxy—the stars and dust clouds we've been exploring—was enclosed in a giant sphere of matter that we cannot see, but whose effects we can observe. The term "dark matter" was quickly applied to this new material. Whatever it is, it does not emit or absorb light or other electromagnetic

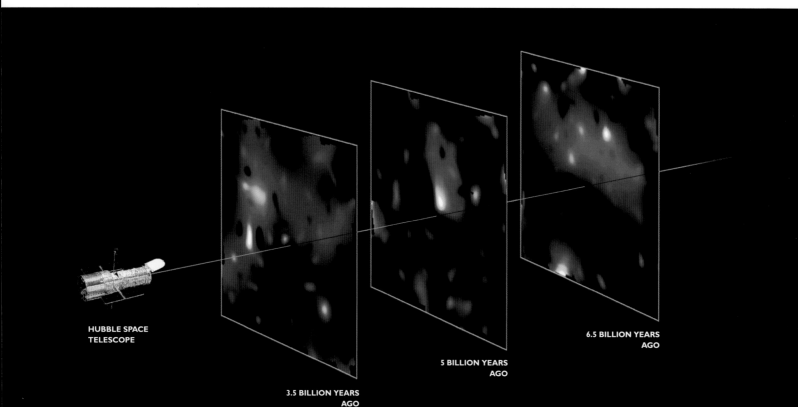

HUBBLE SPACE
TELESCOPE

3.5 BILLION YEARS
AGO

5 BILLION YEARS
AGO

6.5 BILLION YEARS
AGO

radiation, nor does it have any other kind of interaction with ordinary matter except for gravity. This means that while we can detect it by observing its gravitational effect on luminous matter (for example, the effect it has on the rotation curve), we cannot see it directly. Furthermore, calculations showed that for a galaxy like the Milky Way, more than 90 percent of the galaxy's total mass had to be in this same new (and unexpected) form.

Once the idea of dark matter surfaced, evidence for its existence in other venues showed up quickly. For example, there are stellar clusters in which individual stars are moving too fast to be held in by the gravitational attraction of the other stars. In these cases, some extra gravitational oomph is needed, and that, of course, is what dark matter supplies. As we shall see on page 293, today astronomers believe that dark matter makes up 23 percent of the mass of the universe. (For reference, luminous matter like stars makes up a bit less than 5 percent.)

So if there's so much of this stuff around, what is it? Theorists have not been slow in producing hypothetical answers to this question. The most popular candidates for the missing matter are as-yet-undetected objects known as weakly interacting massive particles, or WIMPs. The theories are interesting, but of course the only way we'll know for certain what dark matter is made of is to have someone identify the stuff in a laboratory.

SEARCHING FOR DARK MATTER

There are a number of dark matter searches going on around the world right now. The fact that this new form of matter permeates the galaxy means that the particles of dark matter, whatever they are, have to be passing through us all the time, leaving no record of their passage (remember, they don't interact much with ordinary atoms). In fact, since Earth is constantly moving around the sun, there must be a "dark matter wind" blowing past us constantly, much as one can feel an apparent wind when a car drives through still air.

Attempts to detect dark matter involve trying to see the rare events in which the dark matter wind jostles an atom or two in a detector. Many processes, particularly collisions with cosmic rays, can jostle atoms and mask the very faint dark matter signal. For this reason, dark matter searches tend to be located deep underground in mines or tunnels where the overlying rock provides a shield against this sort of interference. A typical dark matter search experiment is located in an abandoned iron mine in northern Minnesota. The mine, in the town of Soudan, is in the middle of the famous Mesabi Range that supplied much of the iron that built early 20th-century America. (The mine is now a state park and well worth a visit.)

In a chamber over 600 meters (2,000 ft) beneath the surface, a stack of ultrapure germanium and silicon wafers is cooled to within a fraction of a degree of absolute zero. Scientists hope that a few of the particles in the dark matter wind sweeping through the mine will interact with atoms in the stacks, producing vibrations that can be detected by instruments on the crystalline surface. To date no generally accepted dark matter discovery has been made, although some groups have reported tantalizing results. This means that we are in the position of knowing that 90 percent of our galaxy is made of some new material—but we don't know what it is.

A three-dimensional map of dark matter in the universe. Since looking at objects a great distance away is equivalent to looking back in time, we can trace out the evolution of this distribution over billions of years.

HELIX NEBULA

An infrared image of the Helix Nebula, taken by the Spitzer Space Telescope. Located some 700 light-years away, the planetary nebula is the gaseous remains of a dying star, seen as a bright white dwarf in the center of the image.

Our current picture of the universe was born outside of Los Angeles at a newly built telescope atop Mount Wilson. It was the brainchild of one man: Edwin Hubble. Working at the new instrument in the late 1920s, Hubble established that matter in the universe was organized into galaxies like the Milky Way: other "island universes" outside of our own galaxy. More important, he showed that those other galaxies were moving away from us—that the whole universe was expanding.

Given this fact, it's not hard to imagine running the film backward to a time when the entire universe was compacted into a single, unbelievably hot, dense point. This scenario, in which the universe began at a specific time in the past and has been expanding and cooling ever since, is known as the big bang.

THE UN

Confronted with this idea, three questions come to mind: (1) Is this theory correct? (2) How did the big bang begin? (3) How will the universe end?

In this section we will look at these questions, starting with a phenomenon known as the cosmic microwave background, which provides arguably the best evidence for the big bang. The second question will get us into one of the most exciting realizations in science—the idea that to study the largest thing we know about, the universe, we have to study the smallest things we know about, the elementary particles that comprise all of matter.

In recent years we have come to realize that most of the mass of the universe is in the form of a mysterious substance known as dark energy, and the fate of the universe depends on what that dark energy is. At the moment, we have no idea.

NIVERSE

THE UNIVERSE

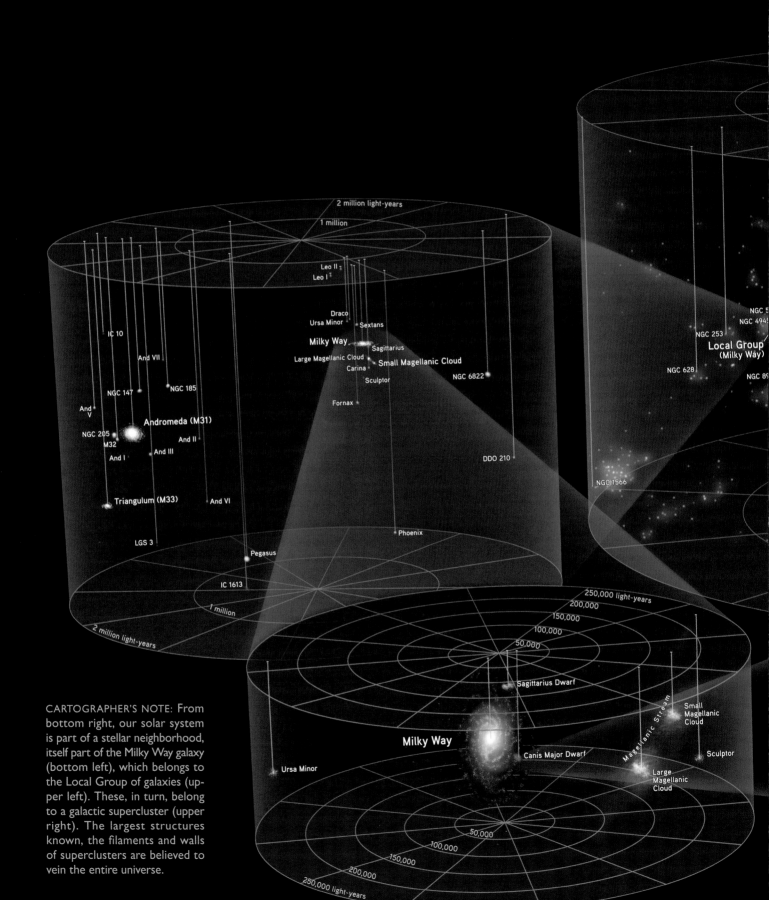

2 million light-years

1 million

Leo II
Leo I

Draco
Ursa Minor   Sextans

Milky Way

Sagittarius

Large Magellanic Cloud   **Small Magellanic Cloud**
Carina
Sculptor
NGC 6822

Fornax

IC 10

And VII

NGC 147   NGC 185

And
V
NGC 205   **Andromeda (M31)**
M32   And II
And I   And III

Triangulum (M33)   And VI

LGS 3

Pegasus

IC 1613

1 million

2 million light-years

DDO 210

Phoenix

NGC 5
NGC 494

NGC 253

Local Group
(Milky Way)

NGC 628

NGC 89

NGC 1566

250,000 light-years
200,000
150,000
100,000
50,000

Sagittarius Dwarf

Small
Magellanic
Cloud

Milky Way

Canis Major Dwarf

Magellanic Stream

Sculptor

Ursa Minor

Large
Magellanic
Cloud

50,000
100,000
150,000
200,000
250,000 light-years

CARTOGRAPHER'S NOTE: From bottom right, our solar system is part of a stellar neighborhood, itself part of the Milky Way galaxy (bottom left), which belongs to the Local Group of galaxies (upper left). These, in turn, belong to a galactic supercluster (upper right). The largest structures known, the filaments and walls of superclusters are believed to vein the entire universe.

Our own solar system and its galaxy are just a small piece in the hierarchical structure of the universe. Gravity binds together galaxies as well as massive galaxy clusters, which can contain thousands of component galaxies. More than 100 billion galaxies altogether are gathered in these clusters throughout the universe—all of them flying apart as the universe is expanding.

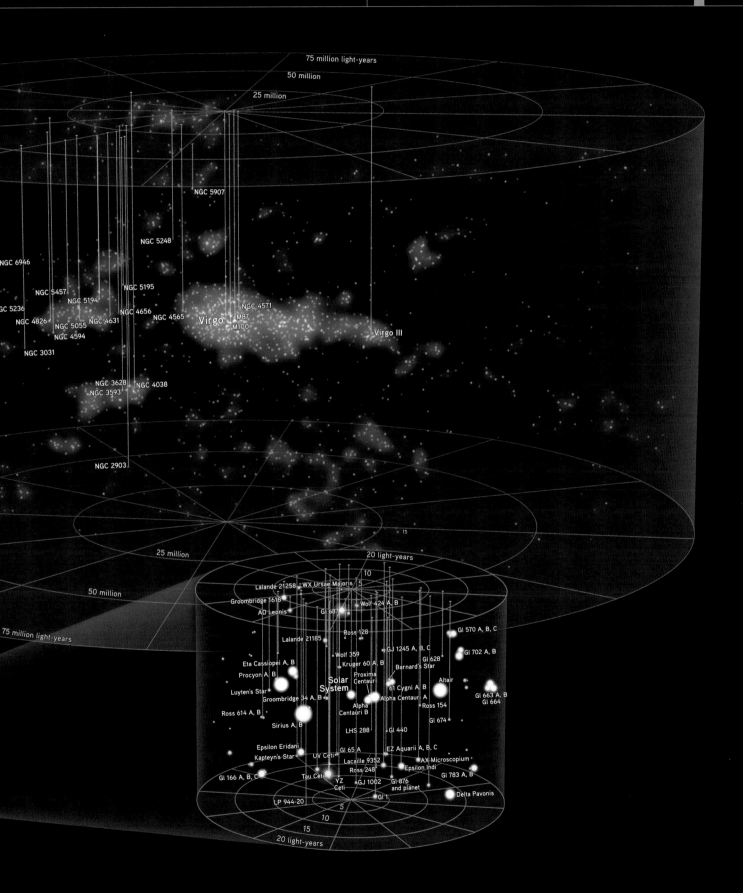

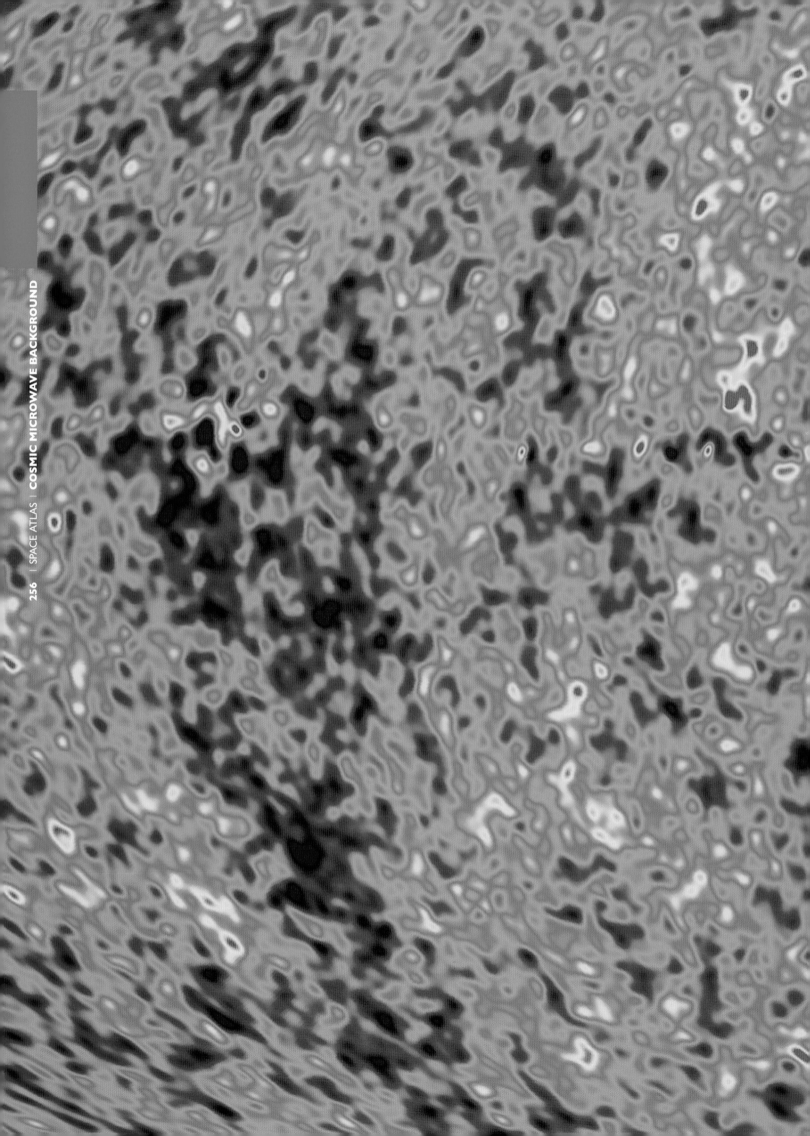

Sometimes the most important discoveries in science happen by accident. The discovery of the cosmic microwave background surely falls into this category. In the early 1960s transcontinental television transmissions were just becoming possible. The technologies were primitive by today's standards—someone who wanted to acquire a TV signal had to point a microwave receiver at the sky. And this raised the question of interference: Was there anything else out there that could also be sending microwaves into the receiver? • In 1964 two scientists at Bell Labs in New Jersey, Arno Penzias and Robert Wilson, began a survey of the microwave sky to resolve this issue. Using an old receiver to scan the sky, they systematically recorded the background microwave radiation that might interfere with TV reception.

COSMIC MICROWAVE BACKGROUND

[MESSAGE FROM THE DAWN OF TIME]

FIRST PREDICTED: 1948, BY RALPH ALPHER AND ROBERT HERMAN

DISCOVERED BY: ARNO PENZIAS AND ROBERT WILSON, 1964

1989: Cosmic Background Explorer (COBE) probe launched

2001: Wilkinson Microwave Anisotropy Probe (WMAP) launched

HOW DISCOVERED: MICROWAVE RADIO TELESCOPE

....

COSMIC BACKGROUND TEMPERATURE: 2.725 KELVINS (-270.425°C/-454.765°F)

VARIATIONS IN COSMIC BACKGROUND: 1 PART IN 14,000

BACKGROUND SHOWS UNIVERSE AT AGE: 380,000 YEARS

CURRENT AGE OF UNIVERSE: 13.7 BILLION YEARS

MILKY WAY'S VELOCITY RELATIVE TO BACKGROUND: 627 KM/SEC (390 MI/SEC)

Detail of microwave map of early universe. (Inset) Launch of the BOOMERANG telescope, which measured the microwave sky

A map of background radiation across the sky reveals tiny differences in the temperatures and density of the early universe.

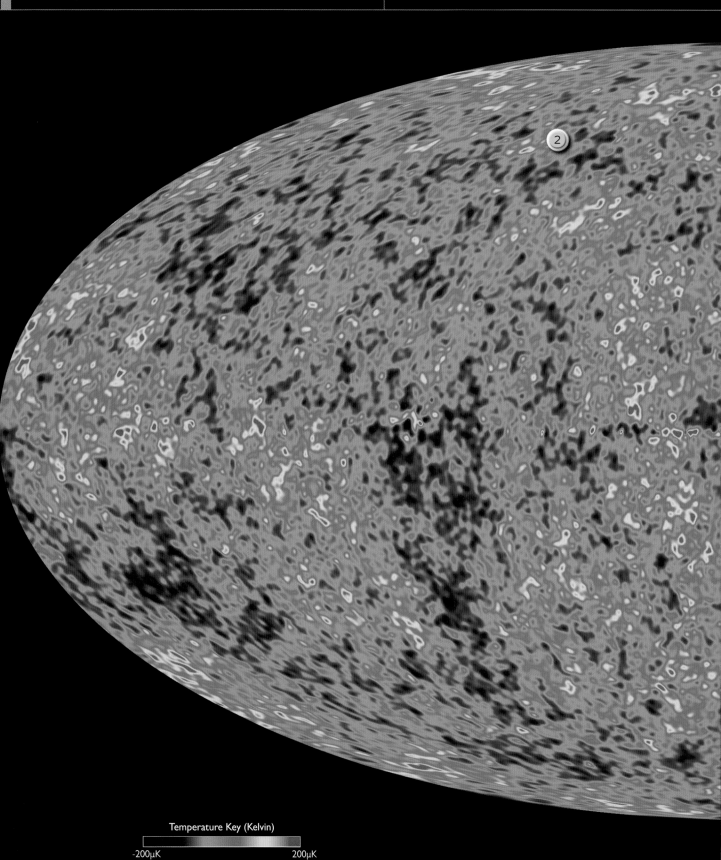

Temperature Key (Kelvin)

-200μK
200μK

TEMPERATURE REGIONS

1. WARM AREA
2. MID RANGE
3. COOLEST AREA

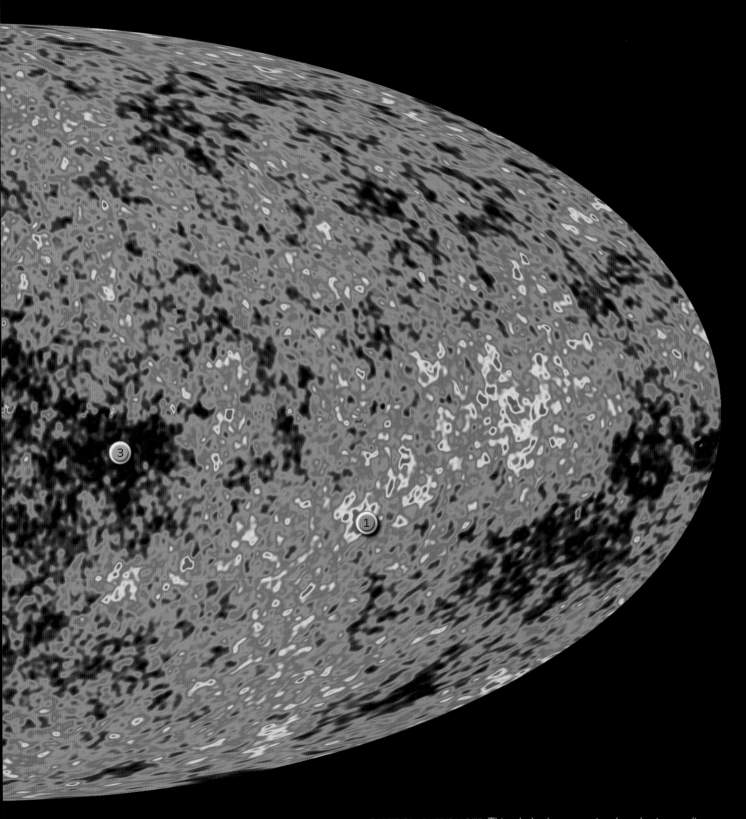

COSMIC MICROWAVE BACKGROUND | SPACE ATLAS | **259**

CARTOGRAPHER'S NOTE: This whole-sky map, using the galactic coordinate system, is presented in a Mollweide projection. Developed from data gathered by NASA's Wilkinson Microwave Anisotropy Probe, it presents differences in the temperature of background radiation using variable color. Warmer areas appear in red, cooler in blue.

As Penzias and Wilson began to scan the sky, a problem quickly arose. No matter which way they pointed their receiver, they detected a faint microwave signal, which showed up as a hiss in the earphones of their apparatus. In a situation like this, scientists always assume that there is something wrong with their electronics, and so Penzias and Wilson began the tedious job of finding the problem. They even evicted some pigeons that had nested in the receiver and had, as they put it delicately, coated part of the interior with a "white dielectric substance." Nothing helped—the hiss just wouldn't go away. Finally someone suggested that they go down the road to Princeton University, where cosmologists were working on the theory of something called the big bang. The theorists were suggesting that there ought to be a universal background of microwave radiation, an echo of the origin of the universe.

Let's take a moment to understand this prediction. If you watch the coals in a fire, you will notice that they change color as time goes by. They are white-hot when the fire is at its fiercest, but as the fire cools they turn red, then orange. The next day, you will feel warmth from the coals, even though they are no longer giving off visible light. From a physicist's point of view, the coals are giving off radiation whose wavelength increases as the temperature drops. In this case, we go from visible light (with a wavelength thousands of atoms across) to infrared radiation that we can feel but not see, but which has a longer wavelength than red light. The Princeton theorists were suggesting that the universe, like the coal in our example, started out very hot, and as it cooled the radiation associated with it went to longer and longer wavelengths. After billions of years, they argued, the radiation would be in the microwave range, with a wavelength of up to a meter. It was this radiation that Penzias and Wilson had discovered—that faint hiss was, in fact,

nothing less than the birth cry of the cosmos! For their work, Penzias and Wilson received the Nobel Prize in physics in 1978.

Earth's atmosphere is transparent to some microwaves—which is why satellite TV works—but it absorbs others. To get a complete picture of the cosmic microwave background, we need to get above the atmosphere. A series of satellites has been launched specifically for this purpose.

EYES IN SPACE

The first of these satellite observatories was the Cosmic Background Explorer (COBE), which was launched in 1989 and operated for four years. It established beyond a doubt that the microwaves were characteristic of an object at a temperature about 3 degrees above absolute zero—2.725 kelvins, if you want to be precise. (This is about minus 270°C/-454°F.) It was also clear that to an accuracy of about one part in a hundred thousand, the microwave radiation is isotropic—that is, it is the same in all directions. The COBE results were the most precise measurement that had ever been made in cosmology. The two men most involved in COBE—John Mather and George Smoot—received the Nobel Prize in physics in 2006 for their work.

Another important result that came out of COBE was the first measurement of the deviations from uniformity in the background radiation. Although the radiation was *almost* the same in all directions, it had tiny

(Upper illustrations) Seeing the microwave radiation from the early universe is like seeing light scattered across clouds. Our instruments see the radiation as it existed when it was scattered by free electrons, before the formation of atoms. (Lower illustration) The electromagnetic spectrum ranges from radio waves, whose wavelengths can be thousands of kilometers long, to gamma rays, whose wavelengths are smaller than the diameter of an atom. Microwaves are high-frequency radio waves whose wavelength can be up to a meter long.

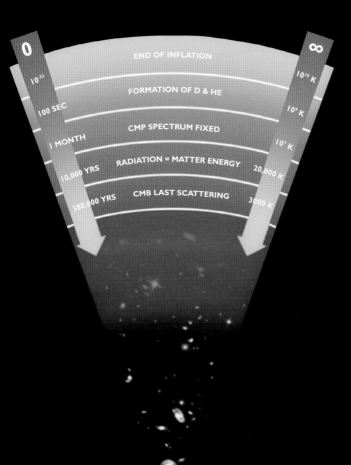

END OF INFLATION

10^{-32}

FORMATION OF D & HE

10^{19} K

100 SEC

CMP SPECTRUM FIXED

10^9 K

1 MONTH

RADIATION = MATTER ENERGY

10^7 K

10,000 YRS

20,000 K

CMB LAST SCATTERING

380,000 YRS

3000 K

SURFACE OF LAST SCATTER

The microwaves we see come from what is called the surface of last scatter, and they have been traveling freely since then.

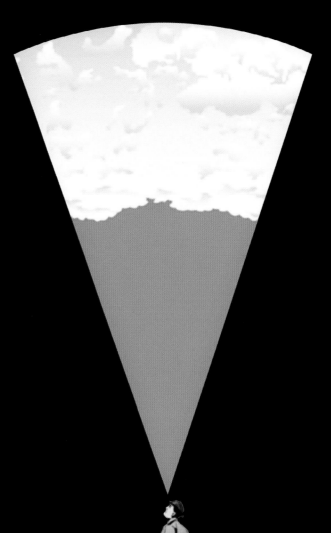

CLOUD SCATTER

Seeing microwaves from the surface of last scatter is like looking at light coming through a cloud—we see light that last scattered from the clouds' water molecules.

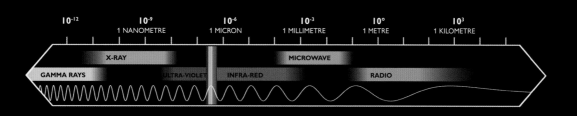

10^{-12}	10^{-9}	10^{-6}	10^{-3}	10^0	10^3
	1 NANOMETRE	1 MICRON	1 MILLIMETRE	1 METRE	1 KILOMETRE

X-RAY

MICROWAVE

GAMMA RAYS ULTRA-VIOLET INFRA-RED RADIO

1965 PENZIAS AND WILSON

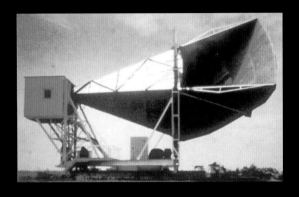

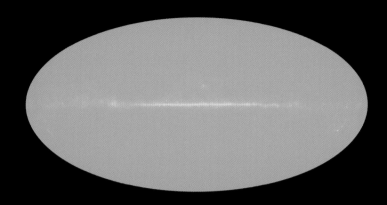

1992 COBE

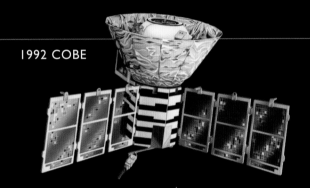

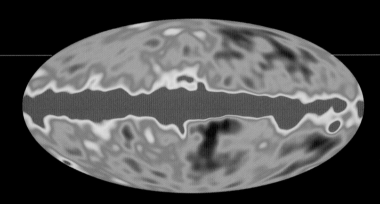

2003 WMAP

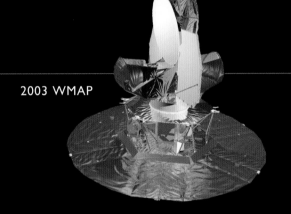

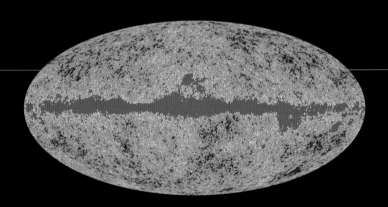

2009 PLANCK

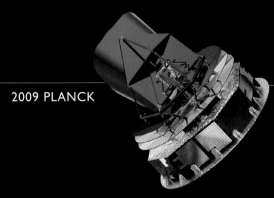

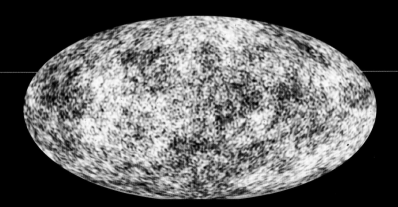

variations. The technical term for these deviations is microwave anisotropy. It turns out that these small differences are extremely interesting: They contain information about the state of the universe in the earliest stages of its existence. As we shall see on page 289, early in its life the universe was a plasma, something like the sun. Electrons and protons whizzed around independently, and if an electron happened to hook onto a proton to form an atom, the next collision would knock it free. Furthermore, there was a constant tug-of-war going on, with the particles trying to gather together under the influence of gravity and radiation breaking up the concentrations. When the universe was a few hundred thousand years old, the temperature dropped to the point where atoms could survive these collisions. At this point the universe became transparent, as will be explained on page 289. The radiation released at that time streamed outward and eventually became the cosmic microwave radiation we see today. Thus, when we see small differences in temperature, we are, in fact, seeing the matter concentrations in the universe as they were when it was a few hundred thousand years old—what one astronomer called the "ripples at the beginning of time." It is these tiny seeds that grew into the large-scale structures we see in the universe today.

21ST-CENTURY SCANS

The next satellite to study the microwave background was launched in 2001. It was named the Wilkinson Microwave Anisotropy Probe (WMAP) after prominent Princeton cosmologist David Wilkinson, who died a few years after launch. WMAP was not placed into orbit around Earth: It was sent on a three-month journey to a spot between Earth and the sun known as a Lagrange point. At these points, the gravitational forces of Earth and the sun balance in such a way that a satellite can remain in stable orbit. Lagrange points are becoming favorite places to park satellites because, in these locations, radiation from Earth does not interfere with the instruments.

WMAP did a much finer survey of the microwave sky. By comparing the precise data from WMAP with the predictions of their theories, cosmologists were able to provide solid evidence for the theories of the big bang that run through this part of the book. Many cosmologists, in fact, consider the ability of these theories to reproduce the uneven background radiation to be the best evidence we have for our picture of the structure and evolution of the universe. In addition, WMAP data allowed cosmologists to pinpoint the age of the universe at 13.7 billion years, with an accuracy of about 100,000 years. The satellite stopped taking data in August 2010, and by 2012 it had released most of the data it had collected.

In May 2009 the European Space Agency launched the next-generation microwave probe. It is named the Planck, after the early 20th-century German scientist Max Planck, one of the founders of quantum mechanics. The probe will provide a more detailed map of the microwave anisotropies, which will give cosmologists a chance to refine their theories. It will also provide measurements of the polarization of the microwaves, which theorists expect will give them deeper insights into the early development of the universe.

Quite a lot of science for what started as an attempt to get better TV pictures!

As instruments have improved, so have maps of the cosmic microwave background. At top left is the large land-based microwave receiver Penzias and Wilson used to discover the radiation. Below are the satellites that have been launched since 1992 to map the radiation in increasing detail. On the right are maps of the microwave sky as seen by each of these instruments; the map associated with Planck is a computer simulation.

We are so used to thinking of our sun as one star among billions in the Milky Way galaxy, and of the Milky Way as one among billions of galaxies in the universe, that it is likely that few readers even noticed when these notions were introduced without explanation in the last section. Yet there was a time when the notion that matter in the universe was bunched up into what we call galaxies was hotly debated. • There are many possible ways a universe might be organized. Matter could be scattered randomly throughout space; one central clump of matter might be surrounded by emptiness; or matter could clump into galaxies, with the galaxies themselves scattered about randomly.

THE GREAT GALAXY DEBATE

[EDWIN HUBBLE AND ISLAND UNIVERSES]

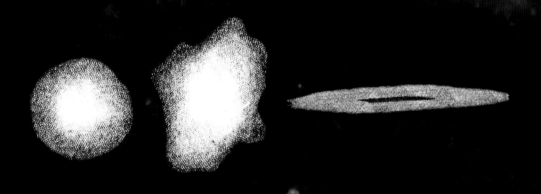

400S B.C.: Greek philosophers claim Milky Way is fiery vapor
1610: Galileo Galilei sees Milky Way is made of stars
1918: Harlow Shapley finds sun not at center of Milky Way
1920S: Edwin Hubble discovers galaxies are made of stars
1929: Hubble observes galaxies are receding
1932: Karl Jansky detects radio noise from center of Milky Way
1936: Hubble develops "tuning fork" galaxy classification
1939: Grote Reber finds radio source Cygnus A
1952: Milky Way's spiral arms are detected in radio
1970S: Vera Rubin discovers dark matter in galaxies
1990S: Hubble Space Telescope surveys deep field

Three galaxies known as ARP 274. (Inset) Three nebulae drawn by astronomer William Herschel

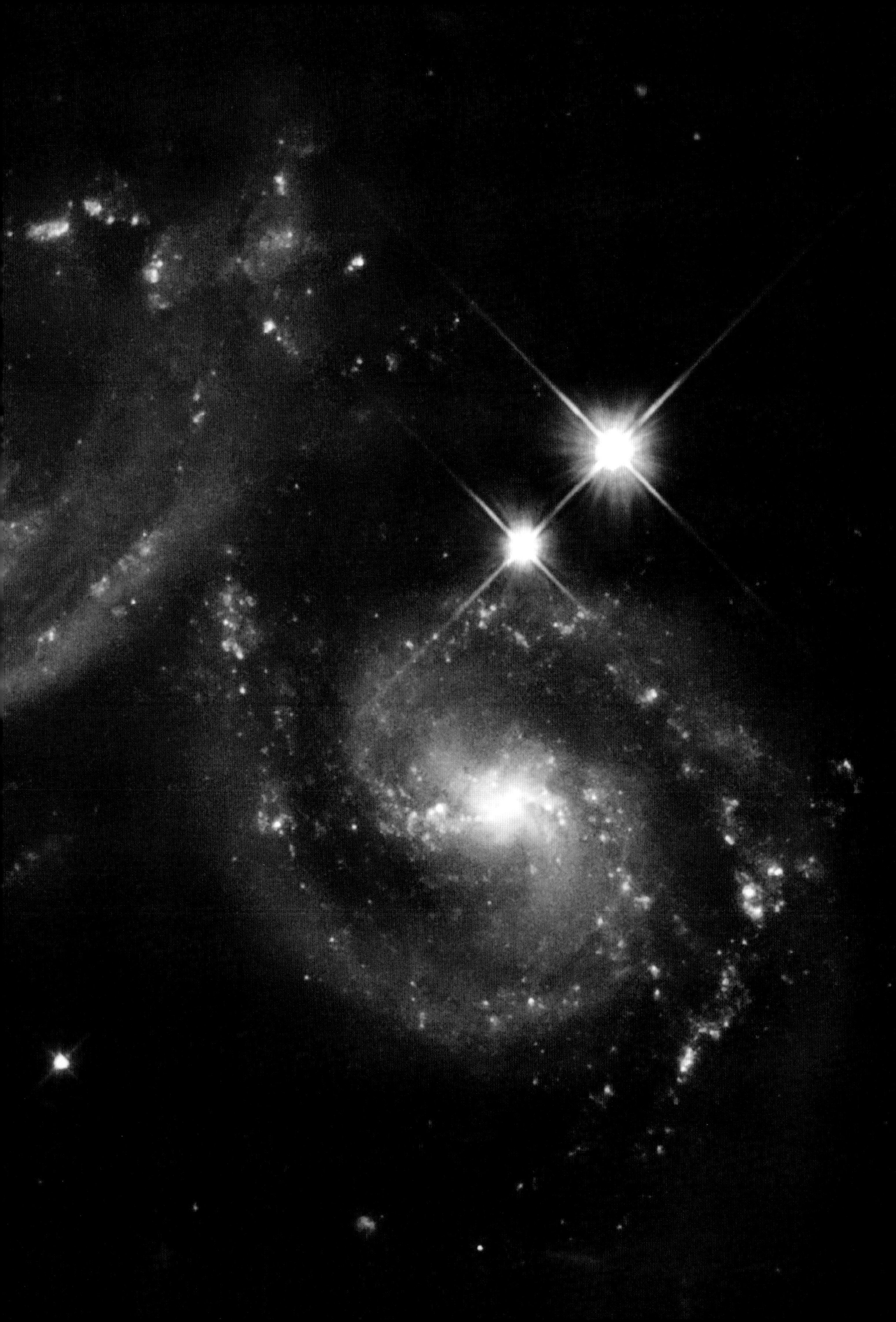

Understanding what different imagined universes might be like is the job of theoretical astrophysicists. On the other hand, we can live in only one of those possible universes, and finding out which one we actually inhabit is the job of observational astronomers.

What I am calling the Great Galaxy Debate was triggered by the presence in the sky of objects called nebulae. Nebula means "cloud" in Latin, and the name comes from the fact that when nebulae were first observed by early astronomers, they looked like smeared patches of light—luminous clouds. The question was simple: Were nebulae just clouds of luminous stuff inside the Milky Way, or were they other "island universes," far beyond our own galaxy? To answer this question, astronomers needed two things: They needed a telescope with a resolution that would allow them to see individual stars within the nebulae, and they needed a way of measuring the distance to those stars.

By the early 20th century, the distance measurement problem had been solved by Henrietta Leavitt (see page 194). Also, although it was far less obvious, the telescope problem was on its way to being solved as well.

A 100-INCH TELESCOPE

To understand the events that led to our current picture of the universe, we have to go back to the middle of the 19th century and meet one of the most remarkable men ever to appear on the American stage—Andrew Carnegie. Coming to America as a ten-year-old, he began his career delivering telegrams in Pittsburgh and wound up being one of the wealthiest men in America. Among other things, for example, he founded the company that eventually became United States Steel. Then, in what was an astonishing turn for a 19th-century robber baron, he wrote an essay known as "The Gospel of Wealth" in which he put forward the idea that once a man had acquired wealth, it was his duty to see that it was spent on solving important social problems. "The man who dies rich," he said, "dies disgraced." You have only to look at charitable trusts like the Bill and Melinda Gates Foundation to see the Carnegie legacy in operation today.

Andrew Carnegie founded the Carnegie Institution of Washington, which is dedicated to scientific research. One project in which he took a great personal interest was the construction of a major astronomical observatory on Mount Wilson, near Los Angeles. In the early years of the 20th century, this observatory housed the largest telescopes in the world. In 1919 Edwin Hubble (see sidebar opposite) joined the staff at Mount Wilson and proceeded to revolutionize our picture of the universe.

Using the observatory's magnificent new instrument, whose 100-inch mirror captured an unprecedented amount of light, Hubble began a systematic study of nebulae. The first thing he did was set up a classification scheme based on their appearance—a scheme still used today. More important, though, the new telescope could pick out individual Cepheid variable stars in nearby nebulae, which meant that Hubble could use Leavitt's standard candle method to find out how far away those nebulae were. The distances he measured turned out to be millions of light-years—much too great to have the nebulae included as parts of the Milky Way. By 1925 Hubble had established that we live in a universe in which matter is organized into galaxies.

Had this been the extent of Hubble's contribution, it would have ensured that his name would be in all the history books. There was another feature of his work, however, that led to the big bang picture of the universe—the picture that continues to dominate modern cosmology. To understand how he came to his theory, we have to make a slight diversion to discuss a phenomenon known as the Doppler effect.

REDSHIFT

You've had firsthand experience of the Doppler effect if you've ever listened to a car horn as the car drove past you on the highway. You probably noticed that the pitch of the horn dropped as the car went by. When the car is standing still, the crests of the sound wave move out uniformly in all directions, and everyone hears the same pitch. If the car is moving, however, the center for each outgoing wave is at the spot where the car was when that particular crest was emitted. This means that someone standing in front of the car will see the crests bunched up (that is, hear a higher pitch), while someone standing in back will see the crests stretched out (that is, hear a lower pitch). This shift in pitch is the Doppler effect, and it will be seen for any type of wave emitted by a moving source.

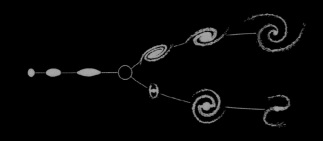

EDWIN HUBBLE

"The history of astronomy is a history of receding horizons."

Born in Missouri to middle-class parents in 1889, Edwin Hubble grew up in Wheaton, Illinois, then as now a railroad suburb of Chicago. At the University of Chicago he was both a top student and a standout athlete, playing on its championship basketball team in 1908. There is a legend in the physics community that Hubble was a high-level amateur boxer and actually had to make a serious decision about whether to stay in school or turn professional. He obviously took the former course, and he won a Rhodes scholarship to Oxford, where he studied law and Spanish. On his return to the United States, he spent a year teaching Spanish at a high school and was admitted to the bar in Kentucky, but he decided to go back to Chicago to study astronomy. While working on his Ph.D., he joined the staff at Yerkes Observatory in Wisconsin and came to the attention of a number of prominent astronomers. He gave his Ph.D. defense in 1917 and volunteered for the Army the next day. When he left the Army as a major in 1919 at the end of World War I, he was offered a position at the new Mount Wilson Observatory and the rest, as they say, is history. Working steadily until his death in 1953, Hubble transformed our understanding of the universe.

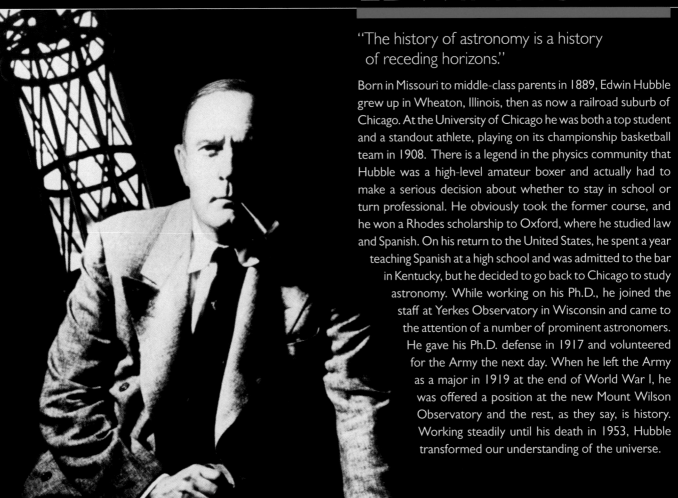

EDWIN HUBBLE IN FRONT OF THE MOUNT WILSON TELESCOPE. (TOP) HUBBLE'S "TUNING FORK" DIAGRAM OF GALAXIES

DISTANT GALAXIES LOOK
REDDER
AS THEY RECEDE FROM US

Astronomers before Hubble had noted that the light from the stars in nebulae was shifted to the longer wavelength (red) end of the spectrum, indicating that the crests were being stretched out. This means that the nebulae in question are moving away from us. Those astronomers didn't know how far away the various nebulae were, however, so they couldn't see any systematic relationships among the observed redshifts. Once Hubble figured out the distances, he was able to note that the larger the redshift (that is, the faster the galaxy was moving away from us), the farther away it was. This is usually expressed in an equation known as Hubble's law:

$$V = H \times D,$$

where V is the speed of the receding galaxy, D the distance to it, and H a number known as Hubble's constant. This equation tells us that if we look at two galaxies and one is twice as far away from us as the other, then the farther galaxy will be moving away from us twice as fast as the closer galaxy.

We'll discuss some of the astonishing consequences of Hubble's discovery in the next section.

The Doppler effect causes light from objects that are receding from us, such as galaxies, to be shifted to longer wavelengths (redshift, middle illustration). Light from approaching objects is shifted to shorter wavelengths (blueshift, bottom illustration).

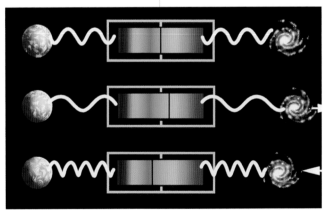

An artist's conception of the cosmic redshift depicts nearby galaxies in white light, but more distant galaxies in progressively redder light as they get farther away (and move more rapidly away from us).

When we look out at the universe, we see many different kinds of galaxies. Most are like the Milky Way—calm, homey places where stars slowly turn the primordial hydrogen of the big bang into the other chemical elements. Only the occasional supernova (see pages 226–31) provides a bit of excitement. Astronomers, following Hubble, classify these normal galaxies by their shapes. With many gradations within each category, these shapes are spiral, elliptical, and irregular. • The Milky Way is one type of spiral galaxy. The best explanation of the spiral appearance of these galaxies states that pressure waves sweep around them, more or less like water sloshing in a bathtub. These waves trigger the formation of bright new stars in rotating arms.

GALACTIC ZOO

[GALAXIES, CLUSTERS, AND SUPERCLUSTERS]

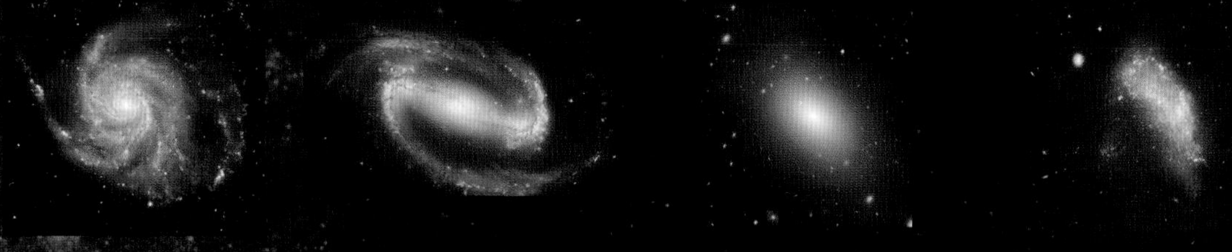

SPIRAL

BARRED SPIRAL

ELLIPTICAL

IRREGULAR

STARS IN GALAXIES: 10 MILLION TO 100 TRILLION

MOST MASSIVE GALAXY: M87, 6 TRILLION SOLAR MASSES

LEAST MASSIVE GALAXY: WILLMAN 1, ABOUT 500,000 SOLAR MASSES

NEAREST GALAXY TO MILKY WAY: CANIS MAJOR DWARF, 25,000 LIGHT-YEARS

GALAXY CLUSTER MILKY WAY BELONGS TO: LOCAL GROUP

NUMBER OF GALAXIES IN LOCAL GROUP: 30–50

LOCAL GROUP LARGEST MEMBERS: MILKY WAY, ANDROMEDA GALAXY

LOCAL GROUP IS PART OF: VIRGO CLUSTER

NUMBER OF GALAXIES IN THE VIRGO CLUSTER: 1,200 TO 2,000

DISTANCE TO THE VIRGO CLUSTER CENTER: 54 MILLION LIGHT-YEARS

MASS OF THE VIRGO CLUSTER (SUN = 1): 1.2 QUADRILLION

Nearby spiral galaxy M74.
(Insets) Various galaxy shapes

Seventy-five nearby galaxies are organized here according to Edwin Hubble's "tuning fork" classification.

KEY TO THE GALAXIES

Unbarred Spirals (**SA**)

Ellipticals (**E**)

E0   E2   E4   E6

The number is derived from a galaxy's eccentricity; zero being round to six more flattened.

S0

Intermediate Spirals (**SAB**)
(Lenticulars)

Stages 0   a   ab   b   bc   c   cd   d   dm   m

Barred Spirals (**SB**)

Irregulars (I)

IA   Unbarred
IB   Barred
IAB Mixed
I0   Non-magellanic
Im   Magellanic

Other Terms

DDO David Dunlop Observatory Catalog of Galaxies
Ho   Holmberg
IC   Index Catalog
M   Messier Catalog
Mrk Markarian

NGC New General Catalog
p   Pecular
Tol Tololo Galaxies
?   Doubtful

CARTOGRAPHER'S NOTE: The Spitzer Space Telescope collected these images in infrared light as part of its Spitzer Infrared Nearby Galaxies Survey. They are placed on the tuning-fork map according to their visible-light attributes.

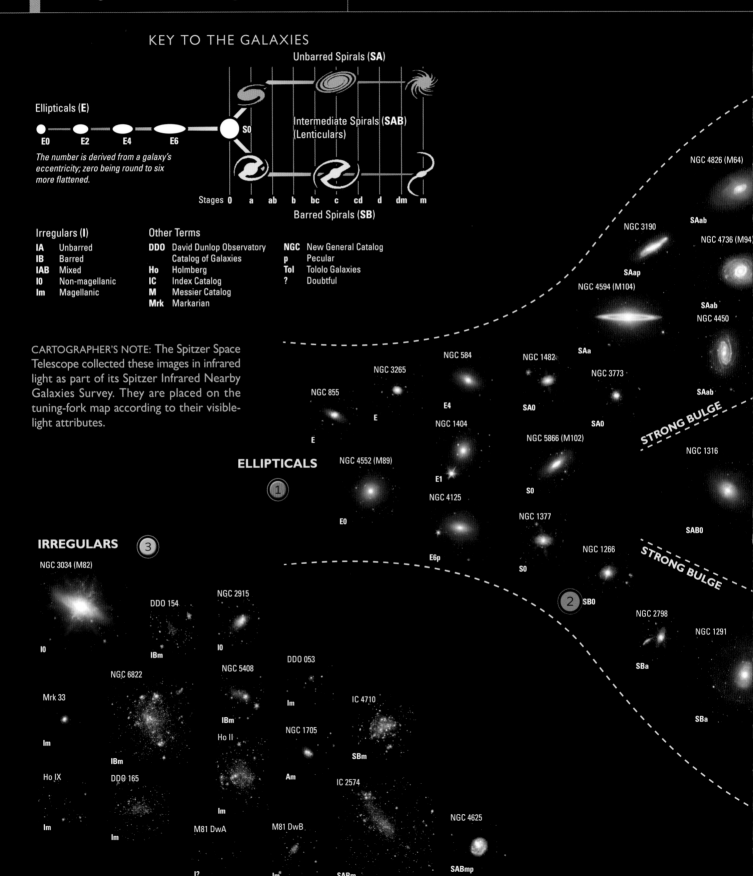

NGC 4826 (M64)

SAab

NGC 4736 (M94)

NGC 3190

SAap

NGC 4594 (M104)

SAab

NGC 4450

NGC 584

NGC 1482

SAa

NGC 3773

NGC 3265

SAab

NGC 855

E4

SA0

STRONG BULGE

E

NGC 1404

SA0

SAab

E

NGC 5866 (M102)

NGC 1316

ELLIPTICALS

①

NGC 4552 (M89)

E1

S0

NGC 4125

NGC 1377

E0

NGC 5866

SAB0

IRREGULARS   ③

E6p

S0

STRONG BULGE

NGC 1266

NGC 3034 (M82)

② SB0

NGC 2798

DDO 154

NGC 2915

NGC 1291

I0

IBm

I0

SBa

NGC 6822

DDO 053

NGC 5408

Mrk 33

Im

IC 4710

SBa

Im

IBm

NGC 1705

Ho II

Ho IX

DDO 165

SBm

Im

Am

IC 2574

Im

NGC 4625

Im

M81 DwA

M81 DwB

I?

Im

SABm

SABmp

KEY FEATURES

(1) ELLIPTICAL GALAXIES: These are made primarily of old stars (blue).

(2) UNBARRED SPIRAL GALAXIES: Red and green colors mark areas of star birth.

(3) IRREGULAR GALAXIES: These are often found in colliding clusters.

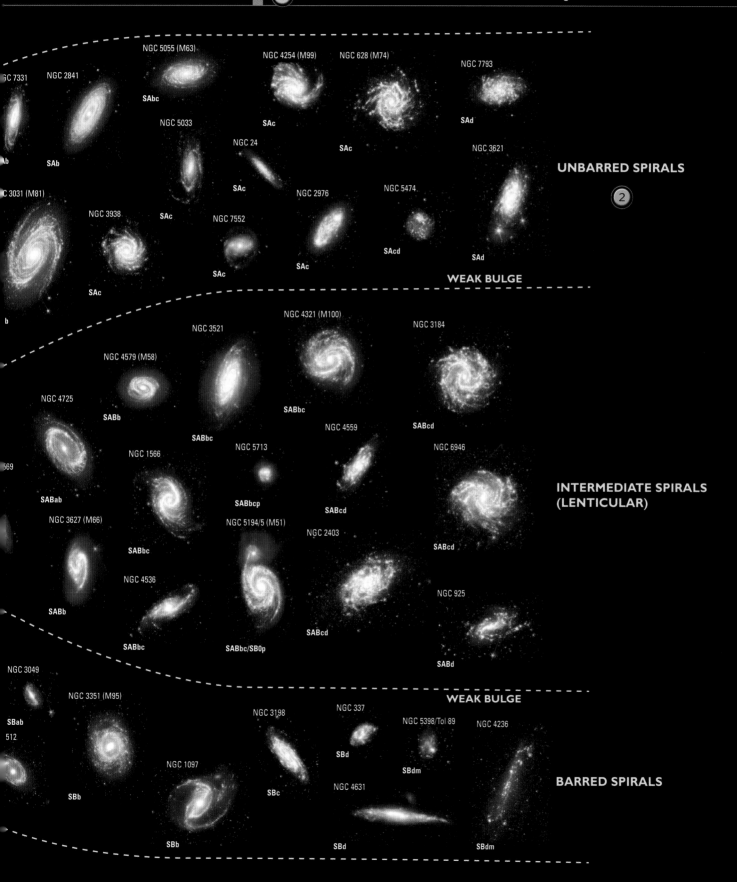

UNBARRED SPIRALS

(2)

NGC 7331

NGC 2841

NGC 5055 (M63)

SAbc

NGC 4254 (M99)

NGC 628 (M74)

NGC 7793

SAd

NGC 5033

SAc

NGC 24

SAc

NGC 3621

Ab

SAb

SAc

NGC 2976

NGC 5474

C 3031 (M81)

NGC 3938

SAc

NGC 7552

SAcd

SAd

b

SAc

SAc

SAc

WEAK BULGE

NGC 4321 (M100)

NGC 3184

NGC 3521

NGC 4579 (M58)

SABbc

NGC 4725

SABb

SABcd

SABbc

NGC 4559

NGC 6946

NGC 1566

NGC 5713

**INTERMEDIATE SPIRALS
(LENTICULAR)**

669

SABab

SABbcp

SABcd

NGC 3627 (M66)

NGC 5194/5 (M51)

NGC 2403

SABbc

NGC 4536

SABcd

SABb

NGC 925

SABbc

SABbc/SB0p

SABd

NGC 3049

WEAK BULGE

NGC 3351 (M95)

NGC 3198

NGC 337

SBab

NGC 5398/Tol 89

NGC 4236

512

SBd

NGC 1097

SBdm

BARRED SPIRALS

SBb

SBc

NGC 4631

SBb

SBd

SBdm

As well as spiral galaxies, the universe holds elliptical galaxies, which as the name implies are basically ovoid blobs of stars. They come in all sizes, from dwarfs to galaxies much larger than the Milky Way. Unlike what we see in spiral galaxies, there appears to be very little star formation in ellipticals.

Finally, irregular galaxies are the leftovers—I think of them as being like the pieces of dough that are left after the cookie cutters have done their work. Irregulars make up most of the galaxies in the universe.

ACTIVE GALAXIES

A small percentage of galaxies are not like our own Milky Way, though. These galaxies are wild and violent places, torn by massive explosions, sometimes ejecting huge jets of hot gases hundreds of light-years into intergalactic space. These are the so-called active galaxies and, like normal galaxies, they come in many shapes and forms. They all appear to have what astronomers call an active galactic nucleus, a small region at the core of the galaxy that is the source of its prodigious energy output. Our best current theory is that each of these galaxies has a large black hole at its center, and that nearby matter falling into the black hole bunches up to form a very hot disk. It is this disk, presumably, that generates both the radiation and the jets in active galaxies.

The most important active galaxies are a class of very energetic objects known as quasars (short for "quasi-stellar radio source"). As their name implies, quasars usually emit a lot of their energy as radio waves and relatively little as visible light. In fact, after they were first discovered in the 1950s, it took over a decade for astronomers to identify a visible object associated with a quasar. When they did, they were surprised to see that the light from the quasar was shifted far to the red, indicating that it was billions of light-years from Earth.

Quasars can be seen at great distances. Scientists believe that, like all active galaxies, quasars are powered by material falling into a central black hole.

CLUSTERS AND SUPERCLUSTERS

Galaxies are not scattered randomly in space, but tend to be bunched together in groups and clusters. The Milky Way, for example, is part of what is called the Local Group. The Local Group includes one other large spiral galaxy, the Andromeda Nebula, and some 30-plus irregular galaxies. It is about ten million light-years across.

The Local Group, in turn, is part of a larger structure known as the Virgo supercluster. This supercluster contains at least a hundred groups and clusters and is about 110 million light-years across. It is one of literally millions of superclusters in the universe. The presence of clusters and superclusters provides evidence for the existence of dark matter (see page 249). If you add up the gravitational force of all of the stars in all of the galaxies, it turns out that the force is too small to hold the clusters and superclusters together. Only if you add in dark matter can the structures keep from flying apart.

So it turns out that the region of the universe in our immediate vicinity is lumpy. The question is whether it remains that way when it's examined on a larger scale. Any attempt to answer this question would involve producing a large-scale, three-dimensional map of the galaxies in the universe—a difficult feat, because the Cepheid variable method Hubble used would be just too cumbersome for this big task. But all is not lost: Hubble's law (see page 269) actually provides a quick way of estimating these measurements. If we measure the redshift of the light from the galaxy, we can find the velocity with which the galaxy is moving away from us and then, from Hubble's law, its distance.

The first of these redshift surveys was completed in 1982 by astronomers Margaret Geller and John Huchra of the Harvard-Smithsonian Center for Astrophysics. Instead of discovering that galactic graininess merged into uniformity, they found a completely unexpected large-scale structure, a finding confirmed by many studies since. The easiest way to visualize the large-scale structure of the universe is to picture a huge mound of soap suds that you cut with a knife: The result would be thin films of soap surrounding empty bubbles. In the same way, the redshift surveys tell us that galactic superclusters are arranged along thin sheets surrounding empty spaces known as voids. The largest structure in the known universe, dubbed the Great Wall, is a filament of superclusters 500 million light-years long, 200 million light-years wide, but only 15 million light-years thick.

Thus, the structure of the universe remains interesting out to the largest scales we can see.

KEEN EYES IN SPACE

With the possible exception of Galileo's telescope, the Hubble Space Telescope is arguably the most important astronomical instrument ever built. Launched in 1990, it resides in a low Earth orbit, a bit more than 160 kilometers (100 mi) up. It is equipped with a telescope whose mirror is 2.4 meters (a little less than 8 feet) across.

The Hubble cannot see farther than other instruments—that distinction will always belong to the largest "light bucket" on Earth. But because the Hubble is above the planet's distorting atmosphere, it can see objects in much more detail. In fact, some writers have suggested that it be thought of as an "astronomical microscope" rather than as a telescope. Additionally, the Hubble can see radiation in the near ultraviolet and near infrared—frequencies that are absorbed in Earth's atmosphere.

The instrument encountered problems soon after launch, when it was discovered that some of its optics had been installed improperly. A repair mission by astronauts fixed the problem, and the Hubble has since then posted an impressive array of achievements. They include the following:

- A refined estimate of the distance to stars that allows us to calculate the age of the universe to an accuracy of 10 percent
- The discovery of the existence of dark matter
- The discovery of galactic black holes

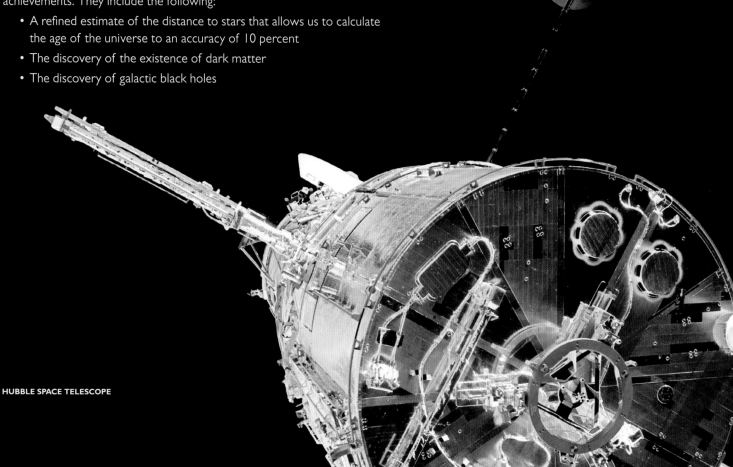

HUBBLE SPACE TELESCOPE

Once Hubble had established the existence of other galaxies and the universe's expansion, it became possible to talk with more confidence about how the universe began. The first point to remember is that the universe's beginning—the big bang—was not an explosion like a bomb blasting out into surrounding air. It was, instead, an expansion of space itself. • An analogy: Imagine that you are making raisin bread with a special transparent dough. If you were standing on any raisin as the dough was rising, you would see the other raisins moving away from you, because the dough between you and each of them is expanding. A raisin that started out twice as far away from you as another would be moving away twice as fast, because there is twice as much dough between you and the farther raisin.

BIG BANG

[THE BEGINNING OF SPACE AND TIME]

BIG BANG

BIG BANG OCCURRED: 13.7 BILLION YEARS AGO
MATTER DOMINATES: FIRST 70,000 YEARS AFTER BIG BANG
HYDROGEN AND HELIUM FORM: 380,000 YEARS AFTER BIG BANG
FIRST STARS: ABOUT 100 MILLION YEARS AFTER BIG BANG
FIRST GALAXIES: 600 MILLION YEARS AFTER BIG BANG
RECESSION RATE (CURRENT): 70.4 KM/SEC/MEGAPARSEC
(43.7 MI/SEC/MEGAPARSEC)
KEY BIG BANG PREDICTION: MICROWAVE BACKGROUND SIGNAL

1912: First evidence seen (but not understood)
1929: Edwin Hubble observes galaxies are receding
1949: "Big Bang" name given by Fred Hoyle
1964: Detection of microwave background signal

Computer simulation of the big bang.
(Inset) Diagram of expanding universe

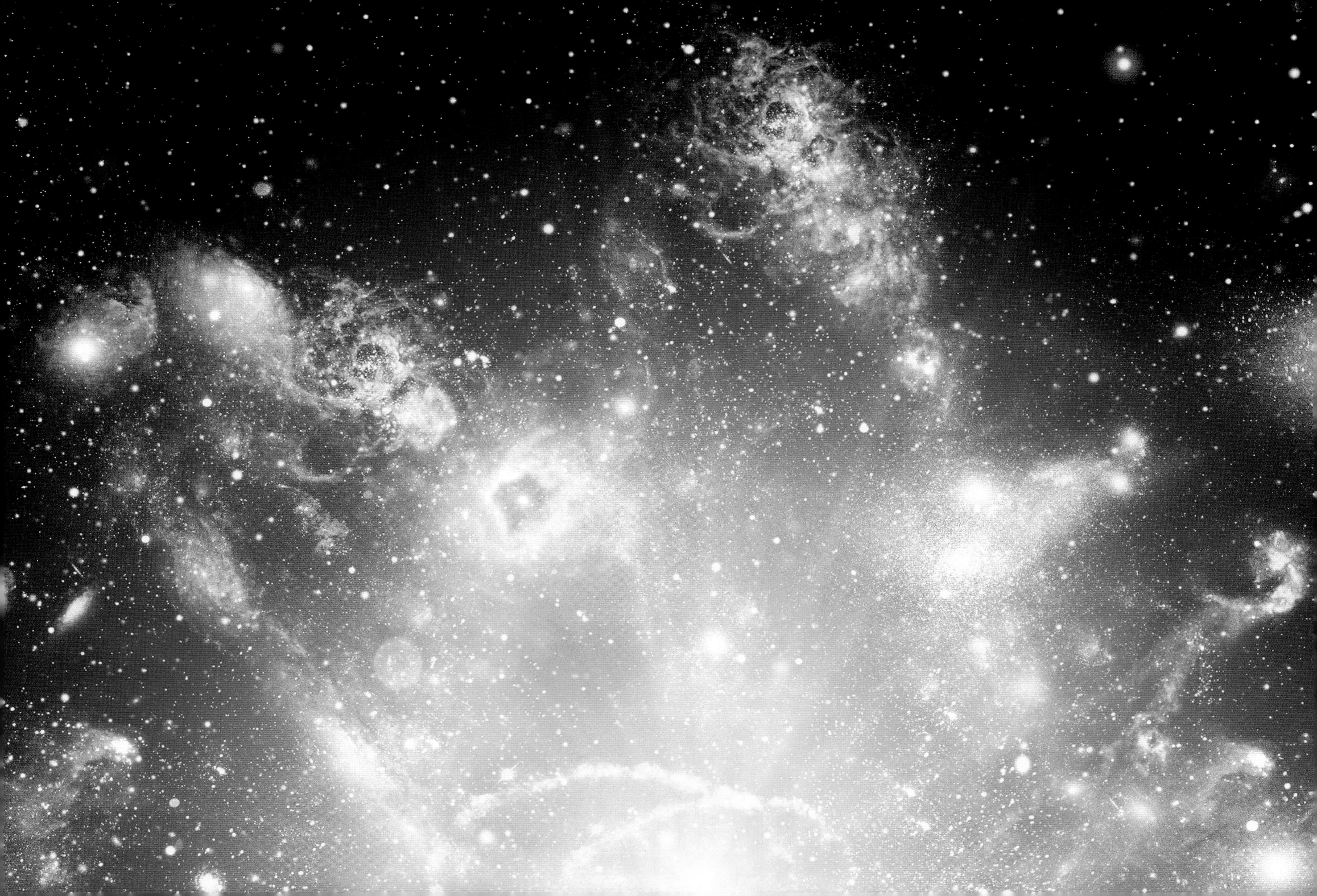

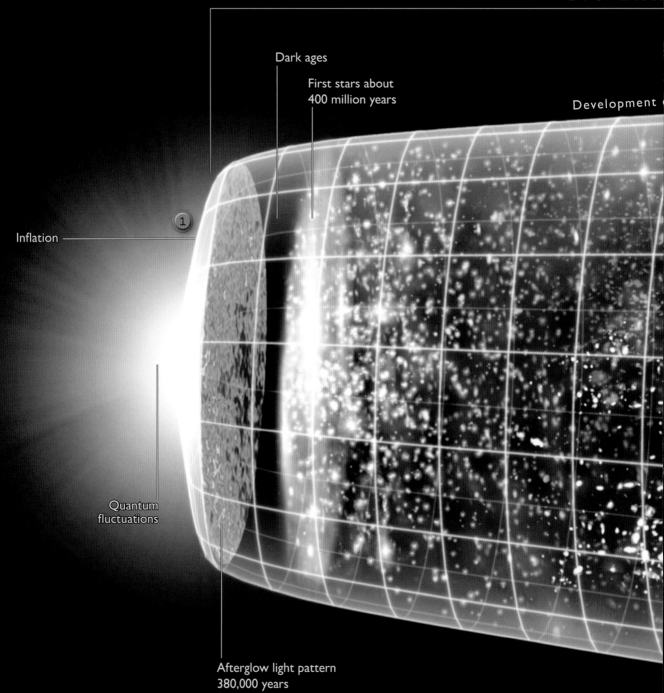

Dark ages

First stars about
400 million years

Development

Inflation

① 1

Quantum
fluctuations

Afterglow light pattern
380,000 years

CARTOGRAPHER'S NOTE: A timeline map of the expansion of the universe
includes data from the Wilkinson Microwave Anisotropy Probe. The verti-
cal grid represents the size of the universe. From a period of rapid inflation
after the big bang, the universe grew steadily until recently, when dark

KEY FEATURES

1. Inflationary period
2. Dark energy speeds up expansion
3. Wilkinson Microwave Anisotropy Probe

XPANSION   **13.7 Billion Years**

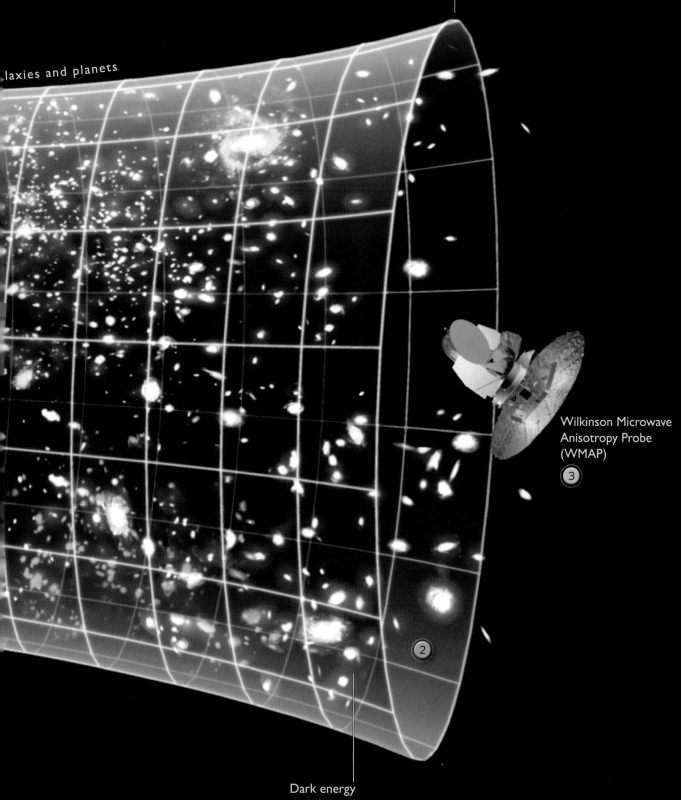

laxies and planets

Wilkinson Microwave
Anisotropy Probe
(WMAP)
3

Dark energy
accelerated
expansion

f you substitute galaxies for raisins in the big bang analogy, you have exactly what Hubble saw—a universal expansion. And just as no raisin is actually moving *through* the dough in our analogy, galaxies in the Hubble expansion are being carried along by the expansion of space itself, not moving through space.

It is easy to believe that the fact that everything is moving away from us means that we must be at the center of the universe, but thinking about our analogy can help clarify that as well. Stand on any raisin in the bread and you will see yourself as being stationary while all of the other raisins move away from you. Every raisin, in other words, sees itself as the center of a universal expansion, which means that while we on Earth may indeed see ourselves as the center of the Hubble expansion, so does everyone else, anywhere in the universe. In the words of the 15th-century theologian Nicolas of Cusa, "The universe has its center everywhere and its edge nowhere."

RUNNING THE FILM BACKWARD

We can learn more about our universe by thinking about the Hubble expansion as a film and then imagining that we run the film backward. In our reversed movie, the universe will keep shrinking down—eventually, in principle, reaching a single point. In other words, our universe had a definite beginning: a moment about 13.7 billion years ago, to be exact.

The fact that the universe began at a specific time in the past has important philosophical consequences. Before we learned about the Hubble expansion, we could imagine that the universe was simply eternal, with no beginning, no end, and no change. Or it could have been cyclic, which would be another kind of eternal universe. Or it might have been linear, with a beginning and an end. The only way to decide which of these options

actually describes the universe is to make observations, which is, of course, exactly what Hubble did. We live in a universe that definitely had a beginning. Deciding what its end will be is a little more complicated, and we'll discuss the current thinking on this subject in the section "End of the Universe" (see pages 290–93).

It is a general property of materials that they tend to heat up when they are compressed and cool when they expand. Thus, we would expect that in its earlier stages the universe was hotter than it is now, simply because it was smaller and more compressed. In other words, we would expect that the universe had a hot beginning and has been cooling ever since. And in fact, we have found evidence for this in the cosmic microwave background (see page 260).

CONDENSATIONS

Before we discuss the beginnings of the universe in the next sections, it's important to understand an important concept: that of transitions, or "condensations." The fact that the universe started out at a high temperature can give us a sense of how it developed in its earliest stages. Here's another analogy that may help you picture this: Imagine that you keep steam at a very high temperature and pressure and then release it suddenly. The steam expands, and cools as it does so, but at 100°C (212°F) something important happens. At this temperature, the steam condenses into water droplets. This pattern—long periods of expansion and cooling with sudden changes in the basic structure of the system—is what we see in the early development of the universe. Let's look at a key condensation—the formation of nuclei at about three minutes—to get a sense of how these transitions worked.

Before the universe was three minutes old, matter existed in the form of free protons and neutrons (the

An artist's two-dimensional conception of the big bang over time ranges from the extreme energies of the beginning, at the white-hot center, to the cooler realms of later millennia, when matter began to condense into stars and galaxies.

particles that make up the nuclei of atoms) and free electrons. If a proton and a neutron came together to form a simple nucleus, the next collision that nucleus experienced would be so violent that it would be torn apart. At three minutes, however, the temperature had fallen to the point that the free protons and neutrons could start to build nuclei. The entire constitution of the universe changed suddenly as a result of this condensation.

On page 230 we pointed out that all of the universe's heavy elements were made in reactions in supernovae. Now that we understand how nuclei were made in the big bang, we can see why this is necessary. Before three minutes, no nuclei can survive. After three minutes, you can start the process of putting protons and neutrons together to make nuclei, but you are doing it in an environment in which the universe is expanding, carrying particles away from each other and making interactions less frequent. This means that there is a narrow window of opportunity, probably less than a minute long, between the time when nuclei can stay together and the time when the density of the universe drops and nuclei no longer form because interactions are too rare. In this window various forms of hydrogen, helium, and lithium appear—everything else, as we saw, is created later in supernovae. The elements that were taken into stars and later forged into the entire periodic table, in other words, were made in a short burst when the universe was only a few minutes old.

The abundance of these light nuclei in the universe today is one of the strongest pieces of evidence we have

ATOMS FORMED WHEN
TEMPERATURES REACHED

3,000

KELVIN

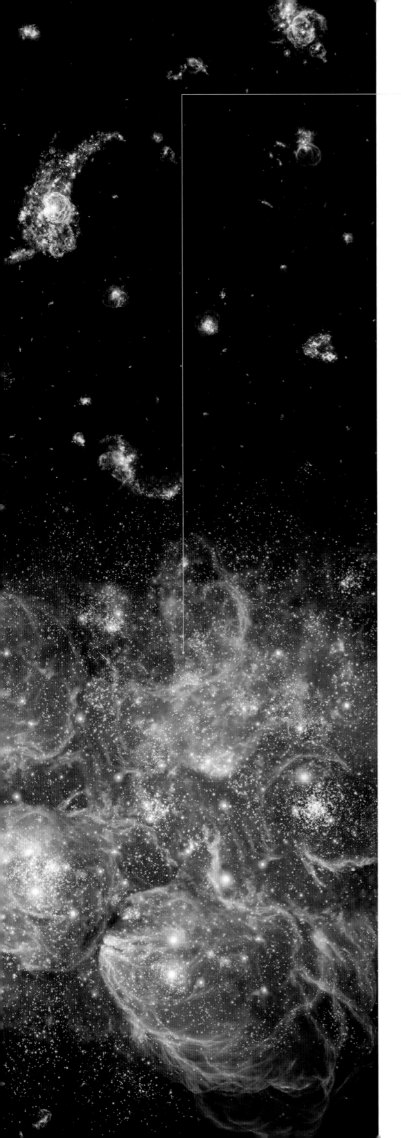

An illustration portrays the early universe, less than one billion years old, when the first stars and galaxies took shape from primordial hydrogen in a burst of star formation, and supernovae exploded across the sky.

for the big bang model. We can reproduce in our laboratories the kinds of energies that particles had when the universe was three minutes old, and our model of the Hubble expansion tells us how often collisions between these particles occurred. Thus, we can make extremely precise and unforgiving predictions of how much of each kind of light element was made in the big bang. The fact that these predictions are borne out by observations provides yet more evidence of the basic correctness of the big bang picture.

THE OBSERVABLE UNIVERSE

One final point: There is a distinction to be made between the universe—by definition, everything there is—and the observable universe—what we can actually see. We can get a rough notion of the size of the observable universe by noting that since the universe is 13.7 billion years old, the farthest objects we can possibly see are 13.7 billion light-years away. In this oversimplified picture, then, the observable universe would be a sphere of radius 13.7 billion light-years centered on Earth, a sphere that grows by 1 light-year every year. (A more detailed calculation would take account of the Hubble expansion and the fact that objects were closer to us when their light was emitted than they are now.)

Many cosmological models situate this sphere inside a much larger universe, like a weak candle in a huge cavern. By definition, of course, we can have no direct knowledge of anything outside of the observable universe—a point to which we will return when we discuss the multiverse.

One of the consequences of Edwin Hubble's discovery of the expanding universe was the realization that at earlier times the universe was smaller and hotter than it is today. One way to think about "hotter" is to note that when an ordinary substance is heated up, its constituent atoms and molecules move faster. This, in turn, means that when those constituents collide, they do so at higher velocities, and the collisions are more violent. We've already seen some examples of how this fact played into the history of the universe. Before the universe was three minutes old, it was too hot for nuclei to exist. Until the universe was several hundred thousand years old, it was too hot, and collisions were too violent, for atoms to exist.

BEGINNING OF THE UNIVERSE

[FROM ENERGY TO MATTER]

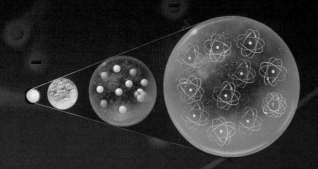

0 TO 10^{-43} SECOND: Planck Era, quantum gravity
10^{-43} TO 10^{-36} SECOND: Electromagnetism, strong, and weak forces unified
UNKNOWN TO 10^{-32} SECOND: Universe inflates exponentially
10^{-36} TO 10^{-12} SECOND: Strong force separates
10^{-12} TO 10^{-6} SECOND: Quark Epoch, weak force separates
10^{-6} TO 1 SECOND: Hadron Epoch, hydrogen nuclei form
1 TO 10 SECONDS: Lepton Epoch, leptons and anti-leptons annihilate
10 SECONDS TO 380,000 YEARS: Plasma of nuclei, electrons, and photons
3 TO 20 MINUTES: Nucleosynthesis, helium nuclei form
150 MILLION TO 1 BILLION YEARS: First stars, quasars, and galaxies
9 BILLION YEARS: Sun and solar system form

Artwork of elementary particles in the first three minutes
after the big bang. (Inset) Formation of the first atoms

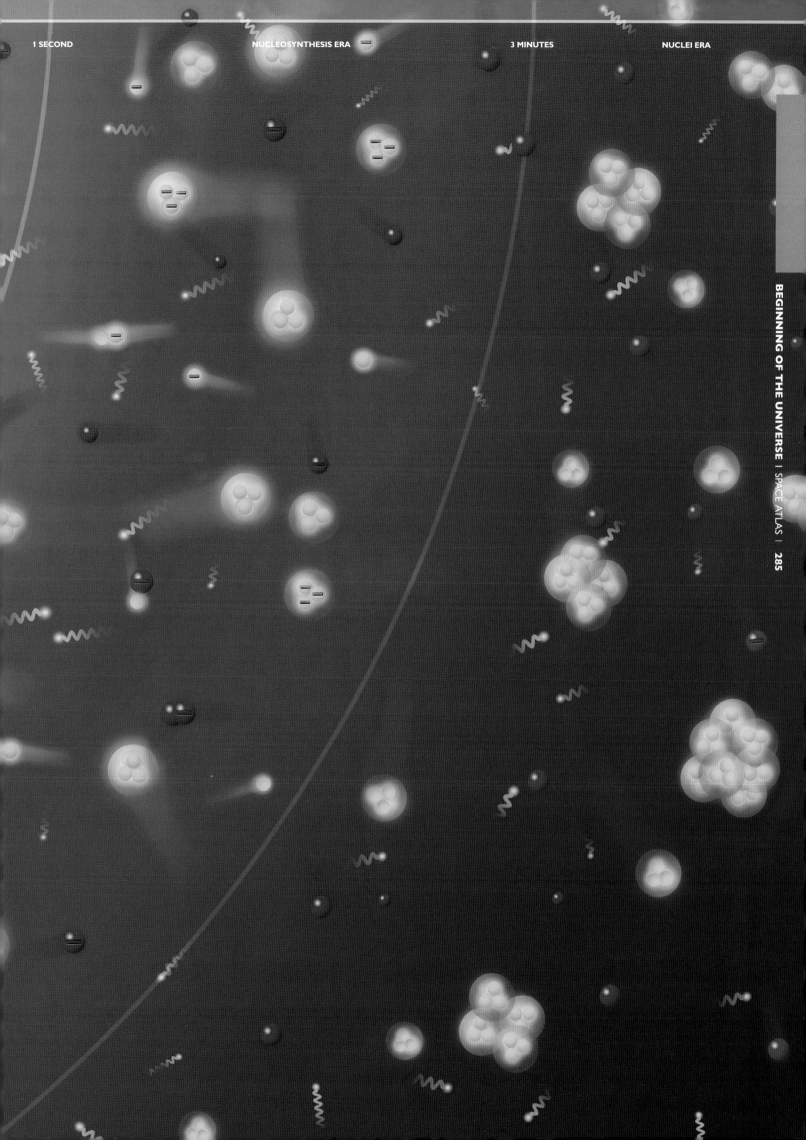

Each milestone in the early minutes and years of the universe marked a kind of "condensation," in which the basic fabric of the universe changed. First, particles came together to form nuclei, and then, much later, nuclei and electrons condensed into atoms. In fact, we will see below that the early history of the universe after the big bang consists of three separations followed by three such condensations.

The first three separations do not involve the basic constituents of matter, but the forces that act between them. Physicists recognize four distinct forces that act in our universe:

1. *Strong force*—the force that holds the nucleus of the atom together

2. *Electromagnetism*—the force that turns on the lights and keeps those notes stuck to your refrigerator door

3. *Weak force*—the force that governs some radio-active decays

4. *Gravity*—the most familiar

In our universe, these forces are distinct and quite different—the forces that hold together the nuclei in the atoms in our bodies are obviously different from the forces that hold the notes on our refrigerators. From the point of view of the theoretical physicist, however, when the temperature gets high enough, they become indistinguishable—the word they use is that the forces become "unified" when the energy is high enough.

THREE SEPARATIONS

Let's start at the very beginning, a point at which we run out of both experiment and theory. We believe that in the instant the universe was born, before it was 10^{-43} second old, all the forces—gravity, electromagnetism, strong, and weak—were unified into a single force. We do not yet have a well-tested theory to back up this belief. But we believe that at 10^{-43} second after the big

bang, gravity separated, or "froze out" from the other unified forces.

The second separation came at the unimaginably small time of 10^{-35} second after the big bang—that's a decimal point followed by 35 zeroes. A well-tested theory, called the standard model, describes what happened then: The strong force separated from the combined electromagnetic and weak, or electroweak. Before 10^{-35} second, in other words, there were only two forces operating in the universe (gravity and the

Recent studies have shown that the universe is much stranger, and more mysterious, than we used to believe. The upper pie chart shows how matter is distributed in its various forms. Note that our familiar world composes only about 5 percent of the total. The lower chart shows the distribution when the universe was 380,000 years old.

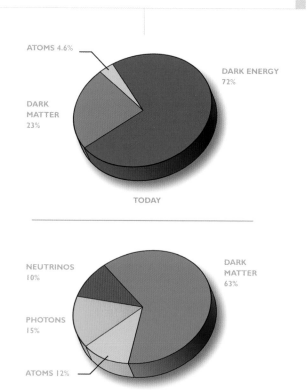

ATOMS 4.6%

DARK ENERGY 72%

DARK MATTER 23%

TODAY

NEUTRINOS 10%

DARK MATTER 63%

PHOTONS 15%

ATOMS 12%

13.7 BILLION YEARS AGO
(UNIVERSE 380,000 YEARS OLD)

An artist's conception tracks the evolution of the universe from the big bang (lower left) to the formation of matter. The yellow indicates the earliest period, the Planck Era when the four forces were unified; orange, the period of rapid inflation; and red, the period when atoms formed.

10⁻⁴³ SEC 10⁻³² SEC 3 MIN 300,000 YEARS

10²⁷ °C 10¹³ °C 10000 °C

An artist's impression of what a quasar might have looked like early in its existence. The supermassive black hole in the center is surrounded by infalling clouds of gas, made luminous by their high level of compression.

unified strong-electroweak), while after this time there were three.

Several other important things happened at this time. The most important of these is that the universe went through a brief but intense expansion called inflation. In the space of about 10^{-35} second, the universe went from something smaller than a proton to something about the size of a grapefruit. (Remember that that this was an expansion *of* space, and not motion *in* space, and this means that it didn't violate Einstein's prohibition against faster-than-light travel.) The brainchild of physicist Alan Guth (now at MIT), the so-called inflationary universe is now a standard part of the story of our universe. And, as you might suspect, the introduction of inflation solved another essential problem in cosmology.

It has to do with the cosmic microwave background. As we pointed out, the radiation is uniform to four decimal places, and this, in turn, means that the different parts of the universe from which that radiation comes are all at the same temperature to the same level of accuracy. The problem was that if you simply extrapolated the Hubble expansion back in time, there was never a time when two parts of the sky 180 degrees apart would have been in contact long enough to establish that kind of uniformity. It would be as if you turned on the hot water tap in your bathtub and all the water in the tub warmed up instantly. What inflation does, in essence, is to say that the observed uniformity in temperature was established when the bathtub was a lot smaller, and that it was simply maintained through the inflationary event.

After inflation, when the universe was a mere tenth of a nanosecond old, the third separation occurred. The unified electroweak force froze out into the electromagnetic force and the weak force. From that point on the universe featured the same four forces we see today.

THREE CONDENSATIONS

We reach the next point when the universe is a mere ten microseconds old. To understand what happened then, we have to talk a little about elementary particles.

Starting in the 1930s, physicists looking at the debris of collisions between cosmic rays and atomic nuclei discovered that there were particles besides the proton and neutron inside the nucleus of the atom. By the 1960s scientists had found that the so-called elementary particles, such as protons, weren't elementary at all: They were combinations of things more elementary still—things that were given the name "quark." (The name has an unusual provenance: There were three kinds of quarks in the original theory, and there was a line in James Joyce's *Finnegan's Wake* that read "Three quarks for Muster Mark.")

Before ten microseconds, the universe consisted of a sea of quarks, these most fundamental constituents of matter. At ten microseconds, the quarks condensed into elementary particles (protons and neutrons are the most familiar). The universe became a hot plasma of atomic nuclei.

At a few hundred thousand years, as the universe cooled, these nuclei captured electrons and became atoms. The formation of atoms was a particularly important event, because it both released the radiation that became the cosmic microwave background and marked the point at which ordinary matter could start to collapse into galaxies.

Physicists say that the universe became "transparent" at this point. Until then, light in the form of photons could not move freely through space, blocked as the photons were by free-floating, energetic electrons. In fact, before the discovery of dark matter (see pages 246–48), this pre-transparent state posed a problem. Photons—electromagnetic radiation—interact and exert a pressure in a plasma, and if ordinary matter tried to condense into galaxies before atoms formed, intense radiation would blow the concentration apart. Furthermore, calculations showed that by the time atoms formed, ordinary matter was spread too thin to be able to collect itself into galaxies. Dark matter, however, doesn't interact with radiation, so it was able to start clumping together before the formation of atoms. When the universe became transparent with the formation of atoms, then, ordinary matter was simply pulled into the concentrations of dark matter that had already formed, creating the galaxies we see today. Far from being a problem, dark matter actually resolved an old problem in cosmology.

The six transitions we've discussed so far can be summarized as follows: Before 10^{-43} second, only one unified force operated in the universe. In the next tiny fractions of a second, first gravity, then the strong force, then the weak and electromagnetic forces split apart. At ten microseconds, quarks condensed into elementary particles, which at three minutes condensed into nuclei and after a few hundred thousand years picked up electrons to become atoms.

Before all of that, of course, there is the ultimate mystery—the creation of the universe itself. And, although scientists are starting to engage in serious speculation about this event, my own sense is that, for the moment at least, we ought to leave this discussion with words from the *Rubaiyat,* written by the poet and astronomer Omar Khayyám:

There was a door to which I found no key
There was a veil past which I could not see

At first glance, predicting the end of the universe seems simpler than tracing its beginnings. The only force acting between far-flung galaxies is gravity, and the only question is whether it's strong enough to reverse the Hubble expansion. The answer to that question depends on a single number—the amount of mass in the universe. • Traditionally, astronomers have distinguished between a situation in which there is not enough mass to halt the expansion—the so-called open universe—and a situation in which the expansion is eventually halted and reversed—a closed universe. The boundary between these two, in which the expansion slows to a halt in an infinite amount of time, is called a flat universe.

END OF THE UNIVERSE

[IT ALL DEPENDS UPON MASS]

TODAY: Universe age, 13.7 billion years
3 BILLION YEARS FROM NOW: Milky Way and Andromeda collide
5 BILLION YEARS FROM NOW: Sun becomes red giant, then white dwarf

IN 2 TRILLION (2×10^{12}) YEARS: Galaxies beyond Local Group supercluster redshift out of sight
IN 100 TRILLION (10^{14}) YEARS: Star formation ceases
IN 10^{34} YEARS TO 10^{40} YEARS: Protons decay and disappear?
IN 10^{40} YEARS TO 10^{100} YEARS: Black holes dominate universe
IN 10^{100} YEARS: Black holes evaporate through Hawking radiation
POSSIBLE FATES: Big chill, big rip, big crunch

Artwork of the "big rip." (Inset) Flat, closed, and open universes

For theoretical as well as observational reasons, astronomers have always assumed that the universe is flat and have tried to find enough mass to bring about this end. That critical amount of mass is said to "close the universe."

Unfortunately, if you count up all visible matter—all the stars and galaxies and nebulae—you get only about 5 percent of the mass you need to close the universe. If you add in dark matter (see page 286), the number goes up to about 28 percent. This was the way things stood in 1998, when some rather astonishing observational results were announced.

For years astronomers had been trying to find a way of determining the end of the universe that doesn't involve adding up masses. When you look at very distant galaxies, you see light that was emitted billions of years ago. By measuring the redshift of that light, you can tell how fast the universe was expanding back then. We would expect that gravity should be slowing that expansion over time. Therefore, this method of tackling the fate of the universe is known as measuring the deceleration parameter.

To measure the deceleration parameter, you need a way of measuring the distance to very distant galaxies, billions of light-years away. Such galaxies are so remote that astronomers cannot make out individual stars, as Edwin Hubble did with his Cepheid variables, so a different standard candle is needed. Enter the Type Ia supernova.

A Type Ia supernova occurs when you have a double-star system in which one partner has gone through its life cycle (see page 230) and become a white dwarf. If the white dwarf pulls material off of its normal-size partner, it can become so massive that nuclear reactions begin and

MICHAEL TURNER

MICHAEL TURNER

"It was probably the most anticipated surprise of all time."

So says University of Chicago cosmologist Michael Turner, speaking of the discovery of the accelerated Hubble expansion in the 1990s. Now director of the Kavli Institute for Cosmological Physics, Turner was one of the organizers of the first cosmology group set up at a particle physics facility—in this case, the Fermi National Accelerator Laboratory outside of Chicago.

Those were exciting times in cosmology. Dark matter had been discovered; the inflationary scenario was explaining the basic geometry of the universe. In spite of all of this, however, a piece of the puzzle seemed to be missing. Turner was among a small group of theoreticians who suggested that that missing piece might be the energy of empty space—the energy of the vacuum itself, often referred to as the cosmological constant. When the accelerated expansion was announced in 1998, everyone recognized that whatever was causing the acceleration was that missing piece.

And what does Michael Turner think dark energy is?

"Half the days of the week I think it's the cosmological constant. The rest of the week I think it's something much more fundamental."

the entire star explodes. The energy released in this sort of event is enormous, and for brief periods the supernova can outshine an entire galaxy. Because all white dwarfs are basically the same size, and because the explosion can be seen from so far away, Type Ia supernovae make excellent standard candles for determining the distance to the farthest galaxies.

By combining the knowledge of the distance to a galaxy (from Type Ia supernovae) with the knowledge of how fast it is receding from us (from the measured redshift), we can deduce the universe's rate of expansion in the distant past. Studying these supernovae in the 1990s, astronomers expected that they would find that the expansion of the universe is slowing down. Instead, to everyone's amazement, just the opposite is true—the universe is expanding faster now than it did billions of years ago. The Hubble expansion is accelerating!

DARK ENERGY

There is only one way to explain this astonishing fact: There must be a force in the universe capable of overcoming the inward pull of gravity. Just as gravity always pulls things together, this new force must push them apart. This new phenomenon was given the name "dark energy," by University of Chicago cosmologist Michael Turner, but it should not be confused with the dark matter we discussed on pages 244–49. However, since energy and mass are equivalent (remember $E = mc^2$), dark energy can be thought of as contributing to the mass of the universe.

The final tally on the mass of the universe, then, is as follows:

Ordinary matter: about 5 percent

Dark matter: about 23 percent

Dark energy: about 73 percent

Altogether, this mass is enough to close the universe.

It is a sobering realization that the stuff we're made of, the stuff we're used to thinking of as the basic fabric of the universe, is actually just a small part of what's out there.

Once Type Ia supernovae were developed as a standard candle, it was possible to trace the history of the Hubble expansion. It appears that for the first five billion years or so, the expansion did indeed slow down. In this era, matter was much more densely packed than it is now, and the inward pull of gravity dominated. As matter became more widely scattered, however, gravity weakened and the outward force of dark energy took over. The expansion started to accelerate, as it has been doing ever since.

Given this new understanding of the universe, what can we say about how it will end? The answer depends on the properties of dark energy, which we do not understand. Nevertheless, we can lay out some possibilities.

BIG CHILL, BIG RIP, OR BIG CRUNCH

In the "big chill" scenario, the Hubble expansion will continue forever. Matter will be spread out farther and farther, and the universe will end as a cold, empty place with occasional bits of matter floating around.

A popular idea about dark energy is that it represents a cost of creating space-time. If so, as the universe expands and the amount of space increases, the amount of dark energy will increase as well. If the amount of dark energy increases with time, the rate of acceleration will also increase until everything—planets, atoms, nuclei—is torn apart in the "big rip." This would be a spectacular (but probably unlikely) ending for the universe.

If the amount of dark energy is fixed, then the expansion will eventually dilute its effects until gravity takes over again at some point in a "big crunch." This would leave us where we started, with a choice between open, closed, and flat, with most evidence pointing toward the last of these.

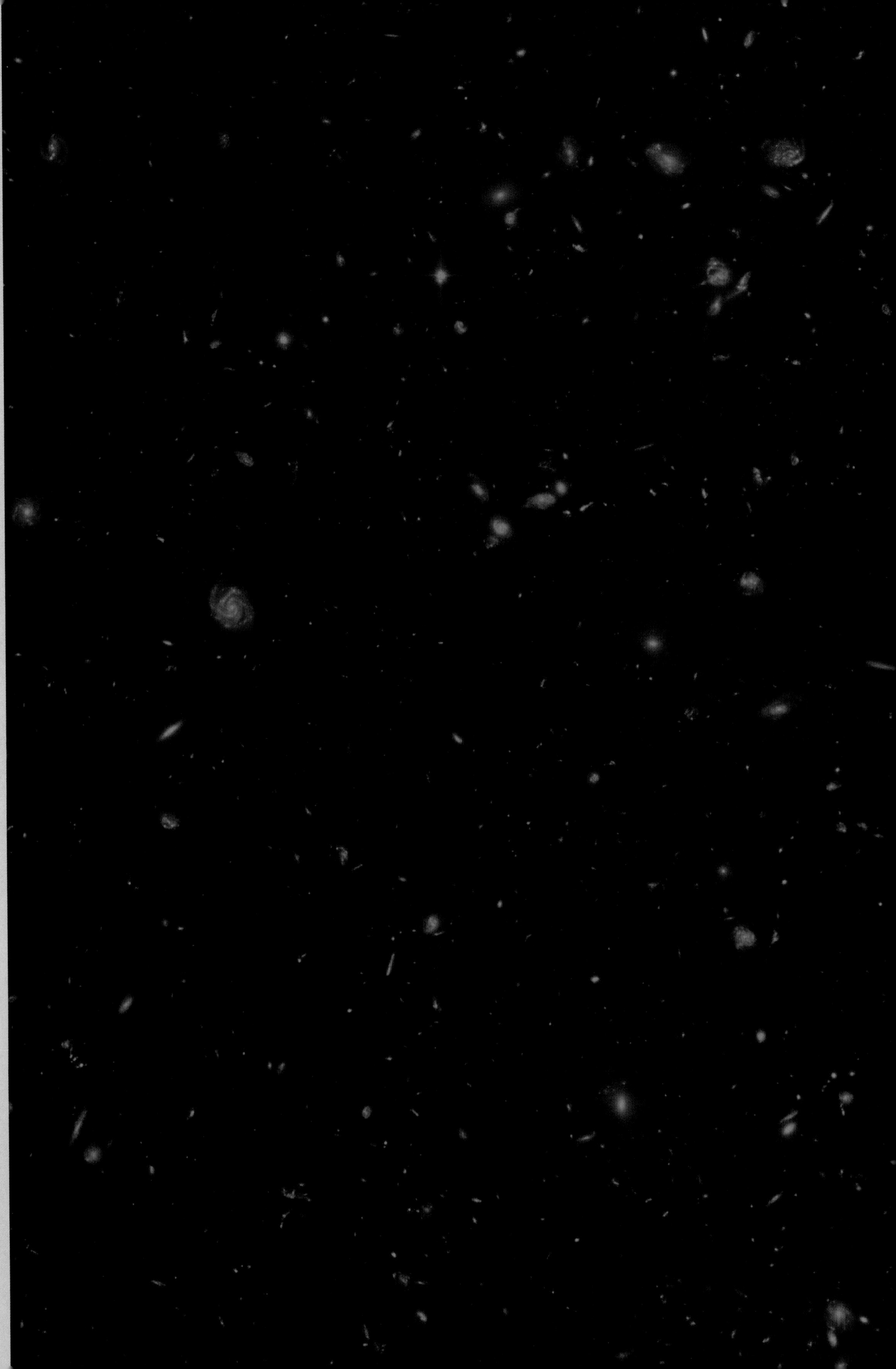

One of the most important developments in our understanding of the heavens was the realization that to study the largest thing we know of—the universe—we have to study the smallest things we know of—the particles that make up the fundamental building blocks of matter. In recent years, the strange and complex world of string theory has led to the equally strange idea of parallel universes. If experiments validate multidimensional string theories, we will end up with a very different view of reality. Just as Copernicus taught us that Earth is not the center of the universe and Hubble taught us that our Milky Way is just one among billions of galaxies, string theorists are telling us that our entire universe may be just one of a huge number of possible universes. The entire collection of universes is termed the multiverse.

MYSTERIES

[EPILOGUE]

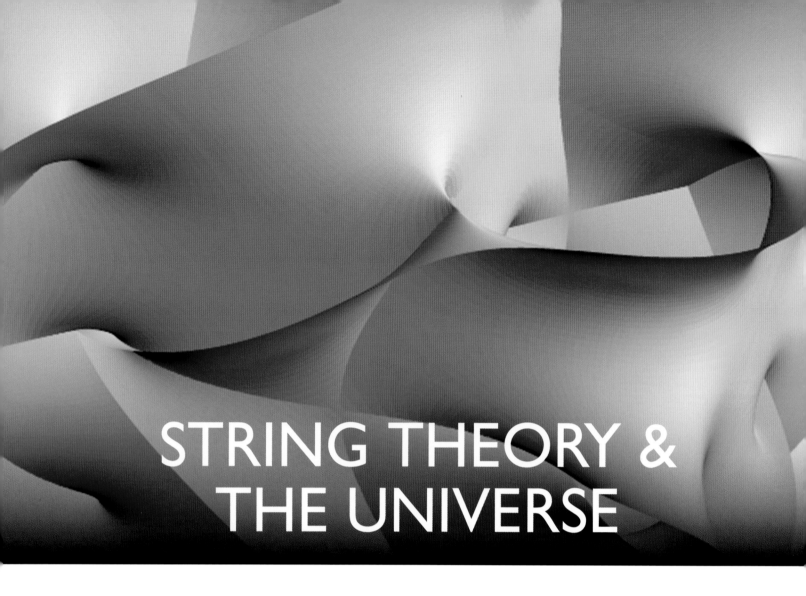

STRING THEORY & THE UNIVERSE

n the last couple of centuries, physicists have delved deeper and deeper into the building blocks of matter. Here is a quick summary of our plunge into the very small:

19th century—we find that matter is made from atoms.

Early 20th century—we find that the atom has a nucleus.

Mid-20th century—we find that the nucleus is made up of elementary particles.

Late 20th century—we find that the elementary particles are made from quarks.

Today—we speculate that the quarks might be made from strange things called "strings."

Physicist Steven Gubser opens his excellent *The Little Book of String Theory* with the statement "String theory is a mystery," and truer words were never spoken. The basic idea of these theories is that quarks are made from tiny

entities called strings. As the name implies, you can think of them as analogous to the strings of a violin or guitar, with different quarks corresponding to different patterns of the strings' vibration. There are many versions of string theories, but for our purposes we can concentrate on a couple features that make strings particularly interesting:

- They unify gravity with the other fundamental forces (see opposite), and thus can describe the earliest stages of the big bang.

- They are difficult mathematically and typically involve vibrations of strings in 10 or 11 dimensions.

MANY DIMENSIONS

This second point is so unfamiliar to most of us that we had better deal with it first. Our everyday world has four dimensions—three in space and one in time. The space dimensions are front and back, up and down, and left

String theory imagines a multidimensional universe composed of string and branes. (Opposite) An artist's conception of how branes might fold together in such a universe

and right. We're less used to thinking of time this way, but "before and after" are familiar concepts.

When physicists started developing string theories, they discovered that the only way to keep their calculations from yielding infinities was to add more dimensions. (In physics jargon, we say that the theories are "renormalizeable" only in this multidimensional world.) So if the theories make sense only in 10 or 11 dimensions, but we live in a world of 4 dimensions, what is to be done?

A simple analogy illustrates the way that string theorists get around this problem. Think about a garden hose lying on a lawn. If you look at the hose from far away, it is a line. If you wanted to move along the hose, you would have only two choices—forward and back. This means that as seen from a distance, the hose is a one-dimensional object. If you get close to the hose, however, you see that it actually has three dimensions—it also has both left and right and up and down. In the same way, string theorists argue, the strings create a four-dimensional world when observed from far away, and the extra dimensions become apparent only when we get close in, something we cannot do with our current technologies. Physicists say the extra dimensions "compactify."

GRAVITY

It is, however, the first characteristic mentioned above—the unification of gravity with other forces—that is of most interest to scientists. It seems that string theories are able to resolve a century-long split in our view of nature. The early 20th century saw two major scientific revolutions. One was Albert Einstein's theory of relativity, which remains our best description of gravity, and the other was quantum mechanics, which is our best description of the subatomic world. The problem is that these two theories look at forces very differently. In relativity the force of gravity arises from the warping of space and time by the presence of mass. Gravity is, in other words, a result of altered geometry. In quantum mechanics, on the other hand, forces arise from the exchange of particles—an essentially dynamic approach. At the moment, the strong, electromagnetic, and weak forces are all described in this way, while gravity is explained by geometry. Reconciling these two competing points of view has been an outstanding problem in theoretical physics for decades.

String theories resolve this long-standing dilemma. In these theories, gravity is the result of the exchange of (as yet undiscovered) particles called gravitons. Thus, gravity is not fundamentally different from the other three forces. We can, in fact, imagine the way that different scientists through the ages would have answered the simple question "Why don't I just float up out of the chair I'm sitting in?"

Isaac Newton: Because Earth and you exert a gravitational force on each other.

Albert Einstein: Because the mass of Earth warps the space-time grid at its surface.

String theorist: Because there is a flood of gravitons being exchanged between you and Earth.

These explanations are not contradictory, but complementary. Apply string theory to massive objects and you get the results of relativity; apply relativity to ordinary objects and you're back to Newton. Mature sciences don't grow by replacing one theory with another, but by incorporating old theories into new ones.

BRANES

Newer versions of string theories involve objects called branes (from "membranes"), which you can think of as sheets flopping around in multidimensional space (imagine moving a string perpendicular to its length so that its path traces out a sheet). Physicists also talk of an ultimate theoretical version of strings, dubbed M theory, that will mark the end of our quest to know the fundamental structure of matter. M theory hasn't been written down yet, although there are a lot of extremely bright people trying to accomplish the task. And oddly enough, given its potential importance, nobody seems to be sure what the M stands for. (My own guess is "membrane.")

Having written this quick introduction to the frontiers of modern theories, I have to say that there is a serious debate in the physics community about whether they really constitute science. The difficulty of the mathematical descriptions has prevented string theorists from making predictions that can be tested by experiment. Skeptics argue that without the traditional interplay between theory and experiment, string theory is just mathematics.

Defenders counter that many of the general features of the theories, such as the prediction that there is a class of so-called supersymmetric particles that has not yet been discovered, are indeed testable. Presumably this debate will be resolved as both our theoretical and experimental abilities improve.

In any case, string theories give us a way of bridging that final gap we discussed, the gap that separates the well-thought-out history of the big bang from the unknowns surrounding the unification of gravity. They allow us to confront the ultimate question: How did the whole thing start? As we shall see in the next section, the theories open some fascinating possibilities that are almost theological in nature.

(Below) British theoretical physicist Michael Green. **(Opposite)** An artist's conception of the multiverse

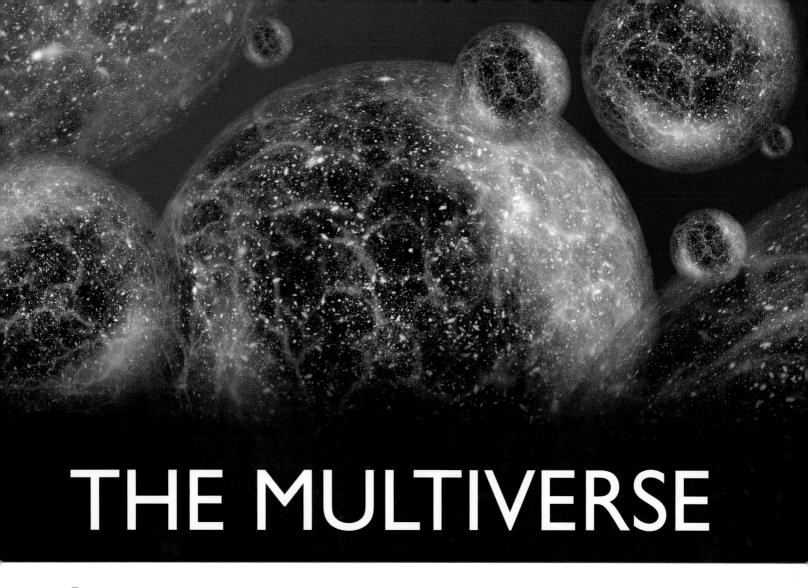

THE MULTIVERSE

Just as there are many different versions of string theory, there are many different versions of the multiverse. Some seem truly bizarre, with multidimensional branes (see pages 297–98) colliding to generate each new universe. The most common picture, however, is most easily visualized as bubbles in boiling water. We can think of each bubble as representing a universe like our own, full of galaxies. Our own bubble is expanding, and in some versions of the multiverse theory a bubble like ours continuously creates small bubbles at its surface, with each small bubble experiencing its own version of the big bang. Our own universe, in other words, might be shedding little baby universes even as you are reading these words.

STRING THEORY LANDSCAPE

String theories predict the existence of the multiverse because of something called the string theory landscape.

Imagine a cosmic pinball game, where balls roll over a surface full of hills and valleys. We know that the ball will eventually settle in one of the valleys—not necessarily the deepest valley on the board, since the ball can get trapped in a shallow valley as well. A valley capable of trapping the ball, even though it is not the deepest valley, is called a false vacuum in the jargon of theoretical physics.

According to string theory, when we map out the surface corresponding to possible energy states in multiple dimensions, we find a huge number of false vacua (valleys)—some 10^{500} of them, in fact. Each of these represents a possible place where the pinball could wind up, or, in string theory, a possible universe. 10^{500} is a huge number—a 1 followed by 500 zeroes! For all intents and purposes, we can say that there are an infinite number of possible universes in the string theory landscape. And, if universes like ours are really shedding baby universes, we

can think of each shedding event as rolling another pinball over the hills and valleys. If we roll enough pinballs, we can expect that eventually most of the false vacua will be filled. In the string theory multiverse, then, any possible universe will eventually appear somewhere in the landscape.

As was the case with string theory, there is a philosophical debate about the multiverse. It centers on the fact that in most versions of the theory there is really no way for one universe to communicate with another, and hence no way to have direct experimental confirmation of the existence of a universe outside of our own. On the other hand, if it turns out that some version of string theory passes the experimental tests to which we can subject it, and if that theory also predicts the existence of the multiverse, then we would have to take that

Artwork depicts the simultaneous creation of multiple universes with differing physical laws. Our universe is in the center, and a universe without matter is at right.

prediction seriously. And this is important, because it may be that the existence of the multiverse could solve a very deep and long-standing problem in our understanding of the universe.

FINE-TUNING PROBLEM

It's called the fine-tuning problem. It can be stated in many ways, but you can get a sense of it by thinking about the gravitational force. If this force were much stronger than it actually is, the big bang would have collapsed soon after it began, simply because the stronger gravity would have ended the expansion before it really got started.

Similarly, if the gravitational force were weaker, it would not have been strong enough to gather matter together into stars or planets. In both cases, the universe would not have produced living creatures capable of asking questions about gravity. So the gravitational force has to be fine-tuned—restricted to certain values—in order for life to develop.

This sort of fine-tuning seems to be a feature of all of nature's constants. Theoretical calculations, for example, indicate that changes of only a few percent in the strong force (the one that holds the nucleus together) or the electromagnetic force (the one that keeps electrons locked in orbits in the atom) would prevent the creation of atoms like carbon and oxygen, eliminating the possibility of life as we know it. In the same way, a number called the cosmological constant, which some theorists believe is involved in the acceleration of the Hubble expansion (see page 280), is measured to be almost (but not quite) zero in our universe. And yet when physicists use quantum mechanics to calculate this number, they are off by 120 orders of magnitude—about as far off from the real value as it is possible to be. (An order of magnitude is one power of ten, so 120 orders of magnitude is a 1 followed by 120 zeroes.) Some as-yet-unknown effect must be almost canceling this large value, but all we can say now is that the cosmological constant seems to be fine-tuned with a vengeance!

This fine-tuning of the forces and constants of nature has always been a problem for scientists. Why should we be in a universe where all of these numbers are exactly as they are? Some theologians have even advanced fine-tuning as proof of the existence of God.

ANTHROPIC PRINCIPLE

The string theory multiverse supports another view, however, related to an old concept called the anthropic principle. Supporters of this view point out that the question "Why are the constants of nature as they are?" is incorrectly posed. The question should really be "Why are the constants of nature as they are, *given that an intelligent living creature is asking the question?*" In a universe incapable of producing life the question would never get asked, so the mere fact that it is being posed is already a statement about the type of universe we're in.

I should point out that there are actually two versions of the anthropic principle: weak and strong. The weak principle is the argument given above—that we must live in a universe capable of producing life because the question is being asked. The strong version asserts that there is some as-yet-undiscovered law that says the universe *must* be such that life can exist. Most scientists prefer the weaker version.

My old statistics professor used to talk about what he called the "golf ball on the fairway" problem. Before a golfer swings, the chance that the ball will land on a particular blade of grass is tiny—but the ball will eventually land on some blade of grass. If it hadn't landed there it would have landed somewhere else equally improbable. In the same way, when we contemplate the string theory landscape it's no use asking why we are in this particular improbable universe, because if we weren't here some version of us would be somewhere else equally improbable.

Looking at things this way leads to some interesting thoughts. The number of possible universes is so large, for example, that the subset capable of producing life is probably large as well. This leads to the standard science-fiction scenario where there is another you reading these words in another universe, except that maybe you have a tail and green scales. The multiverse represents the triumph of the Copernican worldview: the ultimate removal of mankind from the center of existence.

CARINA NEBULA

An infrared image from the European Southern Observatory's Very Large Telescope luminously outlines the structures of the Carina Nebula, a region of intense star formation 7,500 light-years from Earth. At lower left is the unstable star Eta Carinae.

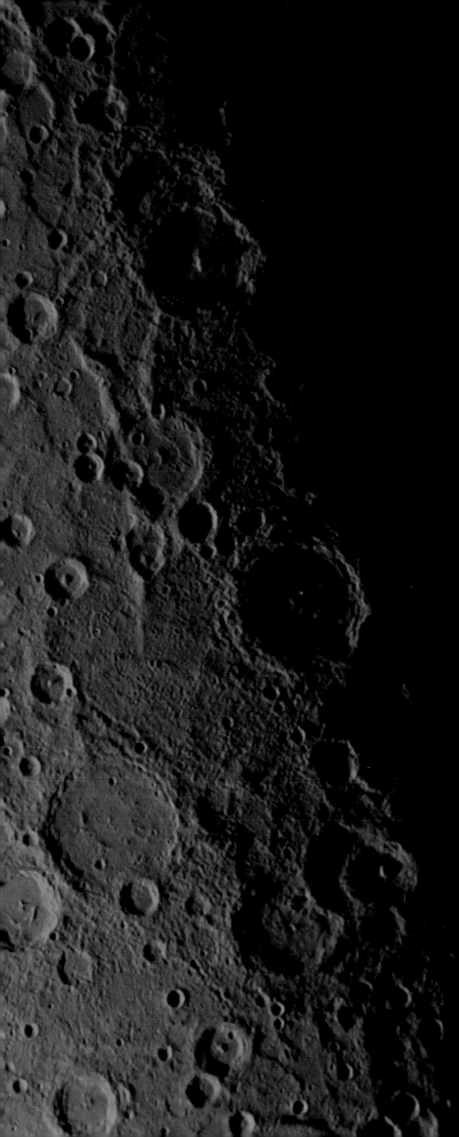

APPENDIX
& INDEXES

OPPOSITE: Earth, its moon, and the sun line up in this composite illustration. Circling the sun on the same orbital plane, Earth and the moon can cause eclipses. THIS PAGE: Craters of Mercury

SOLAR SYSTEM FACTS & FIGURES

SUN

AVERAGE DISTANCE FROM SUN
—

PERIHELION
—

APHELION
—

MASS (EARTH = 1)
332,900

DENSITY
1.409 G/CM^3

EQUATORIAL RADIUS
695,500 KM (432,200 MI)

EQUATORIAL CIRCUMFERENCE
4,379,000 KM (2,715,000 MI)

ORBITAL PERIOD
—

ROTATION PERIOD
25.38 EARTH DAYS (AT 16° LAT)

AXIAL TILT
—

MIN./MAX. SURFACE TEMPERATURE
5500°C (10,000°F)

VOLUME (EARTH = 1)
1,300,000

EQUATORIAL SURFACE GRAVITY (EARTH = 1)
28

SPECTRAL TYPE
G2 V

LUMINOSITY
3.83 × 10^{33} ERGS/SEC

ATMOSPHERE
HYDROGEN, HELIUM

NATURAL SATELLITES
NONE

MERCURY

AVERAGE DISTANCE FROM SUN
57,909,175 KM (35,983,095 MI)

PERIHELION
46,000,000 KM (28,580,000 MI)

APHELION
69,820,000 KM (43,380,000 MI)

MASS (EARTH = 1)
0.055

DENSITY
5.427 G/CM^3

EQUATORIAL RADIUS
2,439.7 KM (1,516 MI)

EQUATORIAL CIRCUMFERENCE
5,329.1 KM (9,525.1 MI)

ORBITAL PERIOD
87.97 EARTH DAYS

ROTATION PERIOD
58.646 EARTH DAYS

AXIAL TILT
0 DEGREES

MIN./MAX. SURFACE TEMPERATURE
173/427°C (-279/801°F)

VOLUME (EARTH = 1)
0.054

EQUATORIAL SURFACE GRAVITY (EARTH = 1)
0.38

ORBITAL ECCENTRICITY
0.20563069

ORBITAL INCLINATION TO ECLIPTIC
7 DEGREES

ATMOSPHERE
TRACE

NATURAL SATELLITES
NONE

VENUS

AVERAGE DISTANCE FROM SUN
108,208,930 KM (67,237,910 MI)

PERIHELION
107,476,000 KM (66,782,000 MI)

APHELION
108,942,000 KM (67,693,000 MI)

MASS (EARTH = 1)
0.815

DENSITY
5.24 G/CM^3

EQUATORIAL RADIUS
6,051.8 KM (3,760.4 MI)

EQUATORIAL CIRCUMFERENCE
38,025 KM (23,627 MI)

ORBITAL PERIOD
224.7 EARTH DAYS

ROTATION PERIOD
243 EARTH DAYS (RETROGRADE)

AXIAL TILT
177.3 DEGREES

MIN./MAX. SURFACE TEMPERATURE
462°C (864°F)

VOLUME (EARTH = 1)
0.88

EQUATORIAL SURFACE GRAVITY (EARTH = 1)
0.91

ORBITAL ECCENTRICITY
0.0068

ORBITAL INCLINATION TO ECLIPTIC
3.39 DEGREES

ATMOSPHERE
CARBON DIOXIDE, NITROGEN

NATURAL SATELLITES
NONE

EARTH

AVERAGE DISTANCE FROM SUN
149,597,890 KM (92,955,820 MI)

PERIHELION
147,100,000 KM (91,400,000 MI)

APHELION
152,100,000 KM (94,500,000 MI)

MASS (EARTH = 1)
1

DENSITY
5.515 G/CM³

EQUATORIAL RADIUS
6,378.14 KM (3,963.19 MI)

EQUATORIAL CIRCUMFERENCE
40,075 KM (24,901 MI)

ORBITAL PERIOD
365.24 EARTH DAYS

ROTATION PERIOD
23.934 HOURS

AXIAL TILT
23.45 DEGREES

MIN./MAX. SURFACE TEMPERATURE
-88/58°C (-126/136°F)

VOLUME
1.0832×10^{12} KM³

EQUATORIAL SURFACE GRAVITY
9.766 M/S²

ORBITAL ECCENTRICITY
0.0167

ORBITAL INCLINATION TO ECLIPTIC
0.00 DEGREE

ATMOSPHERE
NITROGEN, OXYGEN

NATURAL SATELLITES
1

MARS

AVERAGE DISTANCE FROM SUN
227,936,640 KM (141,633,260 MI)

PERIHELION
206,600,000 KM (128,400,000 MI)

APHELION
249,200,000 KM (154,900,000 MI)

MASS (EARTH = 1)
0.10744

DENSITY
3.94 G/CM³

EQUATORIAL RADIUS
3,397 KM (2,111 MI)

EQUATORIAL CIRCUMFERENCE
21,344 KM (13,263 MI)

ORBITAL PERIOD
686.93 EARTH DAYS

ROTATION PERIOD
24.62 HOURS

AXIAL TILT
25.19

MIN./MAX. SURFACE TEMPERATURE
-87/-5°C (-125/23°F)

VOLUME (EARTH = 1)
0.150

EQUATORIAL SURFACE GRAVITY (EARTH = 1)
0.38

ORBITAL ECCENTRICITY
0.0934

ORBITAL INCLINATION TO ECLIPTIC
1.8 DEGREES

ATMOSPHERE
CARBON DIOXIDE, NITROGEN, ARGON

NATURAL SATELLITES
2

JUPITER

AVERAGE DISTANCE FROM SUN
778,412,020 KM (483,682,810 MI)

PERIHELION
740,742,6009 KM (460,276,100 MI)

APHELION
816,081,400 KM (507,089,500 MI)

MASS (EARTH = 1)
317.82

DENSITY
1.33 G/CM³

EQUATORIAL RADIUS
71,492 KM (44,423 MI)

EQUATORIAL CIRCUMFERENCE
449,197 KM (279,118 MI)

ORBITAL PERIOD
11.8565 EARTH YEARS

ROTATION PERIOD
9.925 HOURS

AXIAL TILT
3.12 DEGREES

MIN./MAX. SURFACE TEMPERATURE
-148°C (-234°F)

VOLUME (EARTH = 1)
1,316

EQUATORIAL SURFACE GRAVITY (EARTH = 1)
2.14

ORBITAL ECCENTRICITY
0.04839

ORBITAL INCLINATION TO ECLIPTIC
1.305 DEGREES

ATMOSPHERE
HYDROGEN, HELIUM

NATURAL SATELLITES
65 (INCLUDING 1 UNCONFIRMED)

SATURN

AVERAGE DISTANCE FROM SUN
1,426,725,400 KM (885,904,700 MI)

PERIHELION
1,349,467,000 KM (838,519,000 MI)

APHELION
1,503,983,000 KM (934,530,000 MI)

MASS (EARTH = 1)
95.16

DENSITY
0.70 G/CM³

EQUATORIAL RADIUS
60,268 KM (37,449 MI)

EQUATORIAL CIRCUMFERENCE
378,675 KM (235,298 MI)

ORBITAL PERIOD
29.4 EARTH YEARS

ROTATION PERIOD
10.656 HOURS

AXIAL TILT
26.73 DEGREES

MIN./MAX. SURFACE TEMPERATURE
-178°C (-288°F)

VOLUME (EARTH = 1)
763.6

**EQUATORIAL SURFACE GRAVITY
(EARTH = 1)**
0.91

ORBITAL ECCENTRICITY
0.0541506

ORBITAL INCLINATION TO ECLIPTIC
2.484 DEGREES

ATMOSPHERE
HYDROGEN, HELIUM

NATURAL SATELLITES
62

URANUS

AVERAGE DISTANCE FROM SUN
2,870,972,200 KM (1,783,939,400 MI)

PERIHELION
2,735,560,000 KM (1,699,800,000 MI)

APHELION
3,006,390,000 KM (1,868,080,000 MI)

MASS (EARTH = 1)
14.371

DENSITY
1.30 G/CM³

EQUATORIAL RADIUS
25,559 KM (15,882 MI)

EQUATORIAL CIRCUMFERENCE
160,592 KM (99,787 MI)

ORBITAL PERIOD
84.02 EARTH YEARS

ROTATION PERIOD
17.24 HOURS (RETROGRADE)

AXIAL TILT
97.86 DEGREES

MIN./MAX. SURFACE TEMPERATURE
-216°C (-357°F)

VOLUME (EARTH = 1)
63.1

**EQUATORIAL SURFACE GRAVITY
(EARTH = 1)**
0.86

ORBITAL ECCENTRICITY
0.047168

ORBITAL INCLINATION TO ECLIPTIC
0.770 DEGREE

ATMOSPHERE
HYDROGEN, HELIUM, METHANE

NATURAL SATELLITES
27

NEPTUNE

AVERAGE DISTANCE FROM SUN
4,498,252,900 KM (2,795,084,800 MI)

PERIHELION
4,459,630,000 KM (2,771,087,000 MI)

APHELION
4,536,870,000 KM (2,819,080,000 MI)

MASS (EARTH = 1)
17.147

DENSITY
1.76 G/CM³

EQUATORIAL RADIUS
24,764 KM (15,388 MI)

EQUATORIAL CIRCUMFERENCE
155,597 KM (96,683 MI)

ORBITAL PERIOD
164.79 EARTH YEARS

ROTATION PERIOD
16.11 HOURS

AXIAL TILT
29.58 DEGREES

MIN./MAX. SURFACE TEMPERATURE
-214°C (-353°F)

VOLUME (EARTH = 1)
57.7

**EQUATORIAL SURFACE GRAVITY
(EARTH = 1)**
1.10

ORBITAL ECCENTRICITY
0.00859

ORBITAL INCLINATION TO ECLIPTIC
1.769 DEGREES

ATMOSPHERE
HYDROGEN, HELIUM, METHANE

NATURAL SATELLITES
13

CERES

AVERAGE DISTANCE FROM SUN
2.767 AU

PERIHELION
381,419,582 KM (237,003,140 MI)

APHELION
447,838,164 KM (278,273,734 MI)

MASS (EARTH = 1)
0.00016

EQUATORIAL RADIUS
474 KM (295 MI)

EQUATORIAL CIRCUMFERENCE
~2,900 KM (1,802 MI)

ORBITAL PERIOD
4.60 EARTH YEARS

ROTATION PERIOD
9.075 HOURS

MIN./MAX. SURFACE TEMPERATURE
~167K

**EQUATORIAL SURFACE GRAVITY
(EARTH = 1)**
0.03

ORBITAL ECCENTRICITY
0.0789

ORBITAL INCLINATION TO ECLIPTIC
10.58 DEGREES

ATMOSPHERE
POSSIBLE TRACE

NATURAL SATELLITES
NONE

PLUTO

AVERAGE DISTANCE FROM SUN
5,906,380,000 KM (3,670,050,000 MI)

PERIHELION
4,436,820,000 KM (2,756,902,000 MI)

APHELION
7,375,930,000 KM (4,583,190,000 MI)

MASS (EARTH = 1)
0.0022

EQUATORIAL RADIUS
1,151 KM (715 MI)

EQUATORIAL CIRCUMFERENCE
7,232 KM (4,494 MI)

ORBITAL PERIOD
247.92 EARTH YEARS

ROTATION PERIOD
6.387 EARTH DAYS (RETROGRADE)

MIN./MAX. SURFACE TEMPERATURE
-233/-223°C (-387/-369°F)

**EQUATORIAL SURFACE GRAVITY
(EARTH = 1)**
0.08

ORBITAL ECCENTRICITY
0.249

ORBITAL INCLINATION TO ECLIPTIC
17.14 DEGREES

ATMOSPHERE
TRACE NITROGEN, CARBON MONOX-
IDE, METHANE

NATURAL SATELLITES
4

HAUMEA

AVERAGE DISTANCE FROM SUN
43.34 AU

PERIHELION
—

APHELION
—

MASS (EARTH = 1)
0.00070

EQUATORIAL RADIUS
660–775 KM (410–480 MI)

EQUATORIAL CIRCUMFERENCE
~7,900 KM (4,900 MI)

ORBITAL PERIOD
285.4 EARTH YEARS

ROTATION PERIOD
3.9 HOURS

MIN./MAX. SURFACE TEMPERATURE
~-241°C (-402°F)

**EQUATORIAL SURFACE GRAVITY
(EARTH = 1)**
0.05

ORBITAL ECCENTRICITY
0.195

ORBITAL INCLINATION TO ECLIPTIC
28.22 DEGREES

ATMOSPHERE
—

NATURAL SATELLITES
2

MAKEMAKE

AVERAGE DISTANCE FROM SUN
45.8 AU

PERIHELION
—

APHELION
—

MASS (EARTH = 1)
0.00067

EQUATORIAL RADIUS
~800 KM (500 MI)

EQUATORIAL CIRCUMFERENCE
~4,700 KM (2,900 MI)

ORBITAL PERIOD
309.88 EARTH YEARS

ROTATION PERIOD
—

MIN./MAX. SURFACE TEMPERATURE
-240°C (-400°F)

**EQUATORIAL SURFACE GRAVITY
(EARTH = 1)**
0.05

ORBITAL ECCENTRICITY
0.159

ORBITAL INCLINATION TO ECLIPTIC
28.96 DEGREES

ATMOSPHERE
—

NATURAL SATELLITES
NONE

ERIS

AVERAGE DISTANCE FROM SUN
67.67 AU

PERIHELION
—

APHELION
—

MASS (EARTH = 1)
0.0028

EQUATORIAL RADIUS
1,200 KM (745 MI)

EQUATORIAL CIRCUMFERENCE
~8,200 KM (5,095 MI)

ORBITAL PERIOD
557 EARTH YEARS

ROTATION PERIOD
—

MIN./MAX. SURFACE TEMPERATURE
-230°C (-382°F)

**EQUATORIAL SURFACE GRAVITY
(EARTH = 1)**
0.08

ORBITAL ECCENTRICITY
0.44

ORBITAL INCLINATION TO ECLIPTIC
44.19 DEGREES

ATMOSPHERE
—

NATURAL SATELLITES
1

PLANETARY SATELLITES

NAME	DATE DISCOVERED	DISTANCE FROM PLANET (KM)	RADIUS (KM)
EARTH			
MOON	—	384,400	1,737
MARS			
PHOBOS	1877	9,376	11.1
DEIMOS	1877	23,458	6.2
JUPITER			
IO	1610	421,800	1,821.6
EUROPA	1610	671,100	1,560.8
GANYMEDE	1610	1,070,400	2,631.2
CALLISTO	1610	1,882,700	2,410.3
AMALTHEA	1892	181,400	83.45
HIMALIA	1904	11,461,000	85
ELARA	1905	11,741,000	43
PASIPHAE	1908	23,624,000	30
SINOPE	1914	23,939,000	19
LYSITHEA	1938	11,717,000	18
CARME	1938	23,404,000	23
ANANKE	1951	21,276,000	14
LEDA	1974	11,165,000	10
ADRASTEA	1979	129,000	8.2
THEBE	1980	221,900	49.3
METIS	1980	128,000	21.5
CALLIRRHOE	1999	24,103,000	4.3
THEMISTO	1975/2000	7,284,000	4.0
MEGACLITE	2000	23,493,000	2.7
TAYGETE	2000	23,280,000	2.5
CHALDENE	2000	23,100,000	1.9
HARPALYKE	2000	20,858,000	2.2
KALYKE	2000	23,483,000	2.6
IOCASTE	2000	21,060,000	2.6
ERINOME	2000	23,196,000	1.6
ISONOE	2000	23,155,000	1.9
PRAXIDIKE	2000	20,908,000	3.4
AUTONOE	2001	24,046,000	2.0
THYONE	2001	20,939,000	2.0
HERMIPPE	2001	21,131,000	2.0
AITNE	2001	23,229,000	1.5
EURYDOME	2001	22,865,000	1.5
EUANTHE	2001	20,797,000	1.5
EUPORIE	2001	19,304,000	1.0
ORTHOSIE	2001	20,720,000	1.0
SPONDE	2001	23,487,000	1.0
KALE	2001	23,217,000	1.0
PASITHEE	2001	23,004,000	1.0
HEGEMONE	2003	23,577,000	1.5
MNEME	2003	21,035,000	1.0
AOEDE	2003	23,980,000	2.0
THELXINOE	2003	21,164,000	1.0
ARCHE	2003	23,355,000	1.0
KALLICHORE	2003	23,288,000	2.0
HELIKE	2003	21,069,000	2.0
CARPO	2003	17,058,000	1.5
EUKELADE	2003	23,328,000	2.0
CYLLENE	2003	23,809,000	1.0
KORE	2003	24,543,000	2.0
S/2003 J2	2003	28,455,000	1.0
S/2003 J3	2003	20,224,000	1.0
S/2003 J4	2003	23,933,000	1.0
S/2003 J5	2003	23,498,000	2.0
S/2003 J9	2003	23388,000	0.5
S/2003 J10	2003	23,044,000	1.0
S/2003 J12	2003	17,833,000	0.5

NAME	DATE, DISCOVERED	DISTANCE FROM PLANET (KM)	RADIUS (KM)
S/2003 J15	2003	22,630,000	1.0
S/2003 J16	2003	20,956,000	1.0
S/2003 J18	2003	20,426,000	1.0
S/2003 J19	2003	23,535,000	1.0
S/2003 J23	2003	23,566,000	1.0
S/2010 J1	2010	23,314,335	2.0
S/2010 J2	2010	20,307,150	1.0
S/2011 J1	2011	20,155,290	1.0
S/2011 J2	2011	23,329,710	1.0

SATURN

NAME	DATE, DISCOVERED	DISTANCE FROM PLANET (KM)	RADIUS (KM)
TITAN	1655	1,221,870	2,575.50
IAPETUS	1671	3,560,840	735.60
RHEA	1672	527,070	764.30
TETHYS	1684	294,670	533.00
DIONE	1684	377,420	561.70
MIMAS	1789	185,540	198.20
ENCELADUS	1789	238,040	252.10
HYPERION	1848	1,500,880	135.00
PHOEBE	1898	12,947,780	106.60
JANUS	1966	151,460	89.4
EPIMETHEUS	1980	151,410	56.7
HELENE	1980	377,420	16
TELESTO	1980	294,710	11.8
CALYPSO	1980	294,710	10.7
ATLAS	1980	137,670	15.3
PROMETHEUS	1980	139,380	43.1
PANDORA	1980	141,720	40.3
PAN	1990	133,580	14.8
YMIR	2000	23,040,000	9.0
PAALIAQ	2000	15,200,000	11
TARVOS	2000	17,983,000	7.5
IJIRAQ	2000	11,124,000	6.0
SUTTUNGR	2000	19,459,000	3.5
KIVIUQ	2000	11,110,000	8.0
MUNDILFARI	2000	18,628,000	3.5
ALBIORIX	2000	16,182,000	16
SKATHI	2000	15,540,000	4.0
ERRIAPUS	2000	17,343,000	5.0
SIARNAQ	2000	17,531,000	20
THRYMR	2000	20,314,000	3.5
NARVI	2003	19,007,000	3.5
METHONE	2004	194,440	1.5
PALLENE	2004	212,280	2.0
POLYDEUCES	2004	377,200	2.0
DAPHNIS	2005	136,500	3.5
AEGIR	2005	20,751,000	3.0
BEBHIONN	2005	17,119,000	3.0
BERGELMIR	2005	19,336,000	3.0
BESTLA	2005	20,192,000	3.5
FARBAUTI	2005	20,377,000	2.5
FENRIR	2005	22,454,000	2.0
FORNJOT	2005	25,146,000	3.0
HATI	2005	19,846,000	3.0
HYRROKKIN	2006	18,437,000	4.0
KARI	2006	22,089,000	3.5
LOGE	2006	23,058,000	3.0
SKOLL	2006	17,665,000	3.0
SURTUR	2006	22,704,000	3.0
JARNSAXA	2006	18,811,000	3.0
GREIP	2006	18,206,000	3.0
ANTHE	2007	197,700	0.5
TARQEQ	2007	18,009,000	3.5
AEGAEON	2008	167,500	0.5
S/2004 S7	2005	20,999,000	3.0

NAME	DATE DISCOVERED	DISTANCE FROM PLANET (KM)	RADIUS (KM)
S/2004 S12	2005	19,878,000	2.5
S/2004 S13	2005	18,404,000	3.0
S/2004 S17	2005	19,447,000	3.5
S/2006 S1	2006	18,009,000	3.0
S/2006 S3	2006	16,725,000	3.0
S/2007 S2	2007	16,725,000	3.0
S/2007 S3	2007	18,975,000	2.5
S/2009 S1	2009	117,000	0.15

URANUS

NAME	DATE DISCOVERED	DISTANCE FROM PLANET (KM)	RADIUS (KM)
TITANIA	1787	436,300	788.9
OBERON	1787	583,500	761.4
ARIEL	1851	190,900	578.9
UMBRIEL	1851	266,000	584.7
MIRANDA	1948	129,900	235.8
PUCK	1985	86,000	81
CORDELIA	1986	49,800	20.1
OPHELIA	1986	53,800	21.4
BIANCA	1986	59,200	25.7
CRESSIDA	1986	61,800	39.8
DESDEMONA	1986	62,700	32
JULIET	1986	64,400	46.8
PORTIA	1986	66,100	67.6
ROSALIND	1986	69,900	36
BELINDA	1986	75,300	40.3
CALIBAN	1997	7,231,000	36
SYCORAX	1997	12,179,000	75
PROSPERO	1999	16,256,000	25
SETEBOS	1999	17,418,000	240
STEPHANO	1999	8,004,000	160
PERDITA	1999	76,417	13
TRINCULO	2001	8,504,000	9.0
FRANCISCO	2001	4,276,000	11
FERDINAND	2001	20,901,000	10
MARGARET	2003	14.0345000	10
MAB	2003	97,736	6.0
CUPID	2003	74,392	6.0

NEPTUNE

NAME	DATE DISCOVERED	DISTANCE FROM PLANET (KM)	RADIUS (KM)
TRITON	1846	354,759	1,352.6
NEREID	1949	5,513,787	170
NAIAD	1989	48,227	33
THALASSA	1989	50,074	41
DESPINA	1989	52,526	75
GALATEA	1989	61,953	88
LARISSA	1989	73,548	97
PROTEUS	1989	117,647	210
HALIMEDE	2002	16,611,000	31
SAO	2002	22,228,000	22
LAOMEDEIA	2002	23,567,000	21
NESO	2002	49,285,000	30
PSAMATHE	2003	48,096,000	20

PLUTO

NAME	DATE DISCOVERED	DISTANCE FROM PLANET (KM)	RADIUS (KM)
CHARON	1978	19,600	593
NIX	2005	48,680	44
HYDRA	2005	64,780	36
P4	2011	59,000	~20

ERIS

NAME	DATE DISCOVERED	DISTANCE FROM PLANET (KM)	RADIUS (KM)
DYSNOMIA	2005	33,000	200

HAUMEA

NAME	DATE DISCOVERED	DISTANCE FROM PLANET (KM)	RADIUS (KM)
HI'IAKA	2005	49,500	~310
NAMAKA	2005	39,000	~170

NOTABLE DEEP-SKY OBJECTS

Astronomers know the universe beyond our solar system as the "deep sky": a treasure-house of stars, star clusters, galaxies, nebulae, and more. In the 18th century, French astronomer Charles Messier (to distinguish comets from other fuzzy items in the sky) created a catalog of the most notable deep-sky objects. This Messier Catalog, whose members are designated by "M" numbers, includes many of the most beautiful objects visible today. Later catalogs have added to the lists, notably the New General Catalog (NGC), which was compiled beginning in the 1880s.

Some of these objects have widely accepted common names, but most do not. However, dedicated astronomers know them well. Amateurs with good backyard telescopes can see many of them, though not in great detail. More can be viewed in all their beauty online at such websites as the Hubble Telescope's HubbleSite (http://hubblesite.org/gallery/); the Messier Catalog at SEDS (http://messier.seds.org); or the NGC Catalog Online (http://spider.seds.org/ngc/ngc.html).

GALAXIES

SPIRAL GALAXIES

M31 The Andromeda Galaxy
M33 The Triangulum Galaxy
M51 The Whirlpool Galaxy
M58
M61
M63 The Sunflower Galaxy
M64 The Blackeye Galaxy
M65
M66
M74
M77
M81 Bode's Galaxy
M83 The Southern Pinwheel Galaxy
M88
M90
M91
M94
M95
M96
M98
M99
M100
M101 The Pinwheel Galaxy
M104 The Sombrero Galaxy
M106
M108
M109
NGC 253
NGC 891
NGC 1055
NGC 2403
NGC 2903
NGC 3628)
NGC 4565
NGC 4571
NGC 4631 The Herring or Whale Galaxy
NGC 4656
NGC 5907
NGC 6946
NGC 7331
NGC 7479

LENTICULAR (S0) GALAXIES

M84
M85
M86
M102
NGC 5866 The Spindle Galaxy

ELLIPTICAL GALAXIES

M32
M49
M59
M60
M87
M89
M105
M110
Leo I The Regulus Galaxy
SagDEG
Canis Major Dwarf

IRREGULAR GALAXIES

M82 The Cigar Galaxy
NGC 2976
NGC 3077
NGC 5128
NGC 5195
NGC 6822 Barnard's Galaxy
IC 10
IC 5152
The Large Magellanic Cloud
The Small Magellanic Cloud

NEBULAE

SUPERNOVA REMNANTS

M1 The Crab Nebula
Barnard's Loop

PLANETARY NEBULAE

M27 The Dumbbell Nebula
M57 The Ring Nebula
M76 The Little Dumbbell, Cork, or Butterfly Nebula
M97 The Owl Nebula

NGC 2438
NGC 6543 The Cat's Eye Nebula
NGC 7009 The Saturn Nebula
NGC 7293 The Helix Nebula

STAR-FORMING NEBULAE

M8 The Lagoon Nebula
M16 The Eagle or Star Queen Nebula
M17 The Omega Nebula

M20 The Trifid Nebula
M42 The Orion Nebula
M43 de Mairan's Nebula
M78
NGC 2023
NGC 2237, 2238, 2239, 2246 The Rosette Nebula
NGC 2264 The Cone Nebula
NGC 3372 The Eta Carinae Nebula

NGC 7000 The North America Nebula

DARK NEBULAE

Barnard 33 The Horsehead Nebula
The Coalsack

CLUSTERS

OPEN CLUSTERS

M6 The Butterfly Cluster
M7 Ptolemy's Cluster
M11 The Wild Duck Cluster
M18
M21
M23
M25
M26
M29
M34
M35
M36
M37
M38
M39
M41
M44 Praesepe, the Beehive Cluster
M45 Subaru, the Pleiades (the Seven Sisters)
M46
M47
M48
M50
M52
M67
M93
M103

GLOBULAR CLUSTERS

M2
M3
M4
M5
M9
M10
M12
M13 Great Hercules Globular Cluster
M14
M15
M19
M22
M28
M30
M53
M54
M55
M56
M62
M68
M69
M70
M71
M72
M75
M79
M80
M92
M107

GLOSSARY

Anthropic principle: the idea that the universe contains the necessary conditions to support the development of life, as proven by our existence on Earth

Aphelion: the point in an object's orbit when it is farthest from the sun

Asteroid: small rocky body revolving around the sun. The majority of asteroids orbit between Mars and Jupiter, in the asteroid belt.

Astronomical unit (AU): the average distance from Earth to the sun, 149,597,870 kilometers (92,955,730 mi). Employed as a standard unit of astronomical measurement.

Atmosphere: the gaseous layer surrounding a planet or star that is retained due to the gravitational field and temperature around the object

Big bang model: the widely accepted theory of the origin and evolution of the universe that states that the universe began in an infinitely compact state and is now expanding

Black hole: an object or region of space whose gravitational pull is so strong that nothing, not even light, can escape from it

Brane: Short for membrane, an extended object in string theory that can have any number of dimensions

Celestial sphere: the imaginary surface surrounding Earth upon which all celestial objects appear and which is used to describe the positions of objects in the sky.

Constellation: a shape assigned to an arbitrary arrangement of stars that serves as a tool to differentiate between the stars. There are 88 different shapes identified.

Core: the dense, innermost part of a planet, large moon, asteroid, or star

Corona: the sun's outermost atmosphere; the visible light from the sun, seen around the disk of the moon as it covers the sun during a total eclipse.

Cosmology: the study of the origins and structure of the universe

Crust: the solid, outermost layer of a planet or moon

Dark energy: a property of space that propels the acceleration of the expanding universe

Dark matter: unseen matter in space that exerts a gravitational pull within galaxies

Density: the ratio of the mass of an object to its volume, commonly measured in grams per cubic centimeters (g/cm3) or grams per liter (g/L).

Doppler shift: the observed change in the wavelength of light or sound caused by motion in the observer, the object, or both

Dwarf planet: a body orbiting the sun that has sufficient mass to be nearly round in shape, but which has not cleared its path of other bodies of comparable size

Ecliptic: the apparent path of the sun on the celestial sphere; also, the plane of the Earth's orbit around the sun

Exoplanet: extrasolar planet; a planet orbiting a star other than the sun

Extremophile: an organism that lives in extreme environmental conditions

Fusion: a nuclear reaction that occurs when light nuclei join to produce a heavier nucleus; nuclear fusion

Galaxy: a large collection of stars bound by mutual gravitational attraction

Galilean moons: Jupiter's four largest moons—Io, Europa, Ganymede, and Callisto. Named after the Italian astronomer Galileo Galilei, who is credited with their discovery.

Gas giant: a large, low-density planet composed primarily of hydrogen and helium; in our solar system, Jupiter, Saturn, Uranus, and Neptune are considered gas giants, with Uranus and Neptune also called ice giants

Habitable zone: the orbital distance within which liquid water exists on a planet's surface

Hertzsprung-Russell (H-R) diagram: the graph on which the luminosity and temperatures of stars are plotted

Impact crater: the depression formed as a result of a high-speed solid hitting a rigid surface, such as the circular craters on the surface of the moon.

Infrared: having a wavelength longer than the red end of the visible light spectrum and shorter than that of microwaves

Inner planets: the four small, rocky planets nearest to the sun; also called terrestrial planets

Kuiper belt: (pronounced KI-per) a disk-shaped region located 30 to 50 AU from the sun that is filled with icy bodies

Light-year: the distance light travels in a vacuum in one year; equivalent to about 9.5 trillion kilometers (6 trillion mi)

Mantle: the layer between the crust and the core of a large moon or planet

Milky Way: the large spiral galaxy that is home to the sun; also, the central plane of that galaxy seen in the night sky

Nebula: a cloud of gas and dust in interstellar space

Neutron star: a small, extremely dense star composed almost entirely of neutrons

Nova: a white dwarf in a binary star system that pulls matter from its companion star, periodically brightening in a brilliant burst

Oort cloud: an enormous spherical cloud surrounding our solar system that contains billions of icy objects

Parallax: the apparent change in the position of an object when seen from different vantage points

Planet: a body in orbit around the sun, possessing sufficient mass to form a nearly round shape, and gravitationally dominant, meaning it will have cleared its path of other bodies of comparable size

Planetesimal: small primordial body formed by accretion in the early solar system

Plasma: a gas made up of charged particles

Pulsar: a rapidly rotating neutron star that emits regular pulses of radio waves

Quasar: a contraction of quasi-stellar object; a distant, very luminous, active galactic nucleus emitting radio waves

Radiation: energy emitted in the form of waves or particles

Redshift: the increase in the wavelength of radiation sent out by a celestial body as a result of the Doppler effect, gravitation, or cosmological expansion

Solar mass: the mass of the sun, which is used to measure other celestial objects; one solar mass = $(1.989 \pm 0.004) \times 10^{30}$ kg

Solar nebula: the spinning gaseous cloud that condensed to form the solar system

Standard candle: an object whose absolute magnitude (innate brightness) is well known, and so can serve as an indicator of distance

Supergiant star: a member of a class of the largest and most luminous stars known, with radii between 100 and 1,000 times that of the sun.

Supernova: an enormous stellar explosion; the death of a massive star

Tidal force: the varying gravitational force one massive object exerts on another

MAP TERMS

TERM	PLURAL	DEFINITION
Albedo feature		Geographic area distinguished by amount of reflected light
Arcus	arcūs	Arc-shaped feature
Astrum	astra	Radial-patterned features on Venus
Catena	catenae	Chain of craters
Cavus	cavi	Hollows, irregular steep-sided depressions usually in arrays or clusters
Chaos	chaoses	Distinctive area of broken terrain
Chasma	chasmata	A deep, elongated, steep-sided depression
Collis	colles	Small hills or knobs
Corona	coronae	Ovoid-shaped feature
Crater	craters	A circular depression
Dorsum	dorsa	Ridge
Eruptive center		Active volcanic centers on Io
Facula	faculae	Bright spot
Farrum	farra	Pancake-like structure, or a row of such structures
Flexus	flexūs	A very low ridge with a scalloped pattern
Fluctus	fluctūs	Flow terrain
Flumen	flumina	Channel on Titan that might carry liquid
Fossa	fossae	Long, narrow depression
Insula	insulae	Island, an isolated land area (or group) surrounded by, or nearly surrounded by, a liquid area
Labes	labēs	Landslide
Labyrinthus	labyrinthi	Complex of intersecting valleys or ridges.
Lacuna	lacunae	Irregularly shaped depression on Titan having the appearance of a dry lake bed
Lacus	lacūs	Lake or small plain; on Titan, a "lake" or small, dark plain with discrete, sharp boundaries
Lenticula	lenticulae	Small dark spots on Europa
Linea	lineae	A dark or bright elongate marking, may be curved or straight
Lingula	lingulae	Extension of plateau having rounded lobate or tongue-like boundaries
Macula	maculae	Dark spot, may be irregular
Mare	maria	Sea; large circular plain; on Titan large expanses of dark materials thought to be liquid hydrocarbons
Mensa	mensae	A flat-topped prominence with cliff-like edges
Mons	montes	Mountain
Oceanus	oceani	A very large dark area on the moon
Palus	paludes	Means "swamp"; small plain
Patera	paterae	An irregular crater, or a complex one with scalloped edges
Planitia	planitiae	Low plain
Planum	plana	Plateau or high plain
Plume	plumes	Cryo-volcanic features on Triton
Promontorium	promontoria	Cape; headland
Regio	regiones	A large area marked by reflectivity or color distinctions from adjacent areas or a broad geographic region
Reticulum	reticula	Reticular (netlike) pattern on Venus
Rima	rimae	Fissure
Rupes	rupēs	Scarp
Scopulus	scopuli	Lobate or irregular scarp
Sinus	sinūs	Means "bay"; small plain
Sulcus	sulci	Subparallel furrows and ridges
Terra	terrae	Extensive landmass
Tessera	tesserae	Tile-like, polygonal terrain
Tholus	tholi	Small domical mountain or hill
Unda	undae	Dunes
Vallis	valles	Valley
Vastitas	vastitates	Extensive plain
Virga	virgae	A streak or stripe of color

ACKNOWLEDGMENTS

The author wishes to thank Vera Rubin of the Carnegie Institution and Michael Turner of the University of Chicago for their participation in developing this manuscript. He also wishes to thank editors Susan Tyler Hitchcock and Patricia Daniels, as well as cartographers Carl Mehler and Matthew Chwastyk, for their help is bringing the project to completion.

ABOUT THE AUTHOR

James Trefil is Clarence J. Robinson Professor of Physics at George Mason University in Fairfax, Virginia. His previous books include *Other Worlds* (National Geographic Books, 1999), and he served as Principal Science Consultant for *The Big Idea* (National Geographic Books, 2012). He has received numerous awards for his writing, most recently the Science Writing Award given by the American Institute of Physics, and in 2011 was awarded an honorary degree from the University of Sts. Kiril and Methodius in Skopje, Macedonia. His most recent book is *Science in World History*.

BIBLIOGRAPHY

FURTHER ASTRONOMY READING FROM NATIONAL GEOGRAPHIC

Apt, Jay, et al. *Orbit: NASA Astronauts Photograph the Earth.* Washington, D.C.: National Geographic, 1996.

Daniels, Patricia. *The New Solar System: Ice Worlds, Moons, and Planets Redefined.* Washington, D.C.: National Geographic, 2009.

DeVorkin, David and Robert Smith. *Hubble: Imaging Space and Time.* Washington, D.C.: National Geographic, 2008.

Glover, Linda, et al. *National Geographic Encyclopedia of Space.* Washington, D.C.: National Geographic, 2005.

Gott, J. Richard, and Robert J. Vanderbei. *Sizing Up the Universe.* Washington, D.C.: National Geographic, 2011.

Jones, Tom and Ellen Stofan. *Planetology: Unlocking the Secrets of the Solar System.* Washington, D.C.: National Geographic, 2008.

Raeburn, Paul. *Mars: Uncovering the Secrets of the Red Planet.* Washington, D.C.: National Geographic, 1998.

Schneider, Howard. *National Geographic Backyard Guide to the Night Sky.* Washington, D.C.: National Geographic, 2009.

Trefil, James. *Other Worlds: Images of the Cosmos From Earth and Space.* Washington, D.C.: National Geographic, 1999

ILLUSTRATIONS CREDITS

1, NASA/NOAA/GSFC/Suomi NPP/VIIRS/Norman Kuring; 2-3, NASA, ESA, and the Hubble Heritage Team (STScI/AURA); 4-5, x-ray: NASA/CXC/SAO/J. Hughes et al. Optical: NASA/ESA/Hubble Heritage Team (STScI/AURA); 6-7, NASA/JPL/Space Science Institute; 8-9, NASA, ESA, and J. Maíz Apellániz (Instituto de Astrofísica de Andalucía, Spain); 10-11, E. J. Schreier (STScI) and NASA; 12, NASA Kennedy Space Center; 14-15, NASA/JPL-Caltech/Cornell/Arizona State Univ.; 18-9, NASA; 20, © Visual Language 1996; 21, Jean-Leon Huens/National Geographic Stock; 22, NASA, ESA, and the Hubble SM4 ERO Team; 24, NASA, ESA, J. Merten (Institute for Theoretical Astrophysics, Heidelberg/Astronomical Observatory of Bologna), and D. Coe (STScI); 27 (UP), © Visual Language 1996; 36-7, NASA/JPL-Caltech; 38-9, David Aguilar; 42-3, NASA/JPL-Caltech; 43 (INSET), NASA/JPL-Caltech; 45, NASA/JPL-Caltech/T. Pyle (SSC-Caltech); 46, Ron Miller/Stocktrek Images/Corbis; 50-51, NASA/Johns Hopkins University Applied Physics Laboratory/Arizona State University/Carnegie Institution of Washington. Image reproduced courtesy of Science/AAAS; 51 (INSET), NASA/Johns Hopkins University Applied Physics Laboratory/Carnegie Institution of Washington; 56, Science Source; 57, NASA/Johns Hopkins University Applied Physics Laboratory/Carnegie Institution of Washington; 58, Pierre Mion/National Geographic Stock; 58-9, Rick Sternbach; 60 (INSET), NASA/JPL/USGS; 60-1, NASA/JPL; 66 (UP), ESA/VIRTIS/INAF-IASF/Obs. de Paris-LESIA/Univ. of Oxford; 66 (LO), ESA/VIRTIS/INAF-IASF/Obs. de Paris-LESIA/Univ. of Oxford; 67, Science Source; 68, J. Whatmore/ESA; 70-71, NOAA/NASA GOES Project; 76, Planetary Visions Ltd./Photo Researchers, Inc.; 77, Alain Barbezat/National Geographic My Shot; 79 (UP), NG Maps; 79 (LO), Theophilus Britt Griswold; 80 (INSET), NG Maps; 80-1, Carsten Peter/National Geographic Stock; 86, Christian Darkin/Photo Researchers, Inc.; 87, NASA; 88, Gary Hincks/Photo Researchers, Inc.; 89, Michael Melford/National Geographic Stock; 90 (INSET), NASA/JPL/Malin Space Science Systems; 90-91, NASA/JPL/Cornell; 96, Science Source; 97, NASA/JPL/University of Arizona; 98, Science Source; 99 (UP), NASA/JPL-Caltech; 99 (LO), NASA/JPL/Malin Space Science Systems; 100-101, Mark Garlick/Photo Researchers, Inc.; 101 (INSET), NASA/JPL/USGS; 103, Sanford/Agliolo/CORBIS; 106-107, Science Source; 107, NASA/JPL/University of Arizona; 110, Science Source; 110-11, NASA/JPL; 112, H. Hammel (SSI), WFPC2, HST, NASA; 113, H. Hammel (SSI), WFPC2, HST, NASA; 114 (INSET), NASA/JPL/DLR; 114-5, NASA/JPL/USGS; 124, NASA/JPL; 125, NASA/JPL/University of Arizona; 126, NASA/JPL/University of Arizona; 127, NASA/JPL; 128 (INSET), NASA/JPL/Space Science Institute; 128-9, David Aguilar; 146, Science Source; 147, NASA/JPL/Space Science Institute; 148, Davis Meltzer/National Geographic Stock; 149, NASA/JPL; 150 (INSET), NASA/JPL/Space Science Institute; 150-51, NASA/JPL; 154, Ron Miller; 155, Ludek Pesek/Photo Researchers, Inc.;

156 (INSET), Science Source; 156-7, John R. Foster/Photo Researchers, Inc.; 167, NASA/JPL/Universities Space Research Association/Lunar & Planetary Institute; 167, NASA/JPL/Universities Space Research Association/Lunar & Planetary Institute; 168, Science Source; 169, Mark Garlick/Photo Researchers, Inc.; 170-71, NASA/JPL; 171, Science Source; 172 (INSET), NASA, ESA, H. Weaver (JHUAPL), A. Stern (SwRI), and the HST Pluto Companion Search Team; 172-3, David Aguilar/National Geographic Stock; 174, Bettmann/CORBIS; 175, Science Source; 176-7, Johns Hopkins University Applied Physics Laboratory/Southwest Research Institute (JHUAPL/SwRI); 178-9, Dan Schechter/Getty Images; 179 (INSET), NASA/JPL/UMD; 180, Jonathan Blair/National Geographic Stock; 182 (INSET), David Aguilar/National Geographic Stock; 182-3, Mark Garlick/Photo Researchers, Inc.; 185, NASA/JPL-Caltech/R. Hurt (SSC); 186-7, Viktar Malyshchyts/Shutterstock; 188-9, Hubble image: NASA, ESA, and Q. D. Wang (University of Massachusetts, Amherst). Spitzer image: NASA, Jet Propulsion Laboratory, and S. Stolovy (Spitzer Science Center/Caltech); 192-3, Dr. Fred Espenak/Photo Researchers, Inc.; 193 (INSET), Jon Lomberg/Photo Researchers, Inc.; 194, Harvard College Observatory/Photo Researchers, Inc.; 195, NASA, ESA, and the Hubble Heritage Team (STScI/AURA); 197, NASA/JPL-Caltech; 198-9, Sepdes Sinaga/National Geographic My Shot; 199, SOHO (ESA & NASA); 200-201, SOHO/EIT; 202, David Aguilar/National Geographic Stock; 203, James P. Blair/National Geographic Stock; 204-205, NASA/SDO/GOES-15; 206-207, Robert Stocki/National Geographic My Shot; 207 (INSET), Sebastian Kaulitzki/Shutterstock; 209, O. Louis Mazzatenta/National Geographic Stock; 210 (INSET), NASA/Kepler; 210-11, Dana Berry/National Geographic Stock; 212-13, NASA/Ames/JPL-Caltech; 214-15, NASA/JPL-Caltech; 215, David Aguilar/National Geographic Stock; 216-17, Dr. Verena Tunnicliffe, University of Victoria (UVic); 218 (INSET), NASA; 218-19, SETI; 220, SHNS/SETI Institute/Newscom/File; 222-3, NASA, ESA, and the Hubble SM4 ERO Team; 223 (INSET), Mark Garlick/Photo Researchers, Inc.; 225, Mark Garlick/Photo Researchers, Inc.; 226 (INSET), NASA, ESA, P. Challis and R. Kirshner (Harvard-Smithsonian Center for Astrophysics); 226-7, NASA/JPL-Caltech/O. Krause (Steward Observatory); 228, NASA, ESA, and the Hubble Heritage STScI/AURA)-ESA/Hubble Collaboration. Acknowledgment: Robert A. Fesen (Dartmouth College, USA) and James Long (ESA/Hubble); 228-9, David A. Hardy/www.astroart.org; 230 (UP LE), NASA, ESA, and H. E. Bond (STScI); 230 (UP CTR), NASA, ESA, and H. E. Bond (STScI); 230 (UP RT), NASA and the Hubble Heritage Team (AURA/STScI); 231, NASA, ESA, and the Hubble Heritage Team (STScI/AURA); 232 (INSET), x-ray image: NASA/CXC/ASU/J. Hester et al. Optical Image: NASA/HST/ASU/J. Hester et al.; 232-3, Mark Garlick/Photo Researchers, Inc.; 235, Jonathan Blair/National Geographic Stock; 236-7, NASA, ESA, J. Hester

and A. Loll (Arizona State University); 238 (INSET), Mark Garlick/Photo Researchers, Inc.; 238-9, x-ray: NASA/CXC/CfA/R.Kraft et al.; Submillimeter: MPIfR/ESO/APEX/A.Weiss et al.; optical: ESO/WFI; 241, Detlev Van Ravensway/Photo Researchers, Inc.; 242, Digitized Sky Survey; 242-3, NASA/CXC/M.Weiss; 244-5, x-ray: NASA/CXC/M. Markevitch et al. Optical: NASA/STScI; Magellan/Univ. of Arizona/D. Clowe et al. Lensing map: NASA/STScI; ESO WFI; Magellan/Univ. of Arizona/D. Clowe et al.; 245 (INSET), Volker Springel/Max Planck Institute for Astrophysics/Photo Researchers, Inc.; 246-7, Richard Nowitz/National Geographic Stock; 248, NASA, ESA, and R. Massey (California Institute of Technology); 250-51, NASA/JPL-Caltech/K. Su (Univ. of Arizona); 252-3, NASA/JPL-Caltech/UCLA; 256-7, NASA/WMAP Science Team; 257 (INSET), NASA/NSF; 258-9, NASA/WMAP Science Team; 261 (UP), NASA/WMAP Science Team; 261 (LO), Equinox Graphics/Photo Researchers, Inc.; 262 (UP), NASA/WMAP Science Team; 262 (LO LE), ESA (image by AOES Medialab); 262 (LO RT), ESA; 264 (INSET LE), Royal Astronomical Society/Photo Researchers, Inc.; 264 (INSET CTR), Royal Astronomical Society/Photo Researchers, Inc.; 264 (INSET RT), Royal Astronomical Society/Photo Researchers, Inc.; 264-5, NASA, ESA, M. Livio and the Hubble Heritage Team (STScI/AURA); 267 (UP), Royal Astronomical Society/Photo Researchers, Inc.; 267 (LO), Science Source; 268-9, Chris Butler/Photo Researchers, Inc.; 269, David Parker/Photo Researchers, Inc.; 270-71, NASA, ESA, and the Hubble Heritage Team (STScI/AURA)-ESA/Hubble Collaboration; 271 (INSET LE), NASA, ESA, K. Kuntz (JHU), F. Bresolin (University of Hawaii), J. Trauger (Jet Propulsion Lab), J. Mould (NOAO), Y.-H. Chu (University of Illinois, Urbana), and STScI; 271 (INSET CTR LE), NASA, ESA, and the Hubble Heritage Team (STScI/AURA); 271 (INSET CTR RT), NASA, ESA, and the Hubble Heritage (STScI/AURA)-ESA/Hubble Collaboration; 271 (INSET RT), NASA, ESA, and the Hubble Heritage Team (STScI/AURA); 272-3, NASA/JPL-Caltech/K. Gordon (STScI) and SINGS Team; 275, NASA; 276 (INSET), Ann Feild (STScI); 276-7, Mehau Kulyk/Photo Researchers, Inc.; 278-9, NASA/WMAP Science Team; 281, Don Dixon/cosmographica.com; 282-3, Adolf Schaller for STScI; 284 (INSET), BSIP/Photo Researchers, Inc.; 284-5, Mark Garlick/Photo Researchers, Inc.; 286, NASA/WMAP Science Team; 287, José Antonio Peñas/Photo Researchers, Inc.; 288, NASA/ESA/ESO/Wolfram Freudling et al. (STECF); 290-91, Moonrunner Design Ltd./National Geographic Stock; 291 (INSET), Mark Garlick/Photo Researchers, Inc.; 292, Fermilab; 294-5, NASA, ESA, S. Beckwith (STScI) and the HUDF Team; 296, Wikipedia; 298, Corbin O'Grady Studio/Photo Researchers, Inc.; 299, Detlev Van Ravenswaay/Photo Researchers, Inc.; 300, Mark Garlick/Photo Researchers, Inc.; 302-3, ESO/T. Preibisch; 304, photovideostock/iStockphoto; 305, NASA/Johns Hopkins University Applied Physics Laboratory/Carnegie Institution of Washington.

MAP CREDITS

ALL MAPS

Place names: Gazetteer of Planetary Nomenclature, Planetary Geomatics Group of the USGS (United States Geological Survey) Astrogeology Science Center

Web Page: http://planetarynames.wr.usgs.gov/

IAU (International Astronomical Union)

Web Page: http://iau.org/

NASA (National Aeronautics and Space Administration)

Web Page: http://www.nasa.gov/

SOLAR SYSTEM PP. 40–41, INNER PLANETS PP. 48–49, OUTER PLANETS, PP. 104–105

All images: NASA, JPL (Jet Propulsion Laboratory, California Institute of Technology), Johns Hopkins University Applied Physics Laboratory, Carnegie Institution of Washington

MERCURY PP. 52–55

Global Mosaic: MESSENGER (MErcury Surface, Space ENvironment, GEochemistry, and Ranging), NASA, Johns Hopkins University Applied Physics Laboratory, Carnegie Institution of Washington

VENUS PP. 62–65

Global Mosaic: Magellan Synthetic Aperature Radar Mosaics, NASA, JPL (Jet Propulsion Laboratory, California Institute of Technology)

EARTH PP. 72–75

Surface Satellite Mosaic: NASA Blue Marble, NASA's Earth Observatory

Bathymetry: ETOPO1/Amante and Eakins, 2009

EARTH'S MOON PP. 82–85

Global Mosaic: Lunar Reconnaisance Orbiter, NASA, Arizona State University

MARS PP. 92–95

Global Mosaic: NASA Mars Global Surveyor; National Geographic Society

Moon images: Phobos, Diemos, NASA, JPL (Jet Propulsion Laboratory, California Institute of Technology), University of Arizona

JUPITER PP. 108–109

Global Mosaic: NASA Cassini Spacecraft, NASA, JPL (Jet Propulsion Laboratory, California Institute of Technology), Space Science Institute

MOONS OF JUPITER PP. 116–123

All Global Mosaics: NASA Galileo Orbiter NASA, JPL (Jet Propulsion Laboratory, California Institute of Technology), University of Arizona

SATURN PP. 130–131

Global Mosaic: NASA Cassini Spacecraft NASA, JPL (Jet Propulsion Laboratory, California Institute of Technology)

MOONS OF SATURN PP. 132–145

All Gloabl Mosaics: NASA Cassini Spacecraft NASA, JPL (Jet Propulsion Laboratory, California Institute of Technology) Space Science Institute

SATURN'S RINGS PP. 152–153

NASA Cassini Spacecraft NASA, JPL (Jet Propulsion Laboratory, California Institute of Technology) Space Science Institute

URANUS PP. 158–159, URANUS'S MOONS PP. 160–163, NEPTUNE PP. 164–165, TRITON P. 166

Global imagery: NASA Voyager II, NASA, JPL (Jet Propulsion Laboratory, California Institute of Technology)

THE MILKY WAY PP. 190–191

Artwork: Ken Eward, National Geographic Society

THE SUN PP. 200–201

Artwork: Moonrunner Design, National Geographic Society

THE UNIVERSE PP. 254–255

Artwork: Ken Eward, National Geographic Society

COSMIC MICROWAVE BACKGROUND PP. 258–259

Mosaic: WMAP (Wilkinson Microwave Anisotropy Probe) NASA, WMAP Science Team

GENERAL INDEX

PLACE-NAME INDEX

USING THIS INDEX

The International Astronomical Union regulates the naming of extraterrestrial features. The convention adopted is to use Latin for generic terms (e.g., Utopia Planitia on Mars, *planitia* meaning "low plain"). A listing of those terms appears on page 315. The grid system used to demarcate the location of mapped features is based upon the graticule. Three similar systems are used in this atlas depending on how the maps are presented. Terrestrial planets, Mercury, Venus, Earth, Mars, along with Earth's moon use the first as illustrated below (figure 1). Letter coordinates appear around the perimeter with numbers along the equator. Where both hemispheres of a moon are presented together, as in the satellites of Jupiter and Saturn, a second method of finding features is provided (figure 2). The central meridian of each hemisphere is marked with letters, the equator with numbers that are continuous between the two hemispheres. A final method is used to accommodate bodies in the outer solar system where data is only available in the southern hemisphere. Letters encircle, designating longitude, while numbers represent latitude in rings around the moon (figure 3). The features collected in this index are referenced to these systems. Names without a generic are craters with the exception of those from Earth. Entries with two grid references are features that cross between two mapped hemispheres.

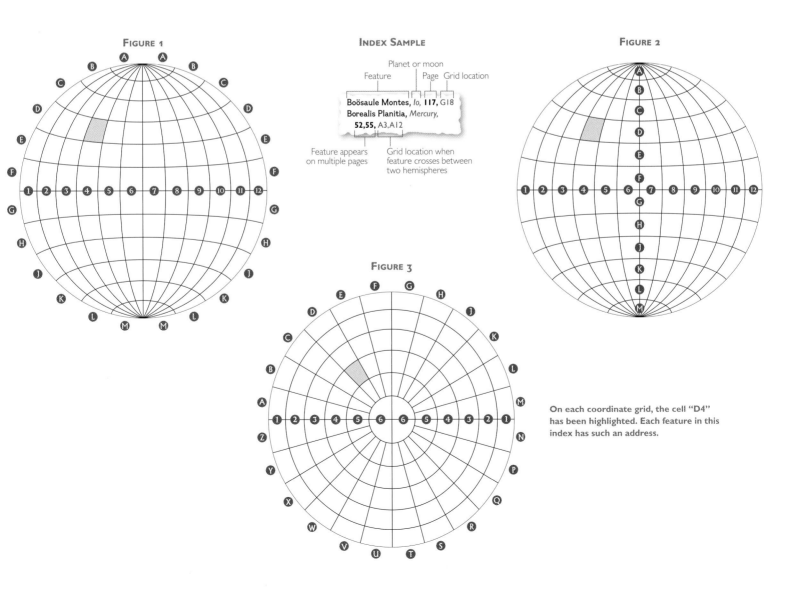

FIGURE 1

INDEX SAMPLE

Planet or moon

Feature   Page   Grid location

Boösaule Montes, *Io,* **117,** G18
Borealis Planitia, *Mercury,*
52,55, A3, A12

Feature appears   Grid location when
on multiple pages   feature crosses between
two hemispheres

FIGURE 2

FIGURE 3

On each coordinate grid, the cell "D4" has been highlighted. Each feature in this index has such an address.

Coc–Fiz

SPACE ATLAS | PLACE-NAME INDEX

328

SPACE ATLAS

James Trefil

PUBLISHED BY THE NATIONAL GEOGRAPHIC SOCIETY

John M. Fahey, Jr., *Chairman of the Board
and Chief Executive Officer*

Timothy T. Kelly, *President*

Declan Moore, *Executive Vice President; President,
Publishing and Digital Media*

Melina Gerosa Bellows, *Executive Vice President;
Chief Creative Officer, Books, Kids, and Family*

PREPARED BY THE BOOK DIVISION

Hector Sierra, *Senior Vice President and General Manager*

Anne Alexander, *Senior Vice President and Editorial Director*

Jonathan Halling, *Design Director, Books and Children's Publishing*

Marianne R. Koszorus, *Design Director, Books*

Susan Tyler Hitchcock, *Senior Editor*

R. Gary Colbert, *Production Director*

Jennifer A. Thornton, *Director of Managing Editorial*

Susan S. Blair, *Director of Photography*

Meredith C. Wilcox, *Director, Administration and Rights Clearance*

STAFF FOR THIS BOOK

Garrett Brown, Patricia Daniels, *Project Editors*

Marty Ittner, *Designer*

Sam Serebin, *Design Consultant*

Carl Mehler, *Director of Maps*

Matthew Chwastyk, *Cartographer*

Madison Doehler, *Map Place-Name
Index Assistant*

Judith Klein, *Production Editor*

Mike Horenstein, *Production Manager*

Galen Young, *Rights Clearance Specialist*

Kate Olsen, *Design Assistant*

MANUFACTURING AND QUALITY MANAGEMENT

Phillip L. Schlosser, *Senior Vice President*

Chris Brown, *Vice President, NGS Book Manufacturing*

George Bounelis, *Vice President, Production Services*

Nicole Elliott, *Manager*

Rachel Faulise, *Manager*

Robert L. Barr, *Manager*

The National Geographic Society is one of the world's largest nonprofit scientific and educational organizations. Founded in 1888 to "increase and diffuse geographic knowledge," the Society's mission is to inspire people to care about the planet. It reaches more than 400 million people worldwide each month through its official journal, *National Geographic,* and other magazines; National Geographic Channel; television documentaries; music; radio; films; books; DVDs; maps; exhibitions; live events; school publishing programs; interactive media; and merchandise. National Geographic has funded more than 10,000 scientific research, conservation and exploration projects and supports an education program promoting geographic literacy. For more information, visit www.nationalgeographic.com.

For more information, please call 1-800-NGS LINE (647-5463) or write to the following address:

National Geographic Society
1145 17th Street N.W.
Washington, D.C. 20036-4688 U.S.A.

Visit us online at www.nationalgeographic.com

For information about special discounts for bulk purchases, please contact National Geographic Books Special Sales: ngspecsales@ngs.org

For rights or permissions inquiries, please contact National Geographic Books Subsidiary Rights: ngbookrights@ngs.org

This 2014 edition printed for Barnes & Noble, Inc. by the National Geographic Society.

ISBN: 978-1-4262-0971-0
ISBN: 978-1-4262-1091-4 (deluxe)
ISBN: 978-1-4351-5411-7 (B&N ed.)

Printed in China

14/RRDS/2